A&P

APPLICATIONS MANUAL

Frederic H. Martini, Ph.D.

Kathleen Welch, M.D.

with

William C. Ober, M.D.
Art Coordinator and Illustrator

Claire E. Ober, R.N.
Illustrator

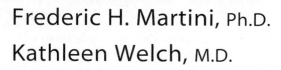

Boston   Columbus   Indianapolis   New York   San Francisco   Upper Saddle River
Amsterdam   Cape Town   Dubai   London   Madrid   Milan   Munich   Paris   Montréal   Toronto
Delhi   Mexico City   São Paulo   Sydney   Hong Kong   Seoul   Singapore   Taipei   Tokyo

Executive Editor: *Leslie Berriman*
Associate Project Editor: *Lisa Damerel*
Assistant Editor: *Cady Owens*
Editorial Assistant: *Sharon Kim*
Director of Development: *Barbara Yien*
Managing Editor: *Mike Early*
Assistant Managing Editor: *Nancy Tabor*
Project Manager: *Dorothy Cox*
Production Management and Composition: *S4Carlisle Publishing Services, Inc.*
Interior and Cover Designer: *tani hasegawa*
Photo Researcher: *Maureen Spuhler*
Senior Procurement Specialist: *Stacey Weinberger*
Senior Marketing Manager: *Allison Rona*

Cover Photo Credit: *MedicalRF.com/Corbis*

Credits and acknowledgments for materials borrowed from other sources and reproduced, with permission, in this textbook appear on page 247.

Copyright © 2015, 2012, 2009 Frederic H. Martini, Inc. Published by Pearson Education, Inc. All rights reserved. Manufactured in the United States of America. This publication is protected by Copyright, and permission should be obtained from the publisher prior to any prohibited reproduction, storage in a retrieval system, or transmission in any form or by any means, electronic, mechanical, photocopying, recording, or likewise. To obtain permission(s) to use material from this work, please submit a written request to Pearson Education, Inc., Permissions Department, One Lake Street, Upper Saddle River, New Jersey 07458. For information regarding permissions, call (847) 486-2635.

Notice: Our knowledge in clinical sciences is constantly changing. The authors and the publisher of this volume have taken care that the information contained herein is accurate and compatible with the standards generally accepted at the time of publication. Nevertheless, it is difficult to ensure that all information given is entirely accurate for all circumstances. The authors and the publisher disclaim any liability, loss, or damage incurred as a consequence, directly or indirectly, of the use and application of any of the contents of this volume.

Many of the designations used by manufacturers and sellers to distinguish their products are claimed as trademarks. Where those designations appear in this book, and the publisher was aware of a trademark claim, the designations have been printed in initial caps or all caps.

Library of Congress Cataloging-in-Publication Data
Martini, Frederic.
 A&P applications manual / Frederic H. Martini, Ph.D., Kathleen Welch, M.D., with William C. Ober, M.D., art coordinator and illustrator, Claire E. Ober, R.N., illustrator.—Tenth edition.
    pages cm
  ISBN 978-0-321-94973-8 — ISBN 0-321-94973-0   1. Physiology, Pathological—Handbooks, manuals, etc.   2. Anatomy, Pathological—Handbooks, manuals, etc.   I. Welch, Kathleen (Kathleen Martini)   II. Title.
  RB113.M32 2015
  612—dc23   2014013336

ISBN 10: **0-321-94973-0**; ISBN 13: **978-0-321-94973-8** (Stand-alone)
ISBN 10: **0-321-97400-X**; ISBN 13: **978-0-321-97400-6** (ValuePack)

www.pearsonhighered.com   2 3 4 5 6 7 8 9 10—RRD—18 17 16 15 14

Contents

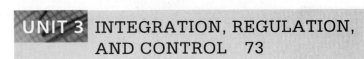

UNIT 3   INTEGRATION, REGULATION, AND CONTROL   73

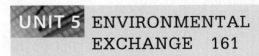

About the Authors

Frederic (Ric) H. Martini, Ph.D.
Author

Dr. Martini received his Ph.D. from Cornell University in comparative and functional anatomy for work on the pathophysiology of stress. In addition to professional publications that include journal articles and contributed chapters, technical reports, and magazine articles, he is the lead author of ten undergraduate texts on anatomy and physiology or anatomy. Dr. Martini is currently affiliated with the University of Hawaii at Manoa and has a long-standing bond with the Shoals Marine Laboratory, a joint venture between Cornell University and the University of New Hampshire. He has been active in the Human Anatomy and Physiology Society (HAPS) for over 20 years and was a member of the committee that established the course curriculum guidelines for A&P. He is now a President Emeritus of HAPS after serving as President-Elect, President, and Past President over 2005–2007. Dr. Martini is also a member of the American Physiological Society, the American Association of Anatomists, the Society for Integrative and Comparative Biology, the Australia/New Zealand Association of Clinical Anatomists, the Hawaii Academy of Science, the American Association for the Advancement of Science, and the International Society of Vertebrate Morphologists.

Kathleen Welch, M.D.
Author

Dr. Welch received her B.A. from the University of Wisconsin–Madison, her M.D. from the University of Washington in Seattle, and did her residency in Family Practice at the University of North Carolina in Chapel Hill. Participating in the Seattle WWAMI rural medical education program, she studied in Fairbanks, Anchorage, and Juneau, Alaska, with time in Boise, Idaho, and Anacortes, Washington, as well. For two years, she served as Director of Maternal and Child Health at the LBJ Tropical Medical Center in American Samoa and subsequently was a member of the Department of Family Practice at the Kaiser Permanente Clinic in Lahaina, Hawaii, and on the staff at Maui Memorial Hospital. She has been in private practice since 1987 and is licensed to practice in Hawaii and Washington State. Dr. Welch is a Fellow of the American Academy of Family Practice and a member of the Maui County Medical Society and the Human Anatomy and Physiology Society (HAPS). With Dr. Martini, she has coauthored a textbook on anatomy and physiology. She and Dr. Martini were married in 1979, and they have one son.

William C. Ober, M.D.
Art Coordinator and Illustrator

Dr. Ober received his undergraduate degree from Washington and Lee University and his M.D. from the University of Virginia. He also studied in the Department of Art as Applied to Medicine at Johns Hopkins University. After graduation, Dr. Ober completed a residency in Family Practice and later was on the faculty at the University of Virginia in the Department of Family Medicine and in the Department of Sports Medicine. He also served as Chief of Medicine of Martha Jefferson Hospital in Charlottesville, VA. He is currently a Visiting Professor of Biology at Washington and Lee University, where he has taught several courses and led student trips to the Galápagos Islands. He was on the Core Faculty at Shoals Marine Laboratory for 24 years, where he taught Biological Illustration every summer. Dr. Ober has collaborated with Dr. Martini on all of his textbooks in every edition.

Claire E. Ober, R.N.
Illustrator

Claire E. Ober, R.N., B.A., practiced family, pediatric, and obstetric nursing before turning to medical illustration as a full-time career. She returned to school at Mary Baldwin College, where she received her degree with distinction in studio art. Following a five-year apprenticeship, she has worked as Dr. Ober's partner in Medical & Scientific Illustration since 1986. She was on the Core Faculty at Shoals Marine Laboratory and co-taught the Biological Illustration course with Dr. Ober for 24 years. The textbooks illustrated by Medical & Scientific Illustration have won numerous design and illustration awards.

Preface

This *A&P Applications Manual* is designed to introduce students of anatomy and physiology to clinical, diagnostic, and other topics of great potential interest. By placing these discussions in a separate applications manual, rather than incorporating them into a traditional anatomy and physiology textbook, the key topics can be considered in greater depth, in a format that facilitates its use as a portable reference.

The organization of the *A&P Applications Manual* parallels that of most introductory textbooks on this subject. The chapters are grouped into six units; with the exception of the introductory unit, the chapters within those units consider specific body systems.

Questions found at the end of each unit are intended to illustrate both how the related systems work together and how disorders affecting any one system can affect others. The Case Studies at the end of the *A&P Applications Manual* provide students with further opportunities to develop their powers of analysis, integration, and problem solving. The studies presented, based on actual clinical histories, draw on material from the entire text (as they would in real life). The questions, keyed to crucial points in the presentation, help students to identify relevant facts and form plausible hypotheses. These case studies can be used as the basis of discussion; alternatively, students can tackle them on their own to sharpen their reasoning skills.

Few instructors will cover all the material in the *A&P Applications Manual*. Because courses differ in their emphases and students differ in their interests and backgrounds, the goal in designing the *A&P Applications Manual* has been to provide maximum flexibility of use. The diversity of applied topics offers instructors many opportunities to integrate the clinical diagnosis and treatment of pathology with normal function, and other health-related topics. Readings from the *A&P Applications Manual* that are not covered in class can be assigned, recommended, or used by individual students for reference. Experience indicates that each student will read those selections that deal with disorders affecting friends or family members, address topics of current interest and concern, or include information relevant to a chosen career path.

With every edition of this manual we have revised numbers and statistics, and sometimes added new diseases to the discussion. We have tried to focus on disorders commonly encountered or of particular interest; as new topics have been added, others have been trimmed to keep the *A&P Applications Manual* of manageable size. However, there are many Internet sources for health and A&P information. Some are accurate and up to date, whereas others present opinion and personal politics as scientific fact. Unfortunately, your search engine will not distinguish among the various types of websites. Evaluating search results can thus be both enlightening and entertaining (as well as time consuming—after all, you still have to study your A&P textbook!).

To save you some time, here are some reliable website sources that provide basic information and links to other reputable sites:

http://medlineplus.gov/
www.cdc.gov
www.who.int/en/
www.cdc.gov/nchs
www.ncbi.nlm.nih.gov/
www.cancer.gov
http://health.nih.gov/

In addition, you will find these resources valuable:

1. the most current edition of your textbook, supplemented by the online resources that contain additional information and exercises,
2. a current medical dictionary,
3. a reference work or text in your field of interest, and
4. an online or print version of a journal in your field of interest

▶ Acknowledgments

We would like to thank Bill Ober, M.D., and Claire Ober, R.N., who prepared the illustrations for this project. We would also like to thank the readers who provided comments and suggestions for improvement of this edition. Finally, we also express our thanks to Leslie Berriman, Lisa Damerel, Cady Owens, Sharon Kim, Dorothy Cox, Mike Rossa, Maureen Spuhler, and Lynn Steines for their work on this project.

Frederic Martini, Ph.D.
martini@pearson.com

Kathleen Welch, M.D.
kwelch@maui.net

Mailing address:
Pearson Education, Applied Sciences
1301 Sansome Street
San Francisco, CA 94111

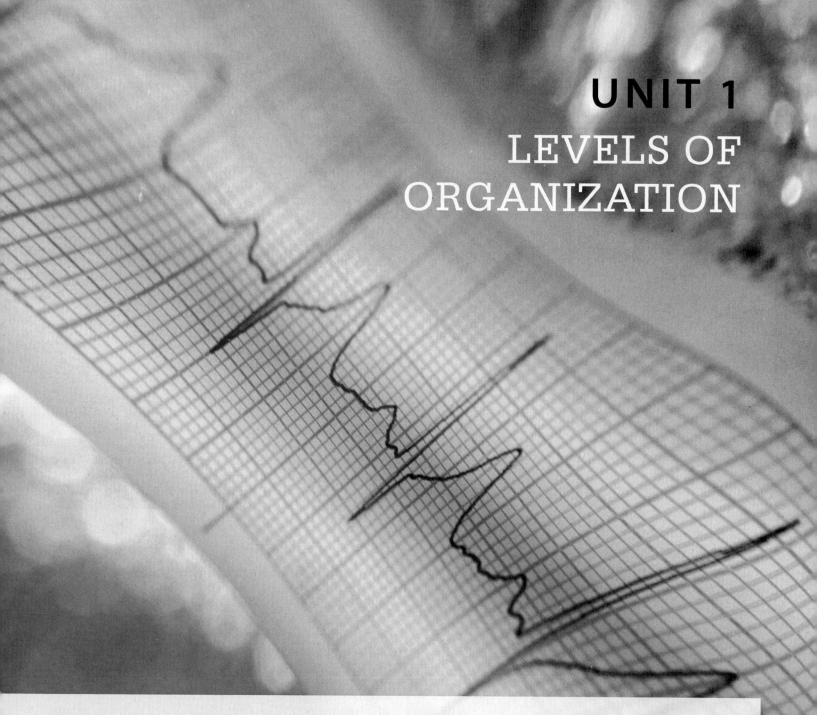

UNIT 1
LEVELS OF ORGANIZATION

The sciences of anatomy and physiology consider the structural and functional characteristics of living things, and this unit introduces basic concepts related to the diagnosis and treatment of human diseases. This unit begins the process by considering the way clinical information is collected and evaluated. It discusses the way disease processes interfere with *homeostasis*, the key to physiological regulation, and how this interference threatens survival in a changing environment. The rest of the unit follows a "levels of organization" theme and relates events and processes at the chemical, cellular, and tissue levels to representative disorders and to clinical procedures useful in diagnosis. One goal is to show the real-world relevance of some of the seemingly abstract concepts presented in an anatomy and physiology textbook.

MAJOR SECTIONS INCLUDED WITHIN THIS UNIT:

An Introduction to Clinical Anatomy and Physiology

The Chemical Level of Organization

The Cellular Level of Organization

The Tissue Level of Organization

1

An Introduction to Clinical Anatomy and Physiology

▶ Disease, Pathology, and Diagnosis

The formal name for the study of disease is **pathology**; the study of functional changes caused by disease processes is called **pathophysiology**. Different diseases typically produce similar signs and symptoms. For example, a person whose lips are paler than normal and who complains of a lack of energy and breathlessness might have (1) respiratory problems that prevent normal oxygen transfer to the blood (as in asthma or COPD), (2) cardiovascular problems that interfere with normal blood circulation to all parts of the body (heart failure), or (3) a reduced oxygen-carrying capacity of the blood (*anemia*). In such cases, doctors must ask questions and collect information to determine the source of the problem. The patient's history and physical exam may be enough for diagnosis in many cases, but laboratory testing and imaging studies such as x-rays are often needed.

A **diagnosis** is a decision about the nature of an illness. The diagnostic process is often a process of elimination, in which several potential causes are evaluated and the most likely one is selected. If tests indicate that, for example, anemia is responsible for the patient's symptoms, then the specific type of anemia must be identified before effective treatment can begin. After all, the treatment for anemia due to a dietary iron deficiency is very different from the treatment for anemia due to internal bleeding. You could not hope to identify the probable type of anemia unless you were already familiar with the physical and chemical structure of red blood cells and with their role in the transport of oxygen. This brings us to a key concept: *All diagnostic procedures presuppose an understanding of the normal structure and function of the human body.*

The Scientific Method

Your course in anatomy and physiology should do more than simply teach you the names and functions of different body parts; it should also provide you with a frame of reference that will enable you to understand new information, draw logical conclusions, and make intelligent decisions. A great deal of confusion and misinformation exists about just how medical science "works," and people make unwise and even dangerous decisions as a result. Nowhere is this more apparent than when a discussion drifts to health, nutrition, or cancer. Whether you are planning to work in a health-related profession or are just trying to make sound decisions about your own life, you will benefit from learning how to organize information, evaluate evidence, and draw logical conclusions.

Logical analysis, a process often called *critical thinking*, does not come naturally; it is too easy to become distracted or misled and then to make a hasty or incorrect decision. Critical thinking is a learned skill that follows rules designed to minimize the chances of error. Critical thinking is important in daily life, but it is absolutely vital in the sciences, especially the medical sciences. In applying critical thinking to scientific investigation, we follow what is called the **scientific method**, a standardized means of organizing and evaluating information to reach valid conclusions.

Forming a Hypothesis

Science involves a lot more than just the collection of information. You could spend the rest of your life carefully observing the world around you, but such a task wouldn't reveal very much, unless you could see some kind of pattern and come up with a **hypothesis**—an idea that explains your observations.

Hypotheses are ideas that may be correct or incorrect. To evaluate a hypothesis, you must have relevant data and a reliable method of analyzing the data. For example, you might propose the hypothesis that *radiation emitted by planet X confers immortality on a living being.* Could anyone prove you wrong? Not likely, particularly if you didn't specify the location of the planet or the type of radiation. Would anyone believe you? If you were a "leading authority" on something (anything), a few people probably would.

That's not as ridiculous as it might seem. For almost 1500 years, "everyone knew" that inhaled air is transported from the lungs through blood vessels to the heart. They "knew" this because the famous Roman physician Galen had said so. But as we now know, Galen and all who agreed with him were quite wrong. Why were Galen's statements about the lungs believed? Because Galen was famous and much of what he said was correct, *all* his statements were accepted as true. To avoid making this kind of error, you must always remember to evaluate the *hypothesis,* not the individual who proposed it!

In evaluating a hypothesis, we must examine it to see if it makes correct predictions about the real world. The steps in this process are diagrammed in ▶ **Figure 1**. A valid hypothesis has three characteristics: It is (1) testable, (2) unbiased, and (3) repeatable.

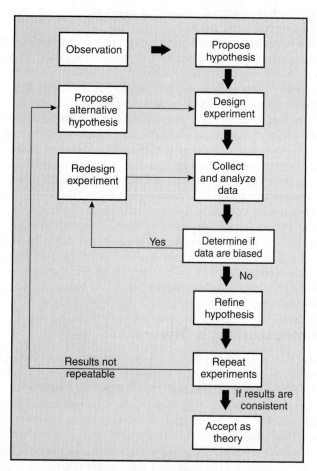

▶ **Figure 1 The Scientific Method.** The basic sequence of steps involved in the development and acceptance of a scientific theory.

A testable hypothesis is a hypothesis that can be studied by experimentation or data collection. Your assertion about planet X qualifies as a hypothesis, but it cannot be tested unless we find the planet and detect the radiation. An example of a testable hypothesis would be "left-handed airplane pilots have fewer crashes than do right-handed pilots." This hypothesis is testable because it makes a prediction about the world that can be checked—in this case, by collecting and analyzing data.

Avoiding Bias

Suppose, then, that you collected information about all the plane crashes in the world and discovered that 80 percent of all airplanes that crashed were flown by right-handed pilots. "Aha!" you might shout, "The hypothesis is correct!" The implications are obvious: Ban all right-handed airline pilots, eliminate four-fifths of all crashes, and sit back and wait for your prize from the Air Traffic Safety Association.

Unfortunately, you would be acting prematurely, because your data collection was biased. To test your hypothesis adequately, you need to know not only how many crashes involved right-handed or left-handed pilots, but also how many right-handed and left-handed pilots were flying. If 90 percent of the pilots were right handed, but they accounted for only 80 percent of the crashes, then left-handed pilots are the dangerous ones! Eliminating bias in this case is relatively easy, but health studies can have all kinds of complicating factors. Because 25 percent of us will probably develop cancer at some point in our lives, we will use cancer studies to exemplify the problems encountered.

Our first example of bias in action concerns cancer statistics, which indicate that cancer rates in the United States and abroad vary by region. For example, although the estimated age-adjusted yearly cancer death rate in the United States was 172.8 per 100,000 population in 2010, the rate in Wyoming was only 172.6 per 100,000, whereas the rate in the District of Columbia was 178.3 per 100,000. It would be very easy to assume that this difference is the direct result of rural versus urban living. But these data alone should not convince you that moving from the District of Columbia to Wyoming would lower your risk of developing cancer. To draw that conclusion, you would have to be sure that the observed rates were the direct result of just a single factor: the difference in physical location. As you will find in later sections, many factors promote the development of cancer. To exclude all possibilities other than geography, you would have to be certain that the populations were alike in all other respects. Here are a few possible sources of variation that could affect that conclusion:

- **Different population profiles.** Cancer rates vary between males and females, among racial groups, and among age groups. Therefore, we need to know how the populations of Wyoming and the District of Columbia differ in each of these respects. The age distribution of the population is so critical for cancer that most statistics are age adjusted, which means that the average age of a state's population is used to adjust the absolute number of cases to arrive at the numbers quoted above.

- **Different lifestyles.** Because tobacco and alcohol use are leading causes of lung and other cancers, we need to know how the populations differ in their patterns of smoking and drinking.

- **Different occupations.** Because chemicals used in the workplace are implicated in many cancers, we need to know how the populations of each region are employed and what occupational hazards they face.

- **Different mobilities.** The region in which a person dies may not be the region in which he or she developed cancer, so we need to know whether people with cancer in Wyoming stay in the state or go elsewhere for critical care and whether people with cancer travel to the District of Columbia to seek treatment.

- **Different health care and habits.** Because cancer death rates reflect differences in patterns of health care, we need to know whether residents of Wyoming pay more attention to preventive health care and have more regular checkups, whether their medical facilities are better, and whether they devote a larger proportion of their annual income to health services than do residents in the District of Columbia.

You can probably think of additional factors, but the point is that avoiding experimental bias can be quite difficult!

A second example of the problem of bias comes from the collection of "miracle cures" that continue to appear and disappear at regular intervals. Pyramid power, coffee enemas, crystals, magnetic-energy fields, and psychic healers come and go in the news. Wonder drugs are equally common, whether they are "secret formulas" or South American plant extracts discovered by colonists from other planets. The proponents of each new procedure or drug report glowing successes with patients who would otherwise have surely succumbed to the disease—and most of these remedies are said to have been suppressed or willfully ignored by the "medical establishment."

Even accepting that the claims aren't exaggerated, does the fact that 1, or 100, or even 1000 patients have been cured prove anything? No, because a list of successes doesn't mean much. To understand why, consider the questions you might pose to an instructor who announced on the first day of class that he or she had given 20 A's last semester. You would want to know how many students were in the class: only 20, or several hundred? You would also want to find out how the rest of the class performed—20 A's and 200 D's might be rather discouraging. You might want to see how the students were selected. If only students with A averages in other courses had enrolled, your opinion should change accordingly. Finally, you might check with the students and compare their grades with those given by other instructors who teach the same course.

With just a couple of modifications, the same questions could be asked about a potential cancer cure:

- How many patients were treated, how many were cured, and how many died?

- How were the patients selected? If selection depended on wealth, degree of illness, or previous exposure to other therapeutic techniques, then the experimental procedure was biased from the start.

- How many might have recovered regardless of the treatment? Even "terminal" cancers sometimes simply disappear for no apparent reason. Such occurrences are rare, but they do happen. Thus, any treatment, however bizarre, will in some cases appear to work. If the frequency of recovery is *lower* than that among other patient groups, the treatment might actually be harmful despite the reported "cures."

- How do the foregoing statistics compare with those of traditional therapies when both are subjected to the same unbiased tests?

The Need for Repeatability

Finally, let's examine the criterion of repeatability. It's not enough to develop a reasonable, testable hypothesis and collect unbiased data. Consider the hypothesis that every time a coin is tossed, it will come up heads. You could build a coin-tossing machine, turn it on, and find that in the first experiment of 10 tosses, the coin came up heads every time. Does this result prove the hypothesis?

No, despite the fact that it was an honest experiment and the data supported the hypothesis. The problem here is one of statistics, sample size, and luck. The odds that a coin will come up heads on any given toss are 50 percent, or 1 in 2—the same as the odds that it will come up tails. The odds that it will come up heads 10 times in a row are about 1 chance in 1×2^{10} (1 in 1024)—small, but certainly not inconceivable. If that coin is tossed 50 times, however, the chance of getting 50 heads drops to 1 in 1×2^{50} (less than 1 chance in a thousand trillion), a figure that most people would accept as vanishingly small. Proving that the hypothesis "a tossed coin always lands heads up" is false requires that the coin come up tails only once. So the truth could be revealed by running the experiment with more coin tosses or by letting other people set up identical experiments and toss their own coins.

For a hypothesis to be correct, anyone and everyone must get the same results when the experiment is performed. If the experiment isn't repeatable, you have to doubt the conclusion even when you have complete confidence in the abilities and integrity of the original investigator.

If a hypothesis satisfies all these criteria—it is testable, unbiased, and repeatable—it can be accepted as a scientific **theory**. The scientific use of this term differs from its use in general conversation. When people discuss "wild-eyed theories," they are usually referring to untested hypotheses. Hypotheses may be true or false, but by definition, theories describe real phenomena and make accurate predictions about the world. Examples of scientific theories include the theory of gravity and the theory of evolution. The "fact" of gravity is not in question, and the theory of gravity accounts for the available data. But this does not mean that theories cannot change over time. Newton's original theory of gravity, though used successfully for more than two centuries, was profoundly modified and extended by Albert Einstein. Similarly, the theory of evolution has been greatly elaborated since it was first proposed by Charles Darwin in the middle of the 19th century. No one theory can tell the whole story, and all theories are continuously being modified and improved as we learn more about our universe.

Homeostasis and Disease

The ability to maintain homeostasis depends on two interacting factors: the status of the physiological systems involved and the nature of the stress imposed. Homeostasis is a balancing act, and each of us is like a tightrope walker. Homeostatic systems must adapt to sudden or gradual changes in our environment, the arrival of pathogens, injuries, and many other factors, just as a tightrope walker must make allowances for gusts of wind, frayed segments of the rope, and thrown popcorn.

The ability to maintain homeostasis varies with one's age, general health, and genetic makeup. The geriatric patient or pregnant patient, or a young infant with the flu is in much greater danger than an otherwise healthy young adult with the same viral infection. If homeostatic mechanisms cannot cope with a particular stress, physiological values will drift outside the normal range. This change can ultimately affect all other systems, with potentially fatal results. After all, a person unable to maintain balance will eventually fall off the tightrope.

Consider a person who is exercising heavily and has a heart rate of 180 beats per minute for several minutes. That would be acceptable in a young, healthy adult, but such a heart rate can be disastrous for an older person with cardiovascular and respiratory problems. If it is allowed to continue, cardiac muscle tissue will be damaged, leading to decreased pumping efficiency and a dangerous drop in blood pressure.

These changes represent a serious threat to homeostasis. Other systems will soon become involved. For example, the drop in blood pressure will suppress kidney function, and waste products will begin accumulating in the blood. The reduced blood flow in other tissues will result in a generalized *hypoxia*, or low tissue oxygen level. Cells throughout the body then begin to suffer from oxygen starvation. The person is now in serious trouble: Unless steps are taken to correct the situation, his or her survival will be threatened.

A failure to maintain homeostatic conditions constitutes **disease**. The disease process may initially affect a specific tissue, an organ, or an organ system, but it will ultimately lead to changes in the function or structure of cells throughout the body. A disease can often be overcome through appropriate, automatic adjustments in physiological systems. In a case of the flu, the disease develops because the immune system cannot defeat the flu virus before that virus has infected cells of the respiratory passageways. For most people, the physiological adjustments made in response to the viral invasion will lead to the elimination of the virus and the restoration of homeostasis. Some diseases, by contrast, cannot easily be overcome. In the case of the person with acute cardiovascular problems, some outside intervention may be necessary to restore homeostasis and prevent fatal complications.

Diseases may result from the following:

- **Pathogens that invade the body.** Examples are the viruses that cause such diseases as flu, mumps, and measles; the bacteria responsible for diseases like anthrax, Lyme disease, and tuberculosis; and the parasites (including protozoans, fungi, and worms) that produce such conditions as malaria, candidiasis ("yeast" infections), and trichinosis. The invasion process is called **infection**. Some parasites do not enter the body, but instead attach themselves to its surface. This process is called **infestation**.

- **Inherited genetic conditions that disrupt normal physiological mechanisms.** These conditions make normal homeostatic control difficult or impossible. Examples include the lysosomal storage diseases, cystic fibrosis, and sickle cell anemia.

- **The loss of normal regulatory control mechanisms.** For example, cancer involves the rapid, unregulated multiplication of abnormal cells. Many cancers have been linked to abnormalities in genes responsible for controlling rates of cell division. A variety of other diseases, called autoimmune disorders, result when regulatory mechanisms of the immune system fail and healthy tissues are attacked.

- **Degenerative changes in vital physiological systems.** Many systems become less adaptable and less efficient as part of the aging process. For example, we experience significant reductions in bone mass, respiratory capacity, cardiac efficiency, and kidney filtration as we age. If the elderly are exposed to stresses that their weakened systems cannot tolerate, disease results.

- **Trauma, toxins, or other environmental hazards.** Accidents can damage organs, impairing their function. Toxins consumed in the diet or absorbed through the skin or lungs can disrupt normal metabolic activities.

- **Nutritional factors.** Diseases can result from diets that are inadequate in proteins, essential amino acids, essential fatty acids, vitamins, minerals, or water. *Kwashiorkor*, a disease caused by protein deficiency, and *scurvy*, caused by vitamin C deficiency, are two examples. Excessive consumption of high-calorie foods, fats, salt, or fat-soluble vitamins can cause or contribute to diseases such as hypertension, coronary artery disease, obesity, and diabetes mellitus.

The Diagnosis of Disease

A person experiencing serious symptoms usually seeks professional help and thereby becomes a patient. The clinician, whether a nurse, a physician, an emergency medical technician, or other health care provider must determine the need for medical care on the basis of observation and assessment of the patient. This is the process of diagnosis: the identification of a pathological process by its characteristic symptoms and signs.

Symptoms and Signs

An accurate diagnosis, or the identification of the disease, is accomplished through the observation and evaluation of symptoms and signs.

A **symptom** is the patient's perception of a change in normal body function. Examples of symptoms include nausea, fatigue, and pain. Symptoms are difficult to measure, and a clinician must ask appropriate questions, such as the following:

"When did you first notice this symptom?"

"Have you ever had this problem before?"

"What does it feel like?"

"Does it come and go, or does it always feel the same?"

"Does anything make it feel better or worse?"

The answers provide information about the location, duration, sensations, recurrence, and triggering mechanisms of the symptoms important to the patient.

Pain, an important symptom of many illnesses, is often an indication of tissue injury. The flowchart in ❱ **Figure 2** indicates the types of pain and introduces important related terminology. We shall consider the control of pain in related sections of the *A&P Applications Manual*.

A **sign** is a physical manifestation of a disease. Unlike symptoms, signs can be measured and observed through sight, hearing, or touch. The yellow color of the skin caused by liver dysfunction and a detectable breast lump are signs of disease. A sign that results from a change in the structure of tissue or cells

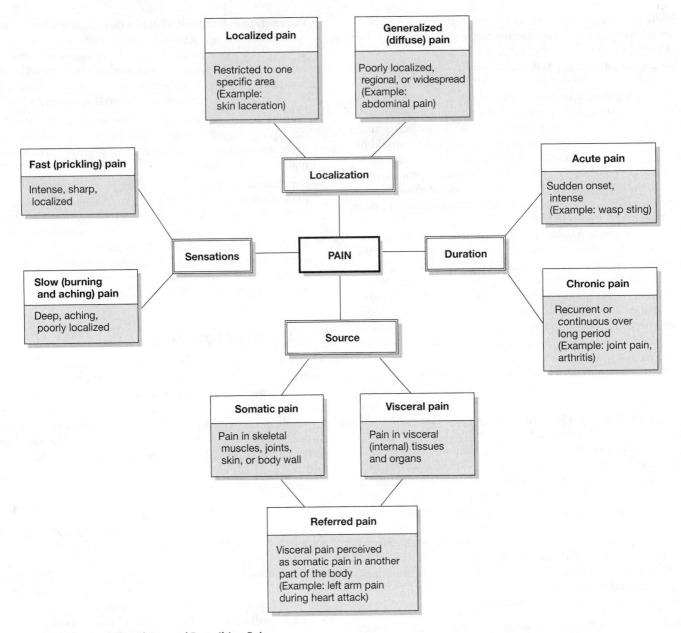

> Figure 2 **Methods of Classifying and Describing Pain.**

is called a **lesion**. We shall consider lesions of the skin in detail in a later section dealing with the integumentary system.

The Steps in Diagnosis

Diagnosis is a lot like assembling a jigsaw puzzle. The more pieces (clues) available, the more complete the picture will be. The process of diagnosis is one of deduction and follows an orderly sequence of steps:

1. **Obtain the patient's medical history.** The medical history is a concise summary of past medical disorders, general factors that may affect the functioning of body systems, and the health of the patient's family. This information provides a framework for considering the individual's current problem. Probably over half of all diagnoses are made from the history, with tests being used chiefly for confirmation.

When taking a history, the examiner gains information about the patient's concerns by asking specific questions and using good observational and listening skills. Starting with "open-ended" questions such as "What problem brings you here today?" or "Tell me about what is bothering you" and then following up with more directed questions to focus on the issues raised. (Physicians in training are often told, "Listen to the patient; she (or he) is trying to tell you what's wrong.") Physical assessment also begins here—this is the time for unspoken questions such as "Is this person moving, speaking, and thinking normally?" The answers will later be integrated with the results of more precise observations.

Other components of the medical history may include the following:

• **Chief complaint.** The patient is asked to specify the primary problem that requires attention. This is recorded as the *chief complaint.* An example would be the entry "Patient complains of pain in the abdomen."

- **History of present illness.** Which areas of the body are affected? What kinds of functional problems have developed? When did the patient first notice the symptoms? The duration and pattern of the disease process is an important factor. For example, an infection may have been present for months, only gradually increasing in severity. This would be called a *chronic infection.* A disease process may have been under way for some time before the person recognizes that a problem exists. Over the initial period, the individual experiences *subclinical symptoms*—symptoms so mild that they are overlooked or ignored. Chronic infections commonly have different causes and treatments than do *acute infections*, which produce sudden, intense symptoms.

- **Review of systems.** The patient is asked questions that focus on the general status of each body system. This process may detect related problems or causative factors. For example, a chief complaint of a headache may be *related* to visual problems (stars, spots, blurs, or blanks seen in the field of vision) or *caused* by visual problems (eyeglasses poorly fitted or made from the wrong prescription).

2. **Perform a physical examination.** The physical examination is a basic and vital part of the diagnostic process. Common techniques used in physical examination are *inspection* (viewing), *palpation* (touching), *percussion* (tapping and listening), and *auscultation* (listening):

- **Inspection** is careful observation. A general inspection involves examining body proportions, posture, and patterns of movements. Local inspection is the examination of sites or regions of suspected disease. Of the four components of the physical exam, inspection is often the most important, because it provides the largest amount of useful information. Many diagnostic conclusions can be made on the basis of inspection alone; most skin conditions, for example, are identified in this way. A number of endocrine problems and inherited metabolic disorders can produce changes in body proportions. Many neurological disorders affect speech and movement in distinctive ways.

- **Palpation** is the clinician's use of the hands and fingers to feel the patient's body. This procedure provides information about skin texture and temperature, the presence and texture of abnormal tissue masses, the pattern of the pulse, and the location of tender spots. Once again, the procedure relies on an understanding of normal anatomy. In one spot, a small, soft, lumpy mass is a salivary gland; in another location, it could be a tumor. A tender spot is important in diagnosis only if the observer knows what organs lie beneath it.

- **Percussion** is tapping with the fingers or hand to obtain information about the densities of underlying tissues. For example, when tapped, the chest normally produces a hollow sound, because the lungs are filled with air. That sound changes in pneumonia, when the lungs contain large amounts of fluid. To get the clearest chest percussions, the fingers must be placed in the right spots.

- **Auscultation** (aws-kul-TĀ-shun; *auscultare*, to listen) is listening to body sounds, typically with a stethoscope.

This technique is particularly useful for checking the condition of the lungs during breathing. The wheezing sound heard in people with asthma is caused by a constriction of the airways, and pneumonia produces a gurgling sound, indicating that fluid has accumulated in the lungs. Auscultation is also important in diagnosing heart conditions. Many cardiac problems affect the sound of the heartbeat or produce abnormal swirling sounds during blood flow.

Every examination also includes measurements of certain vital body functions, such as the body temperature, weight, blood pressure, respiratory rate, and heart (pulse) rate. The results, called **vital signs**, are recorded on the patient's chart. Vital signs can vary over a normal range that differs according to the age, gender, and general condition of the individual. Table 1 indicates the representative ranges of vital signs in infants, children, and adults.

3. **If necessary, perform diagnostic procedures.** The medical history and physical examination may not provide enough information to permit a precise diagnosis. Diagnostic procedures can then be used to focus on abnormalities revealed by the history and physical examination. For example, if the chief complaint is knee pain after a fall, and the examination reveals swelling and localized, acute pain on palpation, the **preliminary diagnosis** may be a torn cartilage. An x-ray, MRI scan, or both may be performed to determine more precisely the extent of the injury and to ensure that there are no other problems, such as broken bones or torn ligaments. With the information the diagnostic procedure provides, the **diagnosis** can be made with reasonable confidence. Diagnostic procedures extend, rather than replace, the physical examination.

Two general categories of diagnostic procedures are performed:

1. **Tests performed on the individual.** Information about representative tests of this type is summarized in Table 2. These procedures allow the clinician to:

- Visualize internal structures (endoscopy; x-rays; scanning procedures such as CT, MRI, and radionucleotide scans; ultrasonography; mammography; see ▶ Figure 3)

- Monitor physiological processes (EEG, ECG, PET, RAIU, pulmonary function tests)

- Assess the patient's homeostatic responses (stress testing, skin tests)

Table 1	Normal Range of Values for Resting Individuals by Age Group		
Vital Sign (at rest)	**Infant (3 months)**	**Child (10 years)**	**Adult**
Blood pressure (mm Hg)	90/50	90–125/60	95/60 to 120/80
Respiratory rate (per minute)	30–50	18–30	8–18
Pulse rate (per minute)	70–170	70–110	50–95

Table 2	Representative Diagnostic Tests, Their Principles, and Their Uses	
Procedure	**Principle**	**Examples of Uses**
Endoscopy	Insertion of fiber-optic tubing into a body opening or through a small incision (laparoscopy and arthroscopy); permits visualization of a body cavity or the interior of an organ; allows direct visualization and biopsy of structures and detection of abnormalities of surrounding soft tissue	*Bronchoscopy:* bronchi and lungs *Laparoscopy:* abdominopelvic organs *Cystoscopy:* urinary bladder *Esophagoscopy:* esophagus *Gastroscopy:* stomach *Colonoscopy:* colon *Arthroscopy:* joint cavity
Standard x-rays	A beam of x-rays passes through the body and then strikes a photographic plate; radiodense tissues block x-ray penetration, leaving unexposed (white) areas on the developed film (Figure 3)	Limb bones: to detect fracture, tumor, growth patterns Chest: to detect tumors, pneumonia, atelectasis, tuberculosis Skull: to detect fractures, sinusitis, metastatic tumors Mammogram: x-rays of each breast taken at different angles for early detection of breast cancer and other masses, such as cysts
Contrast x-rays	X-rays taken after infusion or ingestion of radiodense solutions (Figure 3)	Barium swallow (upper GI): series of x-rays after the ingestion of barium, to detect abnormalities of esophagus, stomach, and duodenum Barium enema: series of x-rays after barium enema, to detect abnormalities of colon IV pyelography: series of x-rays of abdomen after intravenous injection of radiopaque dye filtered by kidneys; reveals anatomical abnormalities of kidneys, ureters, and bladder
Digital subtraction angiography	Produces strikingly clear images of blood vessel distribution by computer analysis of images taken before and after dye infusion (Figure 5)	Analysis of blood flow to the heart, kidneys, and brain to detect blockages and restricted circulation
Computerized tomography (CT or CAT scan)	Produces cross-sectional images of body area viewed; together, all sections can produce a three-dimensional image for detailed examination (Figure 4)	CT scans of the head, abdominal region (liver, pancreas, kidney), chest, and spine, to assess organ size and position, to determine progression of a disease, and to detect abnormal masses
Spiral CT scans	Produce three-dimensional images by computer reconstruction of CT data (Figure 5)	Helps in diagnosis of appendicitis, pulmonary emboli, and early lung cancer
Nuclear scans	Radioisotope ingested, inhaled, or injected into the body becomes concentrated in the organ to be viewed; gamma radiation camera records image on film. Area should appear uniformly shaded; dark or light areas suggest hyperactivity or hypoactivity of the organ	Bone scan: to detect tumors, infections, and degenerative diseases Scans of the brain, heart, thyroid, liver, lung, spleen, and kidney, to assess organ function and the extent of many diseases
Radioactive iodine uptake test (RAIU)	Radioactive labeled iodine compound is given orally; thyroid scans are taken to determine percentage uptake of radioiodine by thyroid gland	Aids in the determination of hyperthyroidism and hypothyroidism and in detection and description of thyroid nodules
Positron emission tomography (PET)	Radioisotopes are given by injection or inhalation; gamma detectors absorb energy and transmit information to computers to generate cross-sectional images (Figure 5)	Used to measure metabolic activity of heart and brain and to analyze blood flow through organs. Helpful in staging some cancers and monitoring response to treatment
Magnetic resonance imaging (MRI)	A magnetic field is produced to align hydrogen protons and is then exposed to radio waves that cause the aligned atoms to absorb energy. The energy is later emitted and captured to produce an image (Figure 4)	Gives excellent contrast of normal and abnormal tissue; reveals extent of tumors, demyelination and other brain and spinal cord abnormalities, obstructions, or aneurysms in arteries, and ligaments and cartilages at joints. No x-ray exposure
Ultrasonography	A transducer contacting the skin or other body surface sends out sound waves and then picks up the echoes (Figure 4)	Avoids x-ray exposure, used to view soft tissues not shielded by bone throughout the body. Used in obstetrics to detect ectopic pregnancy, determine size of fetus, and check fetal rate of growth; abdominal ultrasound detects gallstones, visceral abnormalities, and measures kidneys

(continued)

Procedure	Principle	Examples of Uses
Echocardiography	Ultrasonography of the heart (p. 136)	Used to assess the structure and function of the heart and heart valves
Electrocardiography (ECG)	Graphed record of the electrical activity of the heart, using electrodes on the skin surface	Useful in detection of arrhythmias, such as premature ventricular contractions (PVCs) and fibrillation, and to assess damage after a heart attack
Electroencephalography (EEG)	Graphed record of electrical activity in the brain through the use of electrodes on the surface of the scalp	Analysis of brain wave frequency and amplitude aids in the diagnosis of tumors and seizure disorders
Electromyography (EMG)	Graphed record of electrical activity resulting from skeletal muscle contraction, using electrodes inserted into the muscles	Determination of neural or muscular origin of muscle disorder; aids in the diagnosis of muscular dystrophy, pressure on spinal nerves, and peripheral neuropathies
Pulmonary function tests	Measurement of lung volumes and capacities by a spirometer	Aids in the differentiation between obstructive and restrictive lung diseases; used to test for and monitor asthma
Cytology	Removal of cells for microscopic laboratory analysis	Detects precancerous cells or infections; most often used to assess mucosal cells of cervix (Pap smear)
Stress testing	Monitoring of blood pressure, pulse rate, ECG, and sometimes echocardiography during exercise; may include intravenous injection of radioisotopes to measure cardiac muscle perfusion	May detect and determine the extent of coronary artery disease, which may not be apparent while the individual is at rest
Skin tests	Injection of a substance under the skin, or placement of a substance on the skin surface, to determine the response of the immune system	Tuberculin test: injection of tuberculin protein under skin. Allergen test: injection of allergen or application of a patch containing allergen

Table 2 Representative Diagnostic Tests, Their Principles, and Their Uses (Continued)

2. **Tests performed in a clinical laboratory on tissue samples, body fluids, or other materials collected from the patient.** Table 3 includes details about a representative sample of these tests.

Many of the diagnostic procedures and disorders noted in these tables will be unfamiliar to you now. The main purpose here is to give you an overview; you can refer to the tables as needed throughout the course.

The Purpose of Diagnosis

Two hundred years ago, a physician would arrive at a diagnosis and consider the job virtually done. Once the diagnosis was made and the physician gave a prognosis (likely outcome), the patient and family would know what to expect. In effect, the physician was more of an oracle than a healer. Wounds could be closed, broken bones pulled straight, boils lanced and limbs amputated, but few effective treatment options were available. Less obvious diagnoses than trauma often reflected the culture and beliefs of the era. Curses and bewitching vied

with "unbalanced humours" as explanations of disease. Therapy was often a combination of bleeding (often performed by barbers rather than by surgeons), dietary changes, and herbal medicines (often laxatives). Strong laxatives might have helped in cases of intestinal parasites, but the combination of bleeding and laxatives was potentially dangerous because it reduced both blood volume and blood pressure.

Fortunately, a vast array of treatment options guided by *evidence-based medicine* involving rational, accurate diagnosis are available today. A modern physician addressing a new problem presented by a patient follows the *SOAP* protocol:

S *is for subjective.* The clinician obtains subjective information from the patient and the medical history.

O *is for objective.* The clinician performs the physical examination and obtains objective information about the physical condition of the patient. The examination may include the use of diagnostic procedures.

A *is for assessment.* The clinician arrives at a diagnosis and, if necessary, reviews the literature on the condition. Sometimes the data obtained leads to new questions and

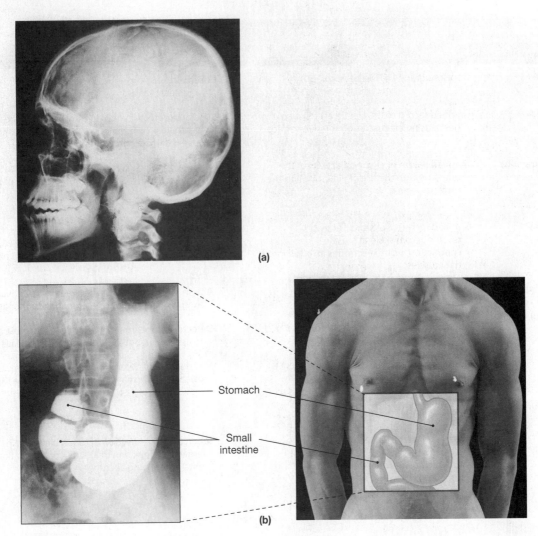

Figure 3 X-rays. (a) An x-ray of the skull, taken from the left side. **X-rays** are a form of high-energy radiation that can penetrate tissues. In the most familiar procedure, a beam of x-rays travels through the body and strikes a photographic plate. Not all of the projected x-rays reach the film; some are absorbed or deflected as they pass through the body. The resistance to x-ray penetration is called **radiodensity**. Radiodensity increases in the following sequence: air, fat, liver, blood, muscle, bone. The usual result is an image with radiodense tissues, such as bone, appearing in white, and less dense tissues in shades of gray to black. **(b)** A **barium contrast x-ray** of a portion of the upper digestive tract. Such an x-ray is produced after a radiodense material is introduced into the body. It is used to provide sharp outlines and contrast and to check the distribution of fluids or the movements of internal organs. In this instance, the patient swallowed a solution of barium, an element that is very dense. The contours of the stomach and intestinal linings are clearly indicated by the white of the barium solution.

additional tests. A preliminary conclusion as to the **prognosis** (probable outcome) is made.

P is for plan. A treatment plan is designed. This can be very simple (lose weight, exercise, and take two aspirin) or highly complex (referral for radiation, chemotherapy, or surgery). If the treatment is complex, one or more treatment options are usually reviewed with the patient and, in many cases, the patient's family. Treatment begins only after informed decisions are made.

The SOAP protocol is both simple to remember and remarkably effective.

The primary goal of an introductory anatomy and physiology course is to provide you with the foundation for other, more specialized courses. In the unit of this manual that deals with body systems, you will be introduced to clinical conditions that demonstrate the relationships between normal and pathological anatomy and physiology. The goal is to acquaint you with the mechanics of the process involved, and show the clinical relevance of even the most basic information taught in Anatomy and Physiology. This knowledge will not enable you to make accurate clinical diagnoses; situations in the real world are much more complicated and variable than the examples provided here. Making an accurate clinical diagnosis is generally a complex process

Table 3   Laboratory Tests Performed on Samples Taken from the Body

Blood Tests:

Serum, plasma, or whole-blood samples can be evaluated. Either venous or arterial blood is taken, depending on the blood constituent or chemical being monitored.

Laboratory Test	Significance	Notes
Complete blood count (CBC): RBC count Hemoglobin (Hb. Hgb) Hematocrit (Hct) WBC and Platelet counts	Data from this test series is usually part of a health checkup Information on the number and percentage of red blood cells and their contents; changes may indicate the presence of disease, hemorrhaging, malnutrition, or other problems Changes may indicate blood loss, infections, or other problems	For more information, see Table 27c, pp. 123–124.
RBC indices (mean corpuscular hemoglobin, MCH, and mean corpuscular hemoglobin concentration, MCHC, among others)	Provide information about the status of hemoglobin production and red blood cell maturation	See Table 27c, pp. 123–124.
WBC count and differential WBC count	The total white blood cell count and the proportions of various cell types reflect the state of the body's immune system and the response to infection. An increased white blood cell count could indicate the presence of infection	See Table 27c, pp. 123–124.
Hemostasis tests: Platelet count Bleeding time Clotting factors assay Plasma fibrinogen Plasma prothrombin time (PT) Plasminogen	A decreased number of platelets could result in uncontrolled bleeding. Other constituents, such as fibrinogen, clotting factors, and prothrombin, also contribute to the clotting process, and these can be tested for separately Frequently used to monitor therapeutic use of anticoagulants	See Table 27c, pp. 123–124.
Electrolytes: Sodium Potassium Chloride Bicarbonate	Sodium, potassium, and chloride levels are levels of electrolytes that function in nerve transmission, skeletal muscle contraction, and cardiac rhythm. Affected by kidney function Abnormal levels of bicarbonate indicate problems with acid–base balance	See Tables 27b, 37, 39, pp. 122, 177–178, 199–200.
Iron	Decreased levels cause iron deficiency anemia; increased levels may cause liver and heart damage	
Arterial blood gases and pH: pH P_{CO_2} P_{O_2}	Respiratory acidosis and alkalosis can be monitored with these values. Decreased oxygen levels occur in respiratory and cardiovascular system dysfunction	See Table 34, pp. 164–165.
Hemoglobin electrophoresis: Hemoglobin A Hemoglobin F Hemoglobin S	Electrophoresis separates the types of hemoglobin for quantitative measurement. Abnormal types of hemoglobin occur in sickle cell anemias and thalassemias	See Table 27c, pp. 123–124.
ABO and Rh typing	Blood typing is critical for correct matching of blood types prior to transfusion. Rh typing during pregnancy is important to determine risk of fetal–maternal Rh incompatibility	See Table 43, p. 229.
Cholesterol	Elevated cholesterol levels increase the risk of atherosclerosis and coronary artery disease	See Table 27b, p. 122.
Lipoproteins: LDL HDL	Electrophoresis is used to separate the LDL fraction of total cholesterol to determine the HDL and LDL levels. High LDL and low HDL are risk factors for coronary artery disease	See Table 27b, p. 122.

(continued)

Table 3 Laboratory Tests Performed on Samples Taken from the Body (*Continued*)

Laboratory Test	Significance	Notes
Enzymes:	Abnormal enzyme levels in the blood are generally due to cellular damage. All enzymes are proteins	See Table 27b, p. 122.
Creatine phosphokinase (CPK and CK)	Present in skeletal and cardiac muscle cells	
Isoenzymes (CK-MM, CK-MB, CK-BB)	CK-MM is useful in the diagnosis of muscle disease; CK-MB can be used in the diagnosis of heart attacks	
Aspartate aminotransferase (AST)	AST levels are important to assess liver damage	
Lactate dehydrogenase (LDH)	Different isoenzymes of LDH can be useful in the detection of heart damage, liver problems, and pulmonary dysfunction	
Troponin	Cardiac-specific protein more sensitive to heart attacks than CK	
Rheumatoid factor	Measures presence of antibodies characteristic of rheumatoid arthritis and (less often) other autoimmune diseases	See Table 16, p. 53.
Hormones	Vary with age and gender; abnormally increased or decreased levels reflect endocrine system disorders	See Table 26, pp. 108–109.
Blood urea nitrogen (BUN) also called Urea	Assesses kidney function, presence of dehydration	See Table 39, pp. 199–200.
Creatinine	Assess kidney function	
Immunoglobulin electrophoresis (IgA, IgG, IgD, IgE, IgM)	Monitors infections and allergic response	See Table 29, p. 147.
Alcohol	Determines level of intoxication	
Human chorionic gonadotropin (hCG)	Detects pregnancy	See Table 43, p. 229.
Phenylalanine	Detects phenylketonuria (PKU), a genetic disorder of amino acid metabolism	
Alpha fetoprotein	Identifies probability of fetal defects or presence of twins; elevated levels with liver tumors	
Glucose tolerance test	Detects hyperglycemic response to glucose ingestion, used to diagnose diabetes mellitus	See Table 26, pp. 108–109.
Blood culture	The presence of intravenous bacterial pathogens occurs in septicemia, pneumonia, and other infectious disorders	See Tables 27c, 29, 37, pp. 123–124, 147, 177–178.
Urine Tests: A single urine sample may be tested, or urine may be collected over a period of time (usually from 2 to 24 hours) and tested. A routine urinalysis aids in the detection of kidney dysfunction, as well as metabolic imbalances and other disorders. The presence of abnormal cellular constituents in urine indicates urinary system disorder, including infection, inflammation, or the existence of a tumor.		
Creatine clearance	Abnormal values indicate reduced kidney function	
Urine electrolytes: Sodium potassium	Abnormal levels reflect fluid or electrolyte imbalances and the effects of hormones on kidney function	See Table 39, pp. 199–200.
Uric acid	Increased levels occur with gout and some kidney disorders	See Table 39, pp. 199–200.
Urine culture	Detects pathogens present in urinary tract infections	

(*continued*)

Table 3	Laboratory Tests Performed on Samples Taken from the Body (*Continued*)	
Laboratory Test	**Significance**	**Notes**
Other Laboratory Tests: Additional laboratory tests can be used to monitor other body fluids, excretory products, or tissues. Here are several examples:		
Cerebrospinal fluid (CSF)	Tested for sugar and protein content, and the presence of antibodies, pathogens, or blood cells	See Table 19, p. 77.
Stool sample	Culturing and microscopic examination of sample to identify pathogenic bacteria or parasites. DNA analysis may detect colon tumors	See Table 37, pp. 177–178.
Semen analysis	Useful in diagnosis of male infertility or in assessing success of vasectomy	See Table 42, pp. 214–215.
Tissue biopsy	The removal of tissue for microscopic examination	See Table 17, p. 65.

that demands a far greater level of experience and training than this course can provide.

For similar reasons, we will not discuss detailed treatment plans; the treatment of serious diseases requires current and specialized training and competence in advanced biochemistry, pharmacology, microbiology, pathology, and other clinical disciplines. However, many of the discussions in later sections include information about the use of specific drugs and other therapeutic procedures in the treatment of disease. These are representative examples intended to show potential treatment strategies, rather than to endorse specific protocols and therapies.

▶ Sectional Anatomy and Clinical Technology

Radiological procedures are used to provide detailed information about internal systems in a living individual. They include (1) scanning techniques using beams of radiation, called x-rays (discovered in 1895), to create a photographic or computer-generated image of internal structures, and (2) methods that involve the administration of radioactive materials. Over the last decade, other imaging techniques such as MRI scans and sophisticated ultrasounds have reduced the reliance on radiation for diagnostic imaging. Physicians who specialize in the performance of these imaging procedures and the analysis of the resulting images are still called *radiologists*.

▶ Figures 3 through 5 compare the views provided by several techniques used. The figures include images produced using *x-rays, computerized tomography (CT) scans, magnetic resonance imaging (MRI) scans, ultrasound procedures, spiral-CT scans, digital subtraction angiography (DSA) techniques,* and *positron emission tomography (PET) scans.* These clinical technologies and more specialized MRI procedures are described and included as figures later in the text.

Whenever you see anatomical diagrams or clinical procedures that present cross-sectional views of the body, remember that each section is oriented as though the observer is standing at the feet of a supine subject (lying face up) and looking toward the head. The section is an inferior view, with anterior at the top and posterior at the bottom, and structures on the left side of the body are seen on the right side of the image.

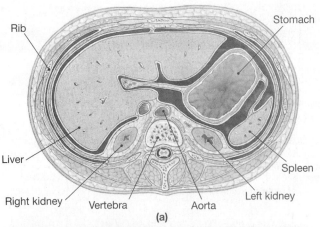

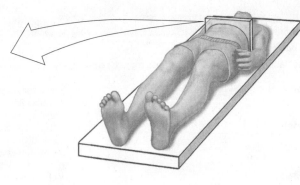

(a)

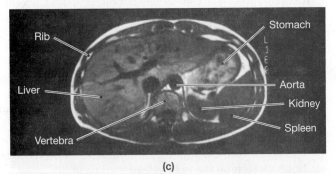

(b)

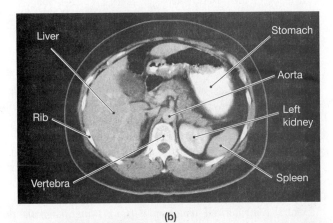

(c)

▶ **Figure 4** **Common Scanning Techniques.**

(a) Drawings of the structures and relative position and orientation of the scans shown in parts (b)–(d).

(b) A CT scan of the abdomen. **C**omputerized **t**omography (**CT**), formerly called **c**omputerized **a**xial **t**omography (**CAT**), uses computers to reconstruct sectional views. A single x-ray source rotates around the body, and the x-ray beam strikes a sensor monitored by the computer. The source completes one revolution around the body every few seconds; it then moves a short distance and repeats the process. The result is usually displayed as a sectional view in black and white, but it can be colorized for visual effect. CT scans show three-dimensional relationships and soft-tissue structure more clearly than do standard x-rays.

(c) An MRI scan of the abdomen. **M**agnetic **r**esonance **i**maging (**MRI**) surrounds part or all of the body with a magnetic field from 3000 to 20,000 times as strong as that of Earth. The MRI field affects protons within atomic nuclei throughout the body. The protons line up along the magnetic lines of force like compass needles in Earth's magnetic field. When struck by a radio wave, a proton absorbs energy. When the wave pulse ends, that energy is released and recorded. Each tissue differs in its response, and very detailed images are obtained by computer analysis. There are no known adverse affects from the MRI scan itself, but metal objects within the body may move in the magnetic field and create problems.

(d) An ultrasound scan of a portion of the abdomen. In **ultrasound** procedures, a small transmitter contacting the skin broadcasts a brief, narrow burst of high-frequency sound and then picks up the echoes. A **sonogram**, or ultrasound picture, can be assembled from the pattern of echoes produced when the sound waves are reflected by internal structures. These images lack the clarity of those produced by other procedures, but no adverse effects have been reported, and fetal development can be monitored without any known risk of birth defects. Special methods of transmission and processing permit analysis of the beating heart, without the complications that can accompany injections of a dye. Note the differences in detail among this image, the CT scan, and the MRI image. Improved technology is reducing this disparity.

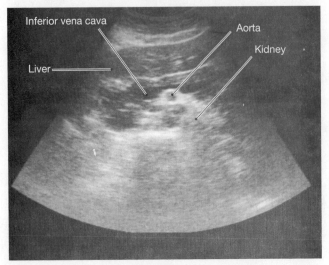

(d)

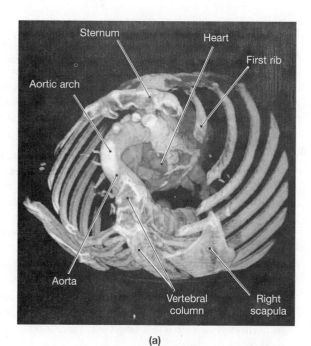

(a)

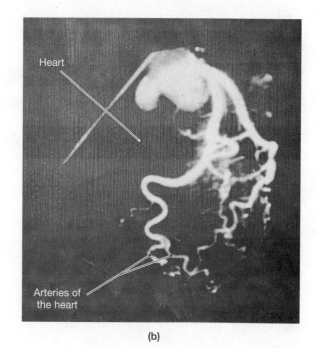

(b)

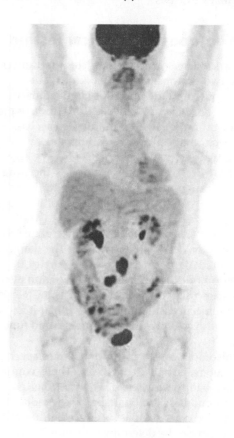

(c)

▶ **Figure 5 Special Scanning Methods. (a)** A *spiral-CT scan* of the chest. Such an image is created by special processing of CT data to permit rapid three-dimensional visualization of internal organs. Spiral-CT scans are becoming increasingly important in clinical settings. **(b)** *Digital subtraction angiography (DSA)* is used to monitor blood flow through specific organs, such as the brain, heart, lungs, or kidneys. X-rays are taken before and after a radiopaque dye is administered into the bloodstream, and a computer "subtracts" details common to both images. The result is a high-contrast image showing the distribution of the dye. **(c)** **P**ositron-**e**mission **t**omography (**PET**) scans rely on the administration of radioactive isotopes that are later detected by gamma-ray detectors and interpreted by computers. A PET scan can reveal the size, shape, position, and importantly, some function (physiology) of organs. MRI and CT scans show anatomy and in some cases blood flow.

The Chemical Level of Organization

▶ Medical Use of Radioisotopes

A **radioisotope**, or radioactive isotope, is an isotope whose nucleus is unstable; that is, the nucleus spontaneously decays, or emits subatomic particles. Many recent technological advances in medicine have involved the use of radioisotopes for the diagnosis and treatment of disease. We will focus on two examples:

1. **The use of radioactive tracers in clinical testing, particularly through the creation of diagnostic images.** Radioisotopes can be attached to organic or inorganic molecules and injected into the body. There, the labeled molecules emit radiation that allows clinicians to monitor their physiological distribution and utilization. The radiation can often be used to create images that provide anatomical information about tissue structure, tumors, blocked or weakened blood vessels, and other abnormalities in the body.

2. **The use of radiopharmaceuticals to destroy abnormal cells and tissues.** If a suitable radioactive compound can be accurately delivered to a target site, radiation can be used to treat many diseases.

Radioisotopes and Clinical Testing

Alpha particles are subatomic particles that consist of a helium nucleus: two protons and two neutrons. These particles generally are emitted by the nuclei of large radioactive atoms, such as uranium. *Beta particles* are electrons, more typically released by radioisotopes of lighter atoms. *Gamma rays* are very-high-energy electromagnetic waves comparable to the x-rays used in clinical diagnosis.

The **half-life** of any radioisotope is the time required for half of a given amount of the isotope to decay. The half-lives of radioisotopes range from fractions of a second to billions of years.

Gamma rays, beta particles, and alpha particles—like x-rays—can damage or destroy living tissues. The danger posed by exposure to radiation varies with the nature of the emission and the duration of exposure. However, radiation also has a variety of beneficial uses in medical research and clinical diagnosis. Weakly radioactive isotopes with short half-lives provide a noninvasive means of checking the structure and functional state of an organ.

Radioisotopes can be incorporated into specific compounds that the body normally processes. These compounds, called *tracers,* are said to be *labeled:* When introduced into the body, labeled compounds can be traced by the radiation they release. After a labeled compound is swallowed, its uptake, distribution, and excretion can be determined by monitoring the radioactivity of samples taken from the digestive tract, body fluids, and waste products, respectively. For example, compounds labeled with radioisotopes of cobalt are used to measure the intestinal absorption of vitamin B_{12}. Normally, cobalt-58, a radioisotope with a half-life of 71 days, is used.

Radioisotopes can also be injected into the blood or other body fluids to provide information about circulatory anatomy and the anatomy and function of specific target organs. In **nuclear imaging**, the radiation emitted by injected radioisotopes creates an image on a special detector. Such a procedure is used to identify regions where particular radioactive materials are concentrated or to check the flow of various substances through vital organs. Radioisotopes can produce pictures of specific organs, such as the liver, spleen, thyroid, or bone, where different labeled compounds are preferentially removed from the bloodstream.

The thyroid gland sits below the larynx (voice box) on the anterior portion of the neck (▶Figure 6a). A normal thyroid gland absorbs iodine, which is then used to produce thyroid hormones. As a result, the thyroid gland will actively absorb and concentrate radioactive iodine. The **thyroid scan** in ▶Figure 6b was taken following the injection of iodine-131, a radioisotope with an 8-day half-life. This procedure, called a thyroid *radioactive iodine uptake measurement*, or **RAIU**, can provide information

about (1) the size and shape of the thyroid gland and (2) the amount of iodine absorption. Comparing the rate of iodine uptake with the level of circulating hormones allows us to evaluate the functional state of the gland.

Radioactive iodine is an obvious choice for imaging the thyroid gland. For most other tissues and organs, a radioactive label must be attached to another compound. *Technetium* (^{99}Tc), a versatile tracer, is the primary radioisotope used in nuclear imaging today. The isotope is artificially produced and has a half-life of 6 hours. This brief half-life significantly reduces the patient's exposure to radiation. Technetium is used in more than 80 percent of all scanning procedures. The nature of the technetium-labeled compound varies with the target organ. Technetium scans are performed to examine the heart, thyroid gland, spleen, liver, kidneys, digestive tract, bone, and other organs. Made from uranium in only a few aging nuclear reactors, a worldwide shortage that started in 2009 may persist until new reactors are built. *PET* (**p**ositron **e**mission **t**omography) scans utilize the same principles as standard radioisotope scans, but the analyses are performed by computer. The scans are much more sensitive, and the computers can reconstruct sections through the body and provide extremely precise localization. This procedure can analyze blood flow through organs and assess the metabolic activity in specific portions of an organ, particularly the brain.

▶Figure 6c is a PET scan of the body showing its activity at a single moment in time. The scan is dynamic, and changing patterns of activity can be followed in real time. PET scans can be used to analyze normal function, as well as to diagnose disorders. For the brain, rapid "functional" MRI scans can be used to monitor small changes in blood flow and tissue activity without the use of radioactive tracers, and are more available than PET scanners.

Radiopharmaceuticals

Nuclear medicine involving injected radioisotopes has been far more successful in producing useful images than in treating specific disorders. The problem is that the doses of radiation must be relatively large to destroy abnormal or cancerous tissues, and it is very difficult to control the distribution of these radioisotopes in the body with sufficient precision. As a result, radiation exposure can damage normal as well as abnormal tissues. It is also difficult to control the radiation dosage administered to the target tissues: Underexposure can have very little effect on the abnormal cells, whereas overexposure can destroy adjacent normal tissues.

Radioactive drugs, or *radiopharmaceuticals,* are effective only if they are delivered precisely and selectively. One success story has been the treatment of *hyperthyroidism* (thyroid oversecretion) and thyroid cancer. The thyroid gland selectively concentrates iodine. Large doses of radioactive iodine (^{131}I) can be administered to treat hyperthyroidism. The radiation that is released destroys the abnormal thyroid tissue and stops the excessive production of thyroid hormones. (Following this treatment, most individuals eventually become *hypothyroid*—deficient in thyroid hormone—but this condition can be easily treated by taking thyroid hormones in tablet form.) The use of

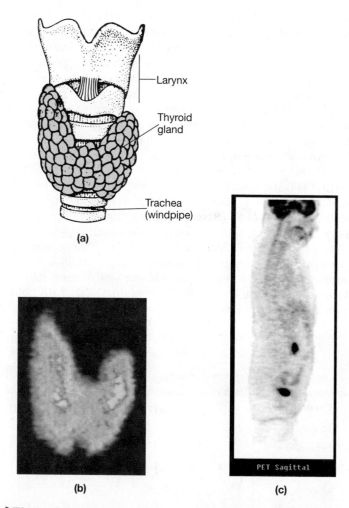

(a)

(b)   (c)

▶**Figure 6 Imaging Techniques. (a)** The position and contours of the normal thyroid gland as seen in dissection. **(b)** After it has been labeled with a radioactive tracer, the thyroid can be examined by special imaging techniques. In this computer-enhanced image, different intensities indicate different concentrations of the radioactive tracer. **(c)** A PET scan of the body. The dark areas indicate where radiolabeled glucose has been absorbed, notably by (1) the brain, (2) an area anterior to the lumbar spine, and (3) the urinary bladder.

Chemical Level

radioactive iodine is now the preferred treatment method, as opposed to surgery or prolonged antithyroid medication, for most adult hyperthyroid patients.

A relatively new application of nuclear medicine involves the attachment of a radioactive isotope to a **monoclonal antibody (mAb)**. Antibodies are proteins produced in the body to provide a selective defense against foreign proteins, toxins, or pathogens. Monoclonal antibodies are produced by immune cells cultured under laboratory conditions. The antibodies these cells produce are then labeled with radioactive materials. Injected into the body, the antibodies will bind to their targets and expose the surrounding tissues to radiation. mAbs specific to certain types of tumor cells have already been approved by the Food and Drug Administration (FDA). For some, the amount of radiation emitted is low, and the procedure is used to produce diagnostic images rather than to treat disease. Radioimmunotherapy with higher doses of radiation is used to target and destroy some forms of lymphoma tumors (p. 149).

▶ Solutions and Concentrations

Physiologists and clinicians pay particular attention to the distribution of ions across membranes and to the electrolyte composition of body fluids. Data must be analyzed from several perspectives, and physiological values can be reported in several ways. One method is to report the concentration of atoms, ions, or molecules in terms of weight per unit volume of solution. Although grams per liter (g/L) can be used, values are most often expressed in grams (g), milligrams (mg), or micrograms (µg) per 100 mL. Because 100 mL = 0.1 liter = 1 deciliter (dL), the abbreviations most often used in this text are **g/dL** (*grams per deciliter*) and **mg/dL** (*milligrams per deciliter*).

Osmotic concentration, or osmolarity, depends on the total number of individual atoms, ions, and molecules in solution, without regard to molecular weight, electrical charge, or molecular identity. As a result, if fluid balance and osmolarity are being monitored, concentrations are usually reported in *moles per liter* (**mol/L**, or **M**) or *millimoles per liter* (**mmol/L**, or **mM**) rather than in g/dL or mg/dL. To convert g/dL to mol/L, multiply by 10 and divide by the atomic weight of the element. For example, a sample of plasma (blood with the cells removed) contains sodium ions at a concentration of roughly 0.32 g/dL (320 mg/dL). We convert this value to mmol/L as follows:

$$\frac{g/dL \times 10}{\text{atomic weight}} = \frac{0.32 \times 10}{22.29} = 0.14 \text{ mol/L} (= 140 \text{ mmol/L})$$

Moles or millimoles per liter can also be used to indicate the concentration of molecules in solution. We can perform the same conversion by substituting molecular weight for atomic weight in the preceding equation. The total solute concentration of a solution can be determined by adding together the concentrations of individual solutes, expressed in moles per liter or millimoles per liter. The resulting value is reported in **milliosmoles**

per liter (mOsm/L). The use of mOsm rather than mmol indicates that multiple solutes are present, each contributing to the total osmolarity.

Because electrolyte concentrations have profound effects on cells, it is often important to know how many positive and negative charges the ions or molecules in a biological solution bear, not just how many ions or molecules are present. For example, a single calcium ion (Ca^{2+}) has twice the electrical charge of a single sodium ion (Na^+), although the two are identical in terms of their effects on osmolarity. One **equivalent (Eq)** is a mole of positive or negative charges. Physiological concentrations are often reported in **milliequivalents per liter (mEq/L)**. You should become familiar with both methods of expression. Fortunately, the conversion from millimoles to milliequivalents is relatively easy to perform. For **monovalent ions**—those with a +1 or −1 charge—millimole and milliequivalent values are identical, so no calculation is needed. For **divalent ions**, with a +2 or −2 charge, the number of charges (mEq) is twice the number of ions (mmol). For an ion with a +3 or −3 charge, the number of milliequivalents is three times the number of millimoles. To convert mEq to mmol, simply divide by the ionic valence (number of charges). **Table 4** compares the methods of reporting the concentration of major electrolytes in plasma in terms of weight, moles, and equivalents.

Physiologists and clinicians surely would benefit from the use of standardized reporting procedures. It can be very frustrating to consult three references and find that the first reports electrolyte concentrations in mg/dL, the second in mmol/L, and the third in mEq/L.

In 1984, the American Medical Association House of Delegates endorsed a plan to standardize clinical test results through the use of **SI (Système Internationale)** units, with a target date of July 1, 1987, for the switchover. Unfortunately, there was no mechanism for enforcing compliance, and the standardization attempt ultimately failed. As of 2013, all scientific and medical journals around the world report data in SI units, but most U.S. clinical laboratories and many journals continue to use their traditional reporting methods, and

Table 4	A Comparison of Methods for Reporting Concentrations of Solutes in Blood*			
Solute	**mg/dL**	**mmol/L**	**mEq/L**	**SI Units**
Electrolytes				
Sodium (Na⁺)	320	140	140	140 mmol/L
Potassium (K⁺)	16.4	4.2	4.2	4.2 mmol/L
Calcium (Ca²⁺)	9.5	2.4	4.8	2.4 mmol/L
Chloride (Cl⁻)	354	100	100	100 mmol/L
Metabolites				
Glucose	90	5	nr	5 mmol/L
Lipids, total	600	nr	nr	0.6 g/L
Proteins, total	7 g/dL	nr	nr	70 g/L

* nr = not reported in these units.

attempts to mandate a synchronous, universal conversion have been abandoned.

The major problem is that the relationships to values currently in use are difficult to remember. Electrolyte concentrations, now most often given in mEq/L, are reported in mmol/L in the SI. Thus, the values for sodium and potassium concentrations remain unchanged, but the normal values for calcium and magnesium are reduced by 50 percent. The situation becomes more confusing in terms of metabolite concentrations. Cholesterol and glucose concentrations are now most often reported in mg/dL, but the SI units are mmol/L. However, total lipid concentrations, also currently reported as mg/dL, and total protein concentrations, now given as g/dL, are reported in terms of g/L under the SI. For these units to be useful in a clinical setting, physicians must not only remember the definition of each SI unit, but must also convert and relearn the normal ranges. As a result, it appears unlikely that the conversion to SI units will be completed in the immediate future. However, with more information technology use and computer reporting of lab test results in a format that includes the normal ranges, these translation issues may fade. Various paper and electronic conversion systems are widely available.

▶ The Pharmaceutical Use of Isomers

A chemical compound is a combination of atoms bonded together in a particular arrangement. The chemical formula specifies the *number* and *types* of atoms that combine to form the compound. The *arrangement* of the atoms, which determines the specific shape of each molecule, is shown by the molecule's structural formula.

Isomers are chemical compounds that have the same chemical formula, but different structural formulas. Isomers called *stereoisomers* are mirror images of each other. Stereoisomers are analogous to the left hand and right hand of the human body. The two hands contain the same palm bones (metacarpals) and finger bones (phalanges), but each hand is a mirror image of the other. A glove designed for the left hand will not fit the right hand, and vice versa. Chemical compounds are also said to be left handed or right handed, depending on their structural configuration. For example, glucose has a left-handed (*levo-*) isomer and a right-handed (*dextro-*) isomer. Just as the right hand cannot fit into a left glove, receptors and enzymes in our cells cannot bind the levo-isomer of glucose. Our cells are therefore unable to metabolize levo-glucose as an energy source.

This pattern is common: Our cells and tissues will typically respond to only one structural form—either levo or dextro—not both. This feature can pose a problem for pharmaceutical chemists, because many of the chemical reactions used to synthesize a drug produce a mixture of levo- and dextro-isomers. In some cases, the inactive isomer is simply ignored; in others, it is removed. For example, the antibiotic *chloramphenicol*

contains both levo- and dextro-isomers, but only the levo form is effective in killing bacterial pathogens. And only the levo form (the active form) of *ephedrine,* a drug that dilates the bronchioles of the lungs, is contained in the popular tablet *Primatene,* which is sometimes used to treat *asthma* attacks. Finally, birth control pills containing the steroid *levonorgestrel,* the levo form of *norgestrel,* are effective at half the dosage of pills containing a mixture of levo- and dextro-isomers.

In some cases, both forms of an isomer are biologically active, but have strikingly different effects, both desired and undesired. The drug *thalidomide* was given to pregnant women in the 1960s to alleviate symptoms of morning sickness. The sedative effect of one isomer was well documented, but the medication sold contained both forms. Unfortunately, the other isomer caused tragic abnormalities in fetal limb development. (We discuss the mechanisms that underlie thalidomide's effects on fetal development on pp. 223–224.)

▶ Artificial Sweeteners

Some people cannot tolerate sugar for medical reasons; others avoid it to comply with recent dietary guidelines that call for reduced sugar consumption or to lose weight. Thus, many people today use artificial sweeteners in their foods and beverages.

Artificial sweeteners are organic molecules that can stimulate taste buds and provide a sweet taste to foods without adding substantial amounts of calories to the diet. These molecules have a much greater effect on the taste receptors than do natural sweeteners, such as fructose or sucrose, so they can be used in minute quantities. For example, *saccharin* is about 300 times as sweet as sucrose. The popularity of this sweetener has declined since it was reported that saccharin can promote bladder cancer in rats. However, the risk is very small, even for rats, and saccharin continues to be used. Several other artificial sweeteners, including *aspartame* (*NutraSweet*), *sucralose,* and *acesulfame potassium* (*Ace-K,* or *Sunette*), are currently available. The market success of an artificial sweetener ultimately depends on its taste and its chemical properties. Stability in high temperatures (as in baking) and resistance to breakdown in an acidic pH (as in carbonated drinks) are important properties of any artificial sweetener.

Molecules of artificial sweeteners do not resemble those of natural sugars. Saccharin, acesulfame potassium, and sucralose cannot be broken down by the body, and they have no nutritional value. Aspartame consists of a pair of amino acids, the building blocks of proteins (as we shall discuss later in this chapter), and they can be broken down in the body to provide energy. However, because aspartame is 200 times as sweet as sucrose, very small quantities are needed, so this artificial sweetener adds few calories to a meal. Because it does not produce the bitter aftertaste sometimes attributed to saccharin, aspartame is used in many diet drinks and low-calorie desserts. People with the rare disease *phenylketonuria* should not use aspartame.

Chemical Level

Fatty Acids and Health

Humans love fatty foods. The smooth, creamy texture of fatty substances, and their appealing taste, make fats a welcome part of our diet. Saturated fats tend to be solid at room temperature, while unsaturated fats are usually liquid. Unfortunately, a diet containing large amounts of saturated fatty acids has been shown to increase the risk of heart disease and other circulatory problems. Saturated fats are found in popular foods like fatty meat and dairy products (including such favorites as butter, cheese, and ice cream).

Some unsaturated fats, by contrast, appear to decrease the risk of heart disease. Most vegetable oils contain a mixture of monounsaturated and polyunsaturated fatty acids. Current research indicates that monounsaturated fats may be more effective than polyunsaturated fats in lowering the risk of heart disease. According to current research, perhaps the healthiest choice is oleic acid, an 18-carbon monounsaturated fatty acid particularly abundant in olive and canola oils. Compounds called *trans* fatty acids, produced during the manufacturing of some margarines and vegetable shortenings (making the product less liquid at room temperature), appear to increase the risk of heart disease. Margarines and prepared foods containing these substances may be no healthier for you than butter. The labels on most packaged foods now show the amount of trans fats (▶ Figure 7).

The Inuit society developed in the Arctic and has lower rates of heart disease than other populations have, even though the traditional Inuit diet is high in fats and cholesterol. Interestingly, the fatty acids in the Inuit diet have an unsaturated bond three carbons before the last, or omega, carbon, a position known as "omega minus 3," or *omega-3*. Fish flesh and fish oils, a substantial portion of the Inuit diet, contain omega-3 fatty acids. Why does the presence of omega-3 fatty acids (or some other unidentified component of fish) in the diet reduce the risks of heart disease, rheumatoid arthritis, and other inflammatory diseases? There is a great deal of interest in this area of research.

Fat Substitutes/Blockers and Weight Loss

The average diet in the United States contains more fat than do the diets of people in many other parts of the world. Diets high in saturated fat and cholesterol have been linked to heart disease, as well as to certain forms of cancer, as has obesity itself. Fat has roughly twice the calories of a similar amount of carbohydrate or protein, so reducing fat intake is one way of reducing calories and hopefully losing weight. Lowering the percentage of calories we derive from fat may also benefit our health. This suggestion has led to an increased interest in the development of fat substitutes, many of which are made from proteins or carbohydrates (including sugars) or bulked up with low-calorie fiber or water.

Fat substitutes try to provide the texture, taste, and cooking properties of natural fats. Two such substitutes, *Simplesse* and *Olestra*, are in widespread use. Simplesse is made from proteins of egg white and skim milk or whey. The heated proteins are mechanically treated to form small spherical masses that have the taste and texture of fats. Simplesse can be used in place of fats in any application other than baking; it is used in low-calorie "ice creams" under the trade name *Simple Pleasures*. These fat substitutes can be broken down in the body, but they provide less energy than do natural fats. For example, ice cream made with Simplesse has half the fat calories of ice cream that contains natural fats.

Olestra is made by chemically combining sucrose and fatty acids. The resulting compounds cannot be used by the body and so contribute no calories. Olestra has been approved as an ingredient in margarines, baked goods, and other snack foods. One of the problems is that Olestra droplets within the digestive tract collect dietary lipids and lipid-soluble materials, including fat-soluble vitamins (A, D, E, and K), and prevent their absorption. In addition, if eaten in large quantities, Olestra can cause diarrhea. To prevent vitamin deficiencies among consumers, manufacturers of snack foods prepared with Olestra now fortify them with fat-soluble vitamins.

One drug, *Orlistat*, is available and is FDA approved for nonprescription use to aid in weight loss. This medication blocks the action of the pancreatic enzymes responsible for fat digestion and prevents the absorption of dietary fats. Side effects of oily diarrhea and potential vitamin deficiencies may occur, but beneficial reductions of weight and blood levels of cholesterol and lipids are possible.

In the 1970s, Dr. Robert C. Atkins, a cardiologist, advocated a high-fat and high-protein but restricted carbohydrate diet. Many

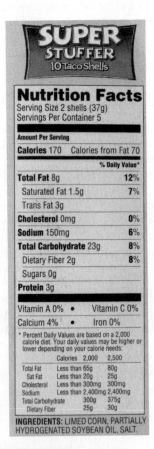

▶ Figure 7 Labeling Fat Content.

heart disease researchers initially thought it would be undesirable because the epidemiologic data linked diets high in fat to heart disease. In recent controlled studies it has been shown to be more effective for initial weight loss than a low-fat and calorie-restricted diet. Some studies have shown equal or better sustained weight loss at one year with a high-fat, low-carbohydrate diet, compared to a low-fat, low-calorie diet. Unfortunately, many dieters are unable to lose weight or maintain weight loss with any diet. However, the studies do show improved blood HDL cholesterol and triglyceride levels in the Atkin's dieters, compared to the low-fat calorie-restricted diets (which did improve LDL lipid levels more than the Atkin's diet). Dr. Atkins theorized that his diet induced a state of ketosis, which contributed to weight loss, and that the reduction in carbohydrates improved blood cholesterol and lipid levels. Researchers (and dieters) suggest the unpalatability of a high-fat/protein, carbohydrate-restricted diet reduced hunger and caloric intake, leading to weight loss. Whatever the mechanism, the high-fat/protein, low-carbohydrate diet has helped some obese people lose weight, with beneficial results. Sustained weight loss, regardless of type of diet, to more normal weights reduces heart disease risk.

▶ Metabolic Anomalies

If enzymes are nonfunctional or are missing, metabolic disorders known as *metabolic anomalies* result. The effects are variable, depending on the enzyme involved, but in severe cases growth and development are impaired and vital tissues are damaged or destroyed. Additional information about many such conditions is given elsewhere in the *Applications Manual*.

Phenylketonuria

Persons diagnosed with **phenylketonuria (PKU)** lack the enzyme that converts the amino acid *phenylalanine* to the amino acid *tyrosine*. Without this enzyme, phenylalanine accumulates in the blood and tissues, and large quantities are excreted in the urine. High levels of phenylalanine damage the developing nervous system. If the condition is not detected shortly after birth, mental retardation can occur. Because milk is a major source of phenylalanine, newborns generally undergo a blood test for PKU 48 hours after nursing begins. Abnormally high levels of phenylalanine in the bloodstream may indicate PKU. Once a diagnosis of PKU is made, the diet is controlled to avoid foods containing high levels of phenylalanine, particularly in the first five years of life when the brain undergoes rapid growth and development, or during pregnancy if the mother herself has PKU. The artificial sweetener aspartame contains phenylalanine and could cause problems for PKU patients. (PKU is discussed further on pp. 187–189.)

Albinism

Albinism is a genetic disorder that results in a lack of pigment in the skin. It is the result of having one defective enzyme out of the four known to be involved in the production of the pigment melanin. The skin of a person afflicted with albinism is white, the hair is pale, and the iris color is light, often showing the red of blood vessels unmasked by color. Among its other functions, melanin helps protect the skin from the effects of ultraviolet (UV) radiation. When outdoors, individuals with albinism must be careful to avoid skin damage from the UV radiation in sunlight.

Hypercholesterolemia

Familial hypercholesterolemia is a genetic disorder resulting in a reduced ability to remove cholesterol from the bloodstream. As circulating levels rise, cholesterol accumulates around tendons, creating yellow deposits called *xanthomas* beneath the skin. The worst aspect of the disorder is the deposition of cholesterol in the walls of blood vessels. This condition, a form of *atherosclerosis*, can restrict the flow of blood through vital organs such as the heart and brain. Atherosclerosis can develop in individuals with normal cholesterol metabolism, but clinical symptoms do not ordinarily appear until age 40 or older. Individuals with congenital hypercholesterolemia may develop acute coronary artery disease or even suffer a heart attack at or before 20 years of age. A group of medicines called the "statins" inhibit an enzyme (HMG–COA reductase) that is a rate-limiting step in cholesterol synthesis. They significantly reduce deaths from atherosclerosis.

Galactosemia

Milk contains the monosaccharide *galactose*, which can be converted to glucose within cells. The genetic disorder **galactosemia** is caused by the absence of the enzyme that catalyzes this reaction. Affected individuals have elevated levels of galactose in the blood and urine. Chronically high levels of galactose during childhood can cause abnormalities in the nervous system, jaundice, liver damage, and cataracts. Preventive treatment involves the early detection of galactosemia and a restriction on the dietary intake of galactose.

The Cellular Level of Organization

The Nature of Pathogens

The presence of a nucleus is the defining characteristic of **eukaryotic cells** (ū-kar-ē-OT-ik; *eu*, true + *karyon*, nucleus). All eukaryotic cells have similar membranes, organelles, and methods of cell division. All multicellular animals, plants, and fungi (plus many single-celled organisms) are composed of eukaryotic cells.

The eukaryotic plan of organization is not the only one in the living world, however. Some organisms do not consist of eukaryotic cells. These organisms are of great interest to us, because they include many of the **pathogens** that are recognized causes of human diseases. Representative pathogens are introduced in ❱ Figure 8.

Bacteria

Prokaryotic cells do not have nuclei or other membranous organelles. Nor do they have a cytoskeleton, and typically, their cell membranes are surrounded by a semirigid cell wall made of carbohydrate and protein.

Bacteria are probably the most familiar prokaryotic cells. They are generally less than 2 μm in diameter. Many bacteria are quite harmless, and many more—including some that live on or within our bodies—are beneficial to us in a variety of ways. Other bacteria are dangerous pathogens that, given the opportunity, will destroy body tissues. These bacteria are dangerous because they absorb nutrients and release enzymes that damage cells and tissues. A few pathogenic bacteria also release toxic chemicals. Bacterial infections are responsible for many serious diseases, as indicated in **Table 5**. We consider these and other bacterial infections in other sections of the *Applications Manual*.

❱ Figure 8a shows the structure of a representative bacterium. ❱ Figure 9 shows the three basic shapes of bacteria: round, rodlike, and spiral. A round bacterium is called a *coccus* (KOK-us; plural, *cocci*, KOK-sē) (example *pneumococcus*, a cause of pneumonia, a lung infection). A rodlike bacterium is a **bacillus** (ba-SIL-us; plural, *bacilli*, ba-SIL-ē) (example *E. coli*, a cause of urine infections). Shapes of spiral bacteria vary, and so do their names. A *vibrio* (VIB-rē-ō) is comma shaped; (example *Vibrio cholerae*, a cause of cholera, a severe diarrhea) and a *spirochete* (SPĪ-rō-kēt) is shaped like a corkscrew (example, *Treponema pallidum*, the cause of syphilis).

Some cocci and bacilli form groupings of cells. The Latin names used to describe these groupings are also used to identify specific bacteria. For instance, pairs of cocci are called *diplococci* (*diplo-*, double). *Streptococci* and *streptobacilli* form twisted chains of cells (*strepto-*, twisted), and *staphylococci* look like a bunch of grapes (*staphylo-*, grapelike). These bacteria cause primarily skin and respiratory infections.

Viruses

Another type of pathogen conforms neither to the prokaryotic nor to the eukaryotic organizational plan. These tiny pathogens, called **viruses**, are not cellular. In fact, when free in the environment, they do not show any of the characteristics of living organisms. They are classified as *infectious agents*—factors that cause infection—because they can enter cells (either prokaryotic or eukaryotic) and replicate themselves.

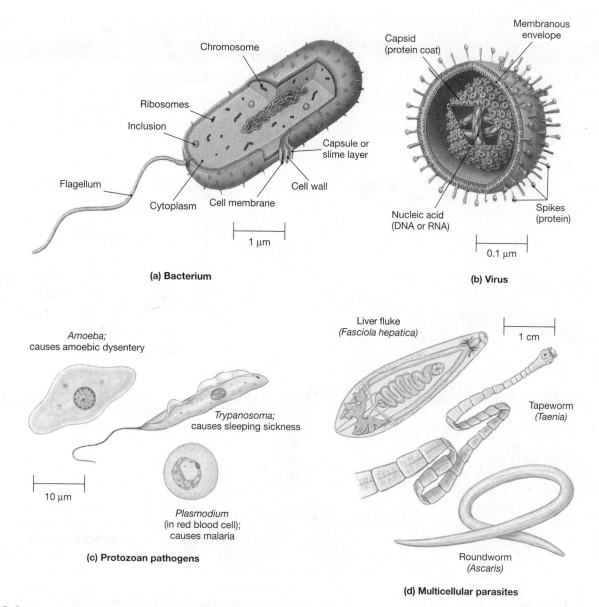

▶ Figure 8 Representative Pathogens. (a) A bacterium, with prokaryotic characteristics indicated. **(b)** A typical virus. Each virus has an inner chamber containing nucleic acid, surrounded by a protein capsid or an inner capsid and an outer membranous envelope. The herpesviruses are enveloped DNA viruses; they cause chicken pox, shingles, and herpes. **(c)** Protozoan pathogens. Protozoa are eukaryotic single-celled organisms, common in soil and water. **(d)** Multicellular parasites. Several groups of organisms are human pathogens, and many have complex life cycles.

Viruses consist of a core of nucleic acid (DNA or RNA) surrounded by a protein coat called a *capsid*. (Some varieties have an *envelope*, a membranous outer covering, as well.) The structure of a representative virus is shown in ▶ **Figures 8b** and **10**. Important viral diseases include influenza (flu), yellow fever, AIDS, hepatitis, polio, measles, mumps, rabies, herpes, and the common cold (**Table 6**).

To enter a cell, a virus must first attach to the plasma membrane. Attachment occurs at one of the normal membrane proteins. Once the virus has penetrated the plasma membrane, the viral nucleic acid takes over the cell's metabolic machinery. In the case of a **DNA virus** (▶ **Figure 11a**), the viral DNA enters the cell nucleus, where transcription begins. The mRNA produced then enters the cytoplasm and is used in translation, in which the cell's ribosomes begin synthesizing viral proteins. The viral DNA replicates in the nucleus, "stealing" the cell's nucleotides. The replicated viral DNA and the new viral proteins then form new viruses that pass out of the cell through the plasma membrane or from cell rupture.

In an **RNA virus**, the situation is somewhat more complicated (▶ **Figure 11b**). In the simplest RNA viruses, the viral RNA entering the cell functions as an mRNA strand that carries the information needed to direct the cell's ribosomes to synthesize viral proteins. These proteins include enzymes essential to the duplication of viral RNA. When the cell is packed with new viruses, the plasma membrane ruptures and the RNA viruses are released into the interstitial fluid.

In **retroviruses**, a group that includes HIV (the virus responsible for AIDS), the replication process is even more complex. These RNA viruses carry an enzyme called *reverse transcriptase*, which directs "reverse transcription"—the assembly of DNA according to

Table 5	Examples of Bacterial Diseases and the Primary Organ Systems Affected	
Organism	**Disease**	**Affected Organ System**
Bacilli		
Bacillus anthracis	Anthrax	Integumentary and respiratory systems
Mycobacterium tuberculosis	Tuberculosis	Respiratory system
Corynebacterium diphtheriae	Diphtheria	Respiratory and integumentary systems
Cocci		
Staphylococcus aureus	Various skin infections	Integumentary system
Streptococcus pyogenes	Pharyngitis (strep throat)	Respiratory and integumentary systems
Neisseria gonorrhoeae	Gonorrhea	Reproductive system
Vibrios		
Vibrio cholerae	Cholera	Digestive system
Spirochetes		
Treponema pallidum	Syphilis	Reproductive and nervous systems
Borrelia burgdorferi	Lyme disease	Skeletal system (joints)
Rickettsias		
Rickettsia prowazekii	Epidemic typhus fever	Cardiovascular and integumentary systems
Coxiella burnetii	Q fever	Respiratory system
Chlamydias		
Chlamydia trachomatis	Trachoma (eye infections)	Integumentary system
	PID (pelvic inflammatory disease)	Reproductive system

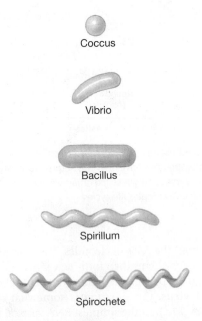

Coccus

Vibrio

Bacillus

Spirillum

Spirochete

▶ **Figure 9 Common Bacterial Shapes.**

the nucleotide sequence of an RNA strand. The DNA created in this way is then inserted into the infected cell's chromosomes. The viral genes become activated, and the cell begins producing RNA by normal transcription. The RNA produced includes viral RNA, mRNA carrying the information for the synthesis of reverse transcriptase, and mRNA controlling the synthesis of viral proteins. These components then combine within the cytoplasm, which

gradually becomes filled with viruses. Finally, the new RNA viruses are shed at the cell surface. Two new anti-influenza medicines, *Relenza* and *Tamiflu*, inhibit a key enzyme involved in the assembly of the virus and its release by infected cells.

Even if the host cell is not destroyed by these events, normal cell function is usually disrupted. In effect, the metabolic activity of the cell is diverted to create viral components, rather than performing tasks needed for cell maintenance and survival. Some viruses, such as *herpes simplex* or *herpes varicella,* can lie dormant within infected cells for long periods before replicating.

Viruses are now becoming important as benefactors as well as adversaries. In *genetic-engineering* procedures (p. 32), viruses whose nucleic acid structure has been intentionally altered can be used to transfer copies of normal human genes into the cells of individuals with inherited enzymatic disorders. This was the method used to insert the gene for the missing enzyme (adenosine deaminase) in persons with severe combined immunodeficiency disease, or SCID (p. 33). The virus integrates the normal gene into the patient's chromosomes, presumably in a random manner. Unfortunately, two of the nine children successfully treated for SCID this way have developed a form of leukemia from the activation of other genes in their chromosomes. (The abnormally activated genes are called *oncogenes.*) Efforts are under way to understand the mechanism responsible and to develop different therapies that will not cause similar problems.

Researchers and clinicians are studying ways to treat cystic fibrosis by gene therapy. **Cystic fibrosis (CF)** is a debilitating genetic defect whose most obvious—and potentially deadly—symptoms involve the respiratory system. The underlying problem is an abnormal gene that carries instructions for a chloride ion

Human liver cell

▶**Figure 10 Viruses.** A variety of viruses, shown with a typical bacterial cell, a human liver cell, and a ribosome for scale.

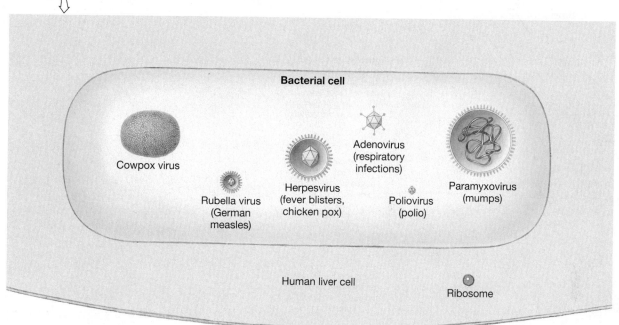

Table 6	Examples of Viral Diseases and the Primary Organ Systems Affected		
Nucleic Acid	**Virus**	**Disease**	**Affected Organ System**
RNA	Influenza A, B, C	Flu	Respiratory system
	Paromyxovirus	Mumps	Digestive and reproductive systems
	Hepatitis A, C, D, E	Infectious hepatitis	Digestive system (liver)
	Rhinovirus	Common cold	Respiratory system
	Human immunodeficiency virus (HIV)	AIDS	Lymphatic system
DNA	Herpesvirus		
	Herpes simplex 1	Cold sore/fever blister	Integumentary system
	Herpes simplex 2	Genital herpes	Reproductive system
	Varicella-zoster	Chicken pox	Integumentary system
	Varicella-zoster	Shingles	Nervous system
	Hepatitis B	Hepatitis	Digestive system (liver)
	Epstein–Barr	Mononucleosis	Respiratory and lymphatic systems

channel that occurs in plasma membranes throughout the body. Researchers have treated CF in laboratory animals by inserting the normal gene into a virus that infects cells lining the respiratory passageways.

Prions

Transmissible spongiform encephalopathic (*TSE*) diseases are caused by a novel class of infectious agents called prions. **Prions** (from "protein infectious 'ions,'" [particles]) are unique among agents of transmissible disease, because they contain no nucleic acids (either DNA or RNA). Rather, they appear to be abnormal three-dimensional forms of the ordinarily harmless protein *PrPc* found in cells. Apparently, an abnormally folded protein can serve as a template for converting normal proteins to the pathogenic form. When present in large quantities, these proteins cause degenerative cellular gaps in brain tissue, which takes on a microscopic "spongy" appearance. Partially metabolized fragments of abnormal prion proteins may form microscopic deposits called amyloid plaques in the brain as well.

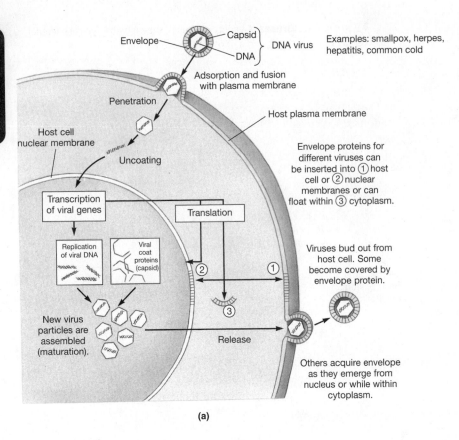

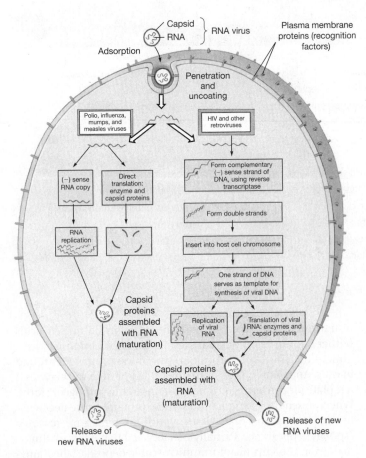

▶ **Figure 11 Viral Replication. (a)** The replication of a DNA virus. **(b)** The replication of RNA viruses.

The same clinical TSE disease can have a genetic, a sporadic, or an infectious origin. Rare genetic variations in the *PrPc* protein cause inheritable forms of TSE. Sporadic cases may come from spontaneous change of the *PrPc* protein to the abnormal shape. In addition, the disease can be transmitted from unrecognized infected donor individuals to recipients of corneal transplants or pituitary hormone extracts. Some are also known to have contracted TSE diseases from exposure to contaminated medical instruments or by eating affected tissues. While TSE diseases are not transmitted by normal household interactions, they can, very rarely, be acquired by contact with the abnormal proteins.

The first recognized human prion disease was *kuru*, a deadly disease affecting members of a society in New Guinea that practiced ritual funeral-related cannibalism. The prions were passed from person to person when uninfected individuals ate infected brains. The infection, which could lead to death within a year, caused half of all childhood and adult deaths in the affected part of New Guinea. Other known prion diseases include inherited, variant and sporadic *Creutzfeldt–Jakob disease* (which usually affects older people) and *fatal familial insomnia*.

Prion infections also occur in domesticated animals. In sheep, the condition is called *scrapie*; in cows, it is called *bovine spongiform encephalopathy* (*BSE*). Infected cows ultimately develop an assortment of strange neurological symptoms (such as pawing at the ground and exhibiting difficulty in walking), giving the condition the common name "mad-cow disease."

In 1995, British and European researchers reported a puzzling variant of Creutzfeldt–Jakob disease among teenagers and young adults. A number of fatal cases in England showed brain changes similar to those of BSE, leading investigators to attribute the outbreak to the consumption of meat products from prion-infected cows. This discovery led to a temporary ban on British beef from the European community, the slaughter and destruction of infected and potentially infected cows, and a change in livestock feeding practices throughout the world. Presumably, many cows became infected between 1980 and 1996 by eating feed containing beef by-products and bonemeal contaminated with prions from the neural tissue of infected animals. The use of such feed was banned in 1996, and greater care is taken when butchering to prevent contact of brain and spinal neural tissue with meat intended for consumption. Since 2003, small clusters of cows in Canada and the United States have been found to have BSE. Public health measures included quarantine and destruction of herds containing the affected cows, and more widespread testing of meat products. Human cases of the variant of Creutzfeldt–Jakob disease in England presumably related to eating beef between 1980 and 1996 peaked in 2000, and the worldwide total was 227 cases as of October 2011. The risk of contracting the disease in England in the 1990s has been estimated at one case per 10 billion servings of beef.

Unicellular and Multicellular Parasites

Bacteria and viruses are the best-known human pathogens, but some pathogens are eukaryotic. Examples of the most important types are shown in ❱ Figure 8c,d. **Protozoa** (❱ Figure 12) are unicellular eukaryotes that are abundant in soil and water. They are responsible for a variety of serious human diseases, including amoebic dysentery and malaria (**Table 7**). Protozoa include (1) flagellates, which use flagella for propulsion; (2) amoeboids, among which are mobile, amoebalike forms that engulf their prey; (3) ciliates, which are covered with cilia; and (4) sporozoans—forms with complex life cycles. **Fungi** (singular, fungus) are eukaryotic organisms that absorb organic materials from the remains of dead cells. Mushrooms are familiar examples of very large fungi. In a mycosis, or fungal infection, a microscopic fungus spreads through living tissues, killing cells and absorbing nutrients. Several relatively common skin conditions (including athlete's foot and candidiasis [diaper rash]) and a few more serious diseases (e.g., histoplasmosis) are the result of fungal infections (**Table 8**).

Larger multicellular organisms, generally referred to as **parasites**, can also invade the human body and cause diseases. The multiplication of these larger parasitic organisms in or on the body is called an **infestation**. Diseases caused by multicellular parasites are listed in **Table 9**. **Helminths** are parasitic worms that can live within the body. They include **flatworms**, such as the flukes and tapeworms, and **roundworms**, or nematodes. These organisms, which range in size from microscopic flukes to tapeworms a meter or more in length, typically cause weakness and discomfort, but do not by themselves kill their host. However, complications resulting from the parasitic infection, such as malnutrition, chronic bleeding, or secondary infections by bacterial or viral pathogens, can ultimately prove fatal.

Arthropods (❱ Figure 13) make up the largest and most diverse group of animals on Earth. The major arthropods that affect humans are the arachnids, including scorpions, spiders, mites, and ticks, and the insects, such as mosquitoes, flies, lice, fleas, and bedbugs. In many cases arthropods serve as vectors, or agents of transmission, of bacterial, viral, and protozoan infectious diseases.

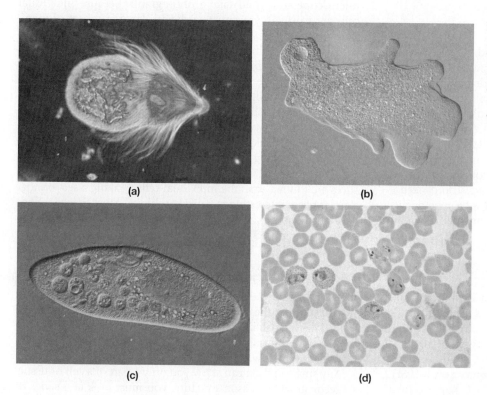

(a)   (b)   (c)   (d)

❱ Figure 12 **Representative Protozoa.**
(a) Trichonympha, a flagellate from a termite gut.
(b) Amoeba proteus, a free-living form found in ponds. (LM × 310) **(c)** Paramecium caudatum, a free-living ciliate. (LM × 160) **(d)** Plasmodium vivax, the parasite that causes malaria, stained within human blood cells. (LM × 125).

Table 7	Examples of Protozoan Diseases and the Primary Organ Systems Affected		
Type of Protozoa	**Name (Genus)**	**Disease**	**Affected Organ System**
Flagellates	*Trypanosoma*	African sleeping sickness	Cardiovascular system
	Leishmania	Leishmaniasis	Lymphatic system
	Giardia	Giardiasis	Digestive system
	Trichomonas	Trichomoniasis	Reproductive system
Amoeboids	*Entamoeba*	Amoebic dysentery	Digestive system
Ciliates	*Balantidium*	Dysentery	Digestive system
Sporozoans	*Plasmodium*	Malaria	Various systems
	Toxoplasma	Toxoplasmosis	Various systems

Table 8	Examples of Fungal Diseases and the Primary Organ Systems Affected	
Organism (Genus)	**Disease**	**Affected Organ System**
Aspergillus	Aspergillosis ("Farmer's lung disease")	Respiratory system
Blastomyces	Blastomycosis	Integumentary system
Histoplasma	Histoplasmosis	Respiratory system
Microsporum, and *Trichophyton*	tinea capitis (scalp)	Integumentary system
	tinea corporis (body)	Integumentary system
	tinea cruris (groin)	Integumentary system
	tinea unguium (nails)	Integumentary system
Candida	Candidiasis	Integumentary system
Coccidioides	Coccidioidomycosis ("San Joaquin valley fever")	Respiratory system

◗ Methods of Microanatomy

Over the last 50 years, our technological gadgetry has improved remarkably, enabling us to view the insides and outsides of cells in new ways. Sophisticated equipment has permitted the detailed analysis of physiological processes within cells. The basic problems facing cytologists stem from the considerable difference in size between the investigator and the object of interest. Cytologists (cell biologists) and histologists (biologists who study tissues) measure intracellular structures in *micrometers* (mm) also known as *microns*. Although the range of cell sizes is considerable, an "average cell" is a cube roughly 10 µm $\times$ 10 µm $\times$ 10 µm. To fill a cubic millimeter, we would need a million cells. Because the human eye cannot recognize details smaller than about 0.1 mm, cytologists rely on special equipment that magnifies cells and their contents.

Light Microscopy

Historically, most information has been provided by *light microscopy*, a method in which a beam of light is passed through the object to be viewed. A light microscope can magnify cellular structures about 1000 times and can show details as fine as 0.25 mm. A camera can be attached to the microscope and used to produce a photograph called a **light micrograph (LM)**. Unfortunately, you cannot simply pick up a cell, slap it onto a microscope slide, and take a photograph. Because individual cells are so small, you must work with large numbers of them. Most tissues have a three-dimensional structure, and small pieces of tissue (biopsies) can be removed for examination. The component cells are prevented from decomposing by first exposing the tissue sample to a poison that will stop metabolic operations, but will not alter cellular structures.

Even then, you still cannot look at the tissue sample through a light microscope, because a cube only 2 mm (0.078 in.) on a slide will contain several million cells. You must slice the sample into thin sections. Living cells are relatively thick, and cellular contents are not transparent. Light can pass through the section only if the slices are thinner than the individual cells. Making a section that slender poses interesting technical problems. Most tissues are not very sturdy, so an attempt to slice a fresh piece will destroy the sample. (To appreciate the problem, try to slice a marshmallow into thin sections.) Thus, before you can make sections, you must preserve the tissue, embed the tissue sample in something that will make it more stable, such as wax, and slice it thin enough to transmit light for microscopic viewing. But you are not done yet: In thin sections, the cell contents are almost transparent; you cannot yet distinguish intracellular details by using an ordinary light microscope. You must first add color to the internal structures by treating the slides with special dyes called *stains*.

Any single section can show you only a part of a cell or tissue. To reconstruct the tissue structure, you must look at a series of

Table 9 Examples of Diseases Caused by Multicellular Parasites and the Primary Organ Systems Affected

Group	Organism	Disease or Condition	Affected Organ System
Helminths			
Roundworms	*Ascaris*	Intestinal infestation	Digestive system
	Enterobius	Pinworm infestation	Digestive system
Flatworms	*Wuchereria*	Elephantiasis	Lymphatic system
Flukes	*Fasciola, Clonorchis* (liver flukes)	Fascioliasis	Digestive system
	Schistosoma (blood fluke)	Schistosomiasis	Cardiovascular, digestive, urinary systems
Tapeworms	*Taenia*	Tapeworm infestation	Digestive system
Arthropods			
Arachnids (eight legs)	Mites	Vectors of bacterial and rickettsial diseases	Various systems
	Ticks	Vectors of bacterial and rickettsial diseases	Various systems
	Spiders, scorpions	Inflammation from bites	Integumentary system
Insects (six legs)	Lice	Vectors of bacterial and rickettsial diseases	Various systems
	Human lice	Pediculosis	Integumentary system
	Mosquitoes	Vectors of bacterial, viral, and parasitic diseases	Various systems
	Flies	Passive carriers of bacterial diseases	Various systems
	Wasps, bees	Inflammation from stings	Various systems

▶**Figure 13** **Representative Disease-Carrying Arthropods.**
(a) *Dermacentor andersoni*, a wood tick.
(b) *Phthirus pubis*, a crab louse, holding onto a human pubic hair. (SEM × 55)
(c) *Musca domestica*, the housefly, which can transport microbes on its body. (×3)
(d) The *Aedes* mosquito, a vector for dengue fever. **(e)** *Ctenocephalides canis*, a common flea.

(a)

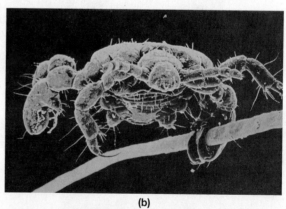

(b)

(c)

(d)

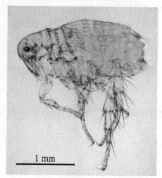

1 mm

(e)

sections made one after the other. After examining dozens or hundreds of sections, you can understand the structure of the cells and the organization of your tissue sample—or can you? Your reconstruction has left you with an understanding of what these cells look like after they have (1) died an unnatural death, (2) been preserved and stabilized, (3) been sliced into thin sections, (4) been stained with various chemicals, and (5) been viewed with the limitations of your equipment. A good cytologist or histologist is extremely careful, cautious, and self-critical and realizes that much of the laboratory preparation is an art as well as a science.

Electron Microscopy

More elaborate procedures can allow for the examination of finer details. In **electron microscopy**, a beam of electrons is passed through or reflected off the surface of a suitably prepared object. In *transmission electron microscopy*, the electrons pass through an ultrathin section. Once through the section, they strike a photographic plate and produce an image known as a **transmission electron micrograph (TEM)**. Transmission electron microscopy can magnify structures up to approximately 500,000 times, revealing details less than a nanometer in size. For instance, with a transmission electron microscope, you can visualize large organic molecules. In *scanning electron microscopy*, a beam of electrons reflects off the surface of an object such as a cell, a broken portion of a cell, or an extracellular structure. (The surfaces are specially coated to enhance reflectivity.) After bouncing off the surface, the electrons strike a photographic plate, producing an image known as a **scanning electron micrograph (SEM)**. ❯Figure 14a,b compares SEM and TEM views of cells that line the intestinal tract, and ❯Figure 14c shows a diagrammatic representation of the intact cell. Scanning electron microscopy can magnify structures about

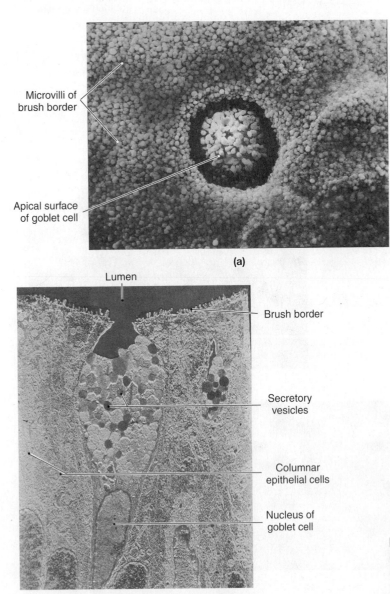

Microvilli of brush border

Apical surface of goblet cell

(a)

Lumen

Brush border

Secretory vesicles

Columnar epithelial cells

Nucleus of goblet cell

(b)

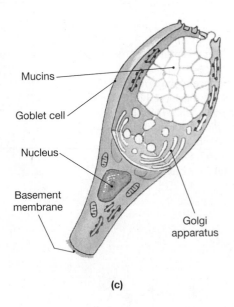

Mucins

Goblet cell

Nucleus

Basement membrane

Golgi apparatus

(c)

❯**Figure 14 A Comparison of Histological Techniques.**
(a) Cell surfaces can be seen with a scanning electron microscope.
(b) Cells similar to those in (a), but viewed with a transmission electron microscope. **(c)** A composite drawing that summarizes the information provided by both scanning and transmission electron microscopy.

50,000 times, but the technique provides a three-dimensional perspective on cellular anatomy that cannot be obtained by other methods.

This level of detail poses problems of its own. At the level of the light microscope, if you were to slice a large cell as you would slice a loaf of bread, you might produce 10 sections from the one cell. You could review the entire series under a light microscope in a few minutes. If you sliced the same cell for examination under an electron microscope, you would have 1000 sections, *each* of which could take several hours to inspect! The illustrations and computer animations associated with this textbook reflect the accumulated data and knowledge of generations of meticulous and patient anatomists!

Lysosomal Storage Diseases

Problems with lysosomal enzyme production cause more than 40 inheritable storage diseases. In each of these conditions, the lack of a specific lysosomal enzyme results in the buildup of materials normally removed and recycled by lysosomes. Eventually, the cell cannot continue to function. Cells that rarely or never divide (such as neurons) accumulate damaging levels earlier, as do macrophages that scavenge damaged cells. Thus the organs most likely to be affected are the nervous system and/or the liver, spleen, and bone marrow. Over time, these conditions are often fatal, so most cases are diagnosed in children. We shall consider three important examples here: Gaucher disease, Tay–Sachs disease, and glycogen storage disease.

Gaucher disease, caused by the buildup of *cerebrosides*— glycolipids in plasma membranes—is probably the most common type of lysosomal storage disease. The disease takes two forms: (1) an infantile form, marked by severe neurological symptoms and ending in death; and (2) a juvenile form, with enlargement of the spleen, anemia, pain, and relatively mild neurological symptoms. Gaucher disease is most common among Ashkenazi Jews, at a frequency of approximately 1 in 1000 births. Regular intravenous treatment with a form of the enzyme *glucocerebrosidase* is partially effective but extremely expensive. In severe cases, a bone marrow transplant may be lifesaving.

Tay–Sachs disease is another hereditary disorder caused by the inability to break down glycolipids. In this case, the glycolipids are *gangliosides,* which are most abundant in neural tissue. Individuals with the condition develop seizures, blindness, and dementia and generally die by age 3–4 years. Like Gaucher disease, Tay–Sachs disease is most common among Ashkenazi Jews, at a frequency of 0.3 per 1000 births.

Glycogen storage disease (**Type II**), also called *Pompe disease,* affects primarily skeletal muscle, cardiac muscle, and liver cells—the cells that synthesize and store glycogen. In people with this condition, a deficiency of the enzyme *alpha-glucosidase* leaves cells unable to mobilize glycogen normally, and large numbers of insoluble glycogen granules accumulate in the cytoplasm. The granules disrupt the organization of the cytoskeleton, interfering with transport operations and the synthesis of materials. In skeletal and heart muscle cells, the buildup leads to muscular weakness and frequently fatal heart problems.

Mitochondrial DNA, Disease, and Evolution

Several inheritable disorders result from abnormal mitochondrial activity. The mitochondria involved have defective enzymes that reduce their ability to generate ATP. Cells throughout the body may be affected, but symptoms involving muscle cells, neurons, and the receptor cells in the eye are most common, because these cells have especially high energy demands normally met by mitochondrial activity. Acquired mitochondrial disorders affecting limited populations of cells may be involved in *Parkinson's disease.* Abnormal mitochondrial DNA has been found in the motor neurons whose degeneration is responsible for the shuffling gait and uncontrollable tremors that are characteristic of this disease.

In other mitochondrial diseases, mitochondria throughout the body are involved. Most inherited mitochondrial disorders are caused by defective nuclear DNA (which codes for most of the building blocks of mitochondria) but some are caused by abnormal inherited mitochondrial DNA (mtDNA). Examples of conditions caused by mitochondrial DNA dysfunction include one class of epilepsies (*myoclonic epilepsy*) and a type of blindness (*Leber's hereditary optic neuropathy*). These are inherited conditions, but the pattern of inheritance is unusual. Although men or women may have the disease, only affected women can pass the condition on to their children. The explanation for this pattern is that the disorder results from an abnormality in the DNA of mitochondria, not in the DNA of cell nuclei. All the mitochondria in the body are produced by the replication of mitochondria present in the fertilized ovum. Few, if any, of those mitochondria were provided by the father; most of the mitochondria of the sperm do not remain intact after fertilization takes place. As a result, children can generally inherit myoclonic epilepsy and Leber's disease only from their mother. This brings us to an interesting concept: Virtually all your mitochondria were inherited from your mother, hers from her mother, and so on back through time. The same is true for every other human. Now, it is known that small changes in DNA nucleotide sequences accumulate over long periods of time. Mitochondrial DNA, or *mtDNA,* can therefore be used to estimate the degree of relatedness between individuals. The greater the difference between the mtDNA of two individuals, the more time has passed since the lifetime of their most recent common ancestor, and the more distant is their relationship. On this basis, it has been estimated that all humans now alive shared a common female ancestor roughly 200,000 years ago. Appropriately, that individual has been called "Mitochondrial Eve." Although other women alive at that time have living descendants, those descendants do not carry the maternal mitochondrial genes of their distant ancestor because an intervening generation consisted of male children only.

Genetic Engineering and Gene Therapy

The study of the synthesis, structure, and function of macromolecules important to life, such as proteins and nucleic acids, is known as **molecular biology**. Among the main goals of this field are deciphering the genomes of living organisms (especially humans), elucidating the mechanisms that control the transcription of genes and the synthesis of proteins, and relating the intricate structure of a protein to its functions.

Research in molecular biology has greatly enhanced our understanding of both normal functions and disease processes. In medicine, molecular biology has prompted a revolution by uncovering a clear biochemical basis for many complex pathologies. For example, in *sickle cell anemia*, red blood cells undergo changes in shape that lead to blocked blood vessels and subsequent tissue damage due to oxygen starvation. This condition results when an individual carries two copies of a defective gene that determines the structure of *hemoglobin*, the oxygen-binding protein in red blood cells. The genetic defect changes just 2 of the 574 amino acids in this protein, but that is enough to alter the functional properties of the hemoglobin molecule, leading to changes in the properties of the red blood cells. This type of disorder is often called a *molecular disease*, because it results from abnormalities at the molecular level of organization.

Many thousand **genetic disorders**, now defined as diseases caused in whole or in part by gene alterations, have been identified. These are grouped into three categories:

- *Single gene disorders* (several thousand are known) are caused by a mutation of a single gene. Cystic fibrosis and sickle cell anemia are examples. They are inherited along standard Mendelian genetics, autosomal dominant, autosomal recessive, and X-linked.

- *Chromosome disorders* are caused by excess or deficiency of genes located on chromosomes or structural changes within chromosomes. Down's syndrome occurs when a person has three copies of chromosome 21 (▶ Figure 15b).

- *Multifactorial/inheritance disorders* are caused by a combination of variations in one or more genes and environmental factors. Heart disease and many cancers are in this group.

Identifying the genetic defect is the vital step toward the development of an effective gene therapy or other treatment. Such development is an important aspect of the field of **genetic engineering**, a general term that encompasses attempts to change the genetic makeup of cells or organisms, including humans.

What are some of the key problems confronting genetic engineers? Genes code for proteins; the makeup of each protein is determined by the sequence of codons (nucleotide triplets) in a stretch of DNA. A human cell has 46 chromosomes, 2 meters of DNA, roughly 1 billion codons and over 3 billion base pairs of nucleotides. Simply finding a particular gene among the approximately 30,000 protein-coding genes that each of us carries is an imposing task. Yet before a specific gene can be studied, its location must be determined. Locating a gene involves preparing a map of the appropriate chromosome.

Mapping the Human Genome

Several techniques can be used to create a map of the chromosomes. **Karyotyping** (KAR-ē-ō-TĪ-ping; *karyon*, nucleus + *typos*, a mark) is the determination of an individual's complement

of chromosomes. ▶ Figure 15a shows a set of 23 paired normal human chromosomes. Each chromosome has characteristic banding patterns, and segments can be stained with special dyes and seen by light microscopy. Unusual banding patterns can indicate structural abnormalities, which are sometimes linked to specific inherited conditions (including one form of leukemia). Locating the relative positions of protein-coding genes on the chromosomes, a process called *gene mapping*, started in the 1990s. By 2006, the Human Genome Project announced the completion of the DNA sequence of the 23 paired human chromosomes.

However, knowing the nucleotide sequence of a gene doesn't provide as much information about the associated protein as we originally expected. It turns out that while there are roughly 30,000 protein-coding genes in the human genome, there are probably 2–3 million different proteins in the body. The "proteome" or group of proteins in a cell varies from minute to minute depending on many different chemical signals within and outside the cell. Each gene can direct the synthesis of a variety of different related proteins, depending on which nucleotide segments are removed from the mRNA strand during RNA processing. It appears likely that much of the DNA previously thought to be "useless" may be responsible for regulating this process.

The genomes of over 3000 organisms have been determined. With automation, 13 different bacteria were sequenced in a single month in 2007. Most of the organisms sequenced have been unicellular organisms (yeast and bacteria), but complete nucleotide sequences are available for some plants, worms, insects, several fishes, and eleven mammals including the laboratory rat and mouse, as well as dogs, cows, humans, and chimpanzees. To paraphrase Winston Churchill, the accurate sequencing of our genome may be "the end of the beginning" in understanding our genetic selves.

Gene Manipulation and Modification

Suppose that the location of a defective gene has been pinpointed. Before attempting to remedy the defect in an individual, a clinician would have to determine the nature of the genetic abnormality. For example, the gene could be inactive or overactive, or it could produce an abnormal protein. It could even be missing. Only after understanding what the problem is could the clinician try to decide how to remedy the defect. Can the gene be turned on, turned off, modified, supplemented, or replaced?

What's the problem? This can be a particularly difficult question to answer. Many of the thousands of inheritable genetic disorders are classified according to general patterns of symptoms, rather than any specific protein or enzyme deficiency. In some cases, the approximate location of the gene has been determined, but the identity of the protein responsible for the clinical symptoms remains a mystery. In cystic fibrosis, many different abnormalities in the gene and the resulting variations in the protein produced result in different patterns of clinical disease. To complicate things further, several genes contribute to many genetic disorders, and environmental factors that

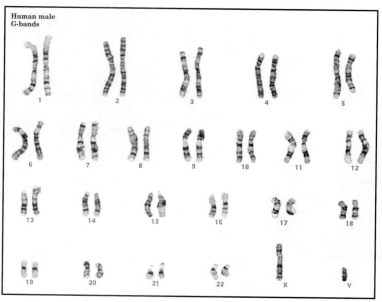

(a) Normal chromosomes

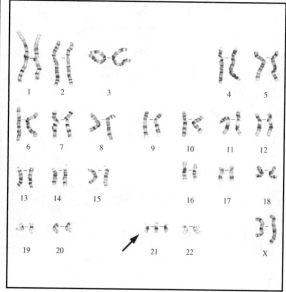

(b) Trisomy 21

▶**Figure 15 Normal and Abnormal Karyotypes.** **(a)** A micrograph of the normal human set of chromosomes; the chromosomes have been arranged in this sequence for ease of comparison. **(b)** The chromosomes of an individual with Down's syndrome. Notice the extra copy of chromosome 21.

influence the timing and amount of gene function also play a role. *What can be done?* If the gene is present, but is overproducing or underproducing a protein, its activity might be controlled by introducing chemical repressors or inducers. Another approach relies on gene manipulation of an organism (usually a bacterium) to produce the missing protein. This process, sometimes called *gene splicing*, creates a **transgenic organism**—an organism of one species into which one or more genes of other species have been incorporated. First the gene is localized and isolated. An accurate copy of the gene is then most commonly spliced into the relatively simple DNA strand of a bacterium, creating *recombinant DNA* (*rDNA*) (▶**Figure 16**). Bacteria grow and reproduce rapidly under laboratory conditions, and before long a colony of identical, transgenic bacteria has formed. All the members of the colony carry the introduced gene and will manufacture the corresponding human protein in large quantities. The protein can then be extracted, concentrated, and administered to individuals who are deficient in the activity of the gene in question. *Hemophilia* (a deficiency of blood-clotting factors) and diabetes caused by insulin deficiency can be treated in this way. Because the gene product is identical to the human protein, it is usually well tolerated. *Human interferon*, an antiviral protein, and human growth hormone are also being produced this way in microbes. Inserting a human gene into mammals that then produce the gene product has been achieved. In 2009, a human blood clotting factor produced in transgenic goats and purified from the goat's milk was approved by the FDA. The medicine, ATRYN, was approved by the European Medicines Agency for marketing in 2006.

The most revolutionary strategies involve "fixing" abnormal cells by giving them copies of normal genes. In general, this method poses significant targeting problems, because the gene must be introduced into the right kind of cell. For example, placing liver enzymes in fingernails would not correct a metabolic disorder, but when the target cells can be removed and isolated, as in the case of bone marrow, the technique is promising. The removal of a defective gene does not appear to be a practical approach, and the focus has been on adding genes that can take over normal functions. Technical difficulties and unexpected problems, such as tendency to develop leukemia following treatment of SCID (p. 24), have slowed progress but several patients appear to have been cured of their disease.

The foregoing procedures attempt to relieve the symptoms of disease by producing supplemental gene products as medicine, or by inserting genes into defective somatic cells for the body to produce the gene products for itself. The new genes do not change the genetic structure of reproductive cells; because oocytes and sperm retain the original genetic pattern, the genetic defect would be passed to future generations. Researchers are much further away from practical methods of changing the genetic characteristics of reproductive cells. Mouse eggs fertilized outside the body have been treated and transplanted into the uterus of a second mouse for development. The gene that was added was one for a growth hormone obtained from a rat, and the large "supermouse" that resulted demonstrated that such manipulations can be performed. Projects are now under way manipulating the characteristics of valuable animal stocks, such as cattle, goats, or chickens. The potential for altering the genetic characteristics of humans is intimidating. Before any clinical variations on this theme are tested, our society will have to come to grips with some difficult ethical issues and safety concerns.

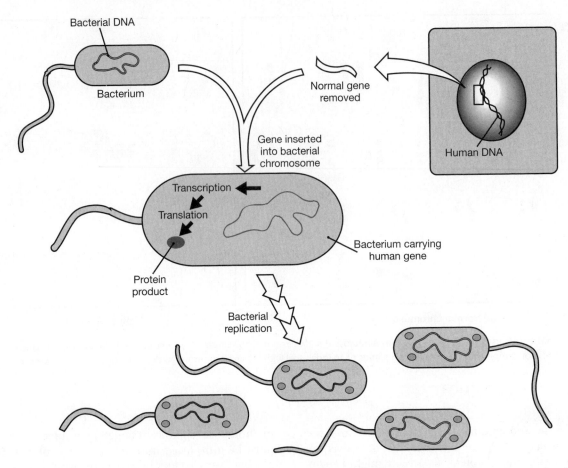

▶ **Figure 16 Gene Splicing.** A gene is removed from a human cell nucleus and is attached to the DNA in a bacterium, where it directs the production of a human protein. Bacterial replication creates a colony of bacteria that share the introduced gene and that can yield large quantities of the protein product.

▶ Drugs and the Plasma Membrane

Many clinically important drugs affect plasma membranes. Although the mechanisms behind the action of general anesthetics, such as *ether, chloroform,* and *nitrous oxide,* have yet to be completely explained, most are lipid-soluble hydrophobic molecules. The potency of most anesthetics is directly correlated with its lipid solubility, which may speed the drug's entry into cells and enhance its ability to block ion channels or change other properties of plasma membranes. The most important clinical result is a reduction in the sensitivity and responsiveness of neurons and muscle cells.

Local anesthetics, such as *procaine* and *lidocaine,* as well as *alcohol* and *barbiturate* drugs, are also lipid soluble. These compounds block sodium channels in the plasma membranes of neurons, reducing or eliminating the responsiveness of the neurons to painful (or any other) stimuli. The very powerful toxin *tetrodotoxin (TTX)* is found in some species of puffers (family Tetraodontidae). Eating the internal organs of these fish causes a severe and potentially fatal form of food poisoning, marked by the disruption of normal neural and muscular activities.

(Nevertheless, the flesh is considered a delicacy in Japan, where it is prepared by specially licensed chefs and served under the name *fugu.*)

Other drugs interfere with membrane receptors for hormones or chemicals that stimulate muscle or nerve cells. *Curare* is a plant extract that interferes with the chemical stimulation of muscle plasma membranes. South American Indians use it to coat their hunting arrows so that wounded prey cannot run away. To prevent reflexive muscle contractions or twitches while surgery is being performed, anesthesiologists may administer a curare derivative (*d-tubocurarine* or a related drug) preoperatively to patients.

▶ Telomeres, Aging, and Cancer

Each **telomere** contains a sequence of roughly 8000 nitrogenous bases, but they are multiple copies of the same base sequence, TTAGGG, repeated over and over again. Telomeres are created by an enzyme called *telomerase.* In most cells, including most tissue stem cells, telomerase is functional early in life, but by adulthood and after cell divisions it becomes inactive. As a result, the telomere segments lost during each mitotic division are not replaced. Eventually, shortening of the telomere reaches a point at which the cell ceases to divide. Embryonic stem cells retain telomerase activity, telomere length, and unlimited potential for cell division.

This mechanism is clearly a major factor in the aging process, since many of the signs of age result from the gradual loss of functional stem cell populations. Experiments are in progress to determine whether activating telomerase (or a suspected alternative repair enzyme) can forestall or reverse the effects of aging. This would seem to be a very promising area of research.

Activate telomerase, and eliminate aging—sounds good, doesn't it? Unfortunately, there's always a catch: In adults, telomerase activation is a key step in the development of cancer. If for some reason a cell with short telomeres does *not* respond normally to repressor genes, it will continue to divide. The result is mechanical damage to the DNA strands, chromosomal abnormalities, and mutations. Interestingly, one of the first consequences of such damage is the abnormal activation of telomerase. Once this occurs, the abnormal cells can continue dividing indefinitely. Telomerase is active in at least 90 percent of all cancer cells. Research is therefore under way to find out how to turn off telomerase that has been improperly activated.

The Tissue Level of Organization

▶ Problems with Serous Membranes

Several clinical conditions, including infection and chronic inflammation, can cause the abnormal buildup of fluid in a body cavity. Other conditions can reduce the amount of lubrication, causing friction between opposing layers of serous membranes. This can promote the formation of **adhesions**—fibrous connections that eliminate the friction by locking the membranes together. Adhesions may also severely restrict the movement of the affected organ or organs and may compress blood vessels or nerves.

Pleuritis, or *pleurisy*, is an inflammation of the pleural cavities. At first the opposing membranes become drier, and scratch against one another, producing a sound known as a *pleural rub*. Adhesions seldom form between the serous membranes of the pleural cavities. More commonly, continued rubbing and inflammation leads to a gradual increase in fluid production to levels well above normal. Fluid then accumulates in the pleural cavities, producing a condition known as *pleural effusions*. Pleural effusions are also caused by heart conditions that elevate the pressure within the pulmonary blood vessels. Fluid then leaks into the alveoli and into the pleural spaces as well, compressing the lungs and making breathing difficult.

Pericarditis is an inflammation of the pericardium. This condition may lead to *pericardial effusion*, an abnormal accumulation of the fluid in the pericardial cavity. When sudden or severe, the fluid buildup can seriously reduce the efficiency of the heart and restrict blood flow through major vessels.

Peritonitis, an inflammation of the peritoneum, can follow infection of, or injury to, the peritoneal lining. Peritonitis is a potential complication of any injury or surgery that penetrates the peritoneal cavity. Appendicitis or stomach ulcers may also perforate the walls of the gastrointestinal tract and lead to peritonitis. Adhesions are common following peritoneal infections and may lead to constriction and blockage of the intestinal tract.

Liver disease, kidney disease, or heart failure can cause an accumulation of fluid in the peritoneal cavity. Called **ascites** (a-SĪ-tēz), this accumulation creates a characteristic abdominal swelling. The pressure and distortion of internal organs by the excess fluid can lead to symptoms such as heartburn, indigestion, shortness of breath, and low-back pain.

▶ Cancer: A Closer Look

In the United States, the current estimated lifetime risk of developing some form of cancer is 45 percent for males and 38 percent for females. In 2010, 575,000 people in the United States died of some form of cancer, making it second only to heart disease as a cause of mortality in the U.S. population. Because both population size and average age are increasing, just comparing the numbers of deaths from year to year can be misleading. The relevant statistics are better presented in terms of the cancer rate per 100,000 population, age adjusted to match the most recent census data (2000). For 2012, this was 186.2 per 100,000 population. From 2000 to 2006, the United States had a decrease in cancer incidence of 6.9 percent, while the cancer death rate over this time decreased 9.5 percent, and the absolute death rate declined 5.1 percent. Much of this improvement has been from the declining incidence of lung cancer due to the decreased incidence of smoking. For those interested in more detailed data analyses, visit the National Center for Health Statistics at *http://www.cdc.gov/nchs*.

Causes of Cancer

Gene mutations can occur in cells during a person's lifetime or can be inherited. A mutation may have little effect or may predispose a person to developing cancer. By definition, inherited cancer involves gene(s) provided by the sperm or oocyte at fertilization; as a result, these genes are in every cell of the individual's body. Such people have a much higher risk of developing a specific cancer than the general population. However, not *everyone* with these genes gets cancer, and this indicates that other genes and/or environmental factors must act as a "trigger." For the general population, it is the interaction of genetic and environmental factors that causes most cancers. Some inherited gene mutations are associated with 5 to 10 percent of all cancers and over 50 hereditary cancer syndromes are known. The *BRCA1* and *BRCA2* genes increase the risk of breast, ovarian, pancreatic, and prostate cancers.

Genetic Factors

An individual born with genes that increase the likelihood of cancer is said to have a *hereditary predisposition* for the disease. Such a person may never develop cancer, but his or her chances are higher than average. Often, there is a family history of cancer at a young age in several close relatives. The inherited genes generally affect the abilities of tissues to metabolize toxins, control mitosis and growth, perform repairs after injury, or identify and destroy abnormal tissue cells. As a result, body cells become sensitive to local or environmental factors that would have less effect on cells from persons lacking these genes.

The majority of cancers result from somatic-cell mutations that modify genes involved in cell growth, differentiation, or mitosis. As a result, an ordinary cell is converted into a cancer cell. The modified genes are called **oncogenes** (ON-kō-jēnz); the normal genes are called *proto-oncogenes. Oncogene activation* occurs by the alteration of normal somatic genes. Because these mutations do not affect reproductive cells, the cancers caused by active oncogenes are not inherited.

A proto-oncogene, like other genes, has a regulatory component that turns the gene "on" and "off" and a structural component that contains the mRNA triplets that determine protein structure. Mutations in either portion of the gene may convert it to an active oncogene. A change of just one nucleotide out of a chain of 5000 can convert a normal proto-oncogene to an active oncogene. In some cases, a viral infection can trigger the activation of an oncogene. For example, a few of the human papilloma viruses (HPV) that cause genital warts are responsible for many cases of cervical cancer. Vaccines against the strains of HPV most implicated in developing cervical cancer were approved in 2006 and have shown over 90 percent efficacy in preventing precancerous changes in cervical (Pap) smear histological testing. If a person with one form of stomach lymphoma (a lymphatic system cancer) is also infected with the bacteria *Heliobacter pylori*, eradication of the bacteria has been followed by the disappearance of the lymphoma.

More than 100 oncogenes have been identified. In addition, a group of anticancer genes has been discovered. These genes, called **tumor-suppressing genes (TSGs)**, or *anti-oncogenes*, suppress division and growth in normal cells, and about 30 have been identified. Mutations that alter TSGs make oncogene activation more likely. Such mutation has been suggested as an important factor in promoting several cancers, including a number of blood cell cancers, breast cancer, colon cancer, and ovarian cancer. Examples of important suppressor genes are the genes *p53* and *p16*. Mutations affecting the *p53* gene are present in the majority of cancers of the colon, breast, and liver. Abnormal *p16* gene activity may be involved in as many as half of all cancer cases.

Environmental Factors

Many cancers can be directly or indirectly attributed to environmental factors called **carcinogens** (kar-SIN-ō-jenz). Carcinogens stimulate the conversion of a normal cell to a cancer cell. Some carcinogens are *mutagens* (MŪ-ta-jenz)—that is, they damage DNA strands and may cause chromosomal breakage. All forms of high-energy radiation, including cosmic rays, x-rays, and UV rays as well as radioisotopes, are mutagens that have carcinogenic effects.

The environment contains many chemical carcinogens. Plants manufacture poisons that protect them from insects and other predators, and although their carcinogenic activities are often relatively weak, many common spices, vegetables, and beverages contain compounds that are carcinogens if consumed in large quantities. Animal tissues may also store or concentrate toxins, and hazardous compounds of many kinds can be swallowed in contaminated food. A variety of laboratory and industrial chemicals, such as coal tar derivatives and synthetic pesticides, have been shown to be carcinogenic. From studies that compared cancer incidence in twins with other data, it has been estimated that 70–80 percent of all cancers are the result of chemical or environmental factors, and 40 percent are due to a single source of carcinogens: cigarette smoke. People have prevented many incidences of cancer by making lifestyle changes, for example, quitting or never starting tobacco use.

Specific carcinogens will affect only those cells capable of responding to that particular physical or chemical stimulus. The responses vary because differentiation produces cell types with specific sensitivities. For example, benzene can produce a cancer of the blood, cigarette smoke a lung cancer, and vinyl chloride a liver cancer. Very few stimuli can produce cancers throughout the body. Radiation is a notable exception. In general, cells undergoing mitosis are most likely to be vulnerable to chemical or radiational carcinogens. As a result, cancer rates are highest in epithelial tissues, like the skin and lining of the intestines, where stem cell divisions occur rapidly, and lowest in nervous and muscle tissues, where divisions do not normally occur.

Detection and Incidence of Cancer

Physicians who specialize in the identification and treatment of cancers are called **oncologists** (on-KOL-ō-jists; *onkos,* mass). Pathologists and oncologists classify cancers according to their cellular appearance and their sites of origin. More than a hundred kinds have been described, but broad categories are used that indicate the location of the primary tumor. A **tumor** is defined as a "new growth" resulting from uncontrolled cell division. A tumor can be *malignant* or *benign* and may *metastasize* (spread to other areas of the body) rapidly or very slowly. Only malignant tumors are called cancers. **Table 10** summarizes information about benign and malignant tumors (cancers) associated with the major tissues of the body.

A statistical profile of cancer incidences and survival rates in the United States is shown in **Table 11**. The numbers from other countries are different. For example, *bladder cancer* is common in Egypt, *stomach cancer* in Japan, and *liver cancer* in Africa. Variations in the combination of genetic factors and dietary, infectious, and other environmental factors are thought to be responsible for these differences.

Clinical Staging and Tumor Grading

The detection of a cancer often begins during a screening physical examination, when the physician notices an abnormal lump or growth. Many laboratory and diagnostic tests are necessary for the correct diagnosis of cancer. Information is usually obtained by the histological examination of a tissue sample, or *biopsy*, typically supplemented by medical imaging and blood studies. A biopsy is one of the most significant diagnostic procedures, because it permits a direct look at the tumor cells. Not only do

malignant cells have an abnormally high rate of mitosis, but they are also structurally distinct from healthy body cells.

If the tissue appears cancerous, other important questions must be answered, including the following: What is the measurable size of the primary tumor? Has the tumor invaded surrounding tissues? Has the cancer already metastasized to develop secondary tumors? Are any regional lymph nodes affected? The answers to these questions are combined with observations from the physical exam, the biopsy results, and information from any imaging procedures to arrive at an accurate diagnosis and prognosis.

In an attempt to develop a standard system, national and international cancer organizations have devised the **TNM system** for staging (i.e., identifying the stage of progression of) cancers. The letters stand for *tumor* (*T*) size and invasion, *lymph node* (*N*) involvement, and degree of *metastasis* (*M*):

- Tumor size is graded on a scale of 0 to 4. T0 indicates the absence of a primary tumor, and the largest dimensions and greatest amount of invasion are categorized as T4.

- Lymph nodes filter the tissue fluids from nearby capillary beds. The fluid, called *lymph*, then returns to the general lymphatic circulation. Once cancer cells have entered the lymphatic system, they can spread very quickly throughout the body. Lymph node involvement is graded on a scale of 0 to 3. A designation of N0 indicates that no lymph nodes have been invaded by cancer cells. A classification of N1 to N3 indicates the involvement of increasing numbers of lymph nodes:

  N1 indicates the involvement of a single lymph node less than 3 cm in diameter.

  N2 includes one medium-sized (3–6 cm) node or multiple nodes smaller than 6 cm.

  N3 indicates the presence of a single lymph node larger than 6 cm in diameter, regardless of whether other nodes are involved.

- Metastasis is graded on a scale of 0 to 1. M0 indicates that there is no evidence of metastasis, whereas M1 indicates that the cancer cells have produced secondary tumors in other portions of the body.

This grading system provides a general overview of the progression of the disease. For example, a tumor classified as T1N1M0 has a better prognosis than one classified as T4N2M1. The latter tumor would be much more difficult to treat. The grading system alone does not provide all the information needed to plan treatment, however, because different types of cancer progress in different ways. Therapies must vary accordingly. Thus, *leukemia,* a cancer of the blood-forming tissues, is treated differently than colon cancer. We will consider specific treatments in discussions dealing with cancers that affect individual body systems; the next section provides a general overview of the strategies used to treat cancer.

Cancer Treatment

It is unfortunate that the media tend to describe cancer as though it were one disease rather than many. This simplistic perspective fosters the belief that some dietary change, air ionizer, or wonder drug will be found that can prevent or cure the affliction. No single, universally effective cure for cancer is likely, because there are too many separate causes, underlying mechanisms, and individual differences.

The goal of cancer treatment is to achieve remission. A tumor in **remission** either ceases to grow or decreases in size. The treatment

Table 10	Benign and Malignant Tumors in the Major Tissue Types
Tissue	**Description**
Epithelia	
Carcinomas	Any cancer of epithelial origin
Adenocarcinomas	Cancers of glandular epithelia
Angiosarcomas	Cancers of endothelial (vascular) cells
Mesotheliomas	Cancers of mesothelial cells
Connective tissues	
Fibromas	Benign tumors of fibroblast origin
Lipomas	Benign tumors of adipose tissue
Liposarcomas	Cancers of adipose tissue
Leukemias	Cancers of blood-forming tissues
Lymphomas	Cancers of lymphoid tissues
Chondromas	Benign tumors in cartilage
Chondrosarcomas	Cancers of cartilage
Osteomas	Benign tumors in bone
Osteosarcomas	Cancers of bone
Muscle tissues	
Myxomas	Benign muscle tumors
Myosarcomas	Cancers of skeletal muscle tissue
Cardiac sarcomas	Cancers of cardiac muscle tissue
Leiomyomas	Benign tumors of smooth muscle tissue
Leiomyosarcomas	Cancers of smooth muscle tissue
Neural tissues	
Gliomas	Cancers of neuroglial origin
Neuroblastomas	Cancers of neuronal origin

			Five-Year Survival Rates	
			Diagnosis Date	
Site	Estimated New Cases (2013)	Estimated Deaths (2013)	1974–76	2003–2009
Digestive tract				
Esophagus	17,990	15,280	5%	19.3%
Stomach	21,600	10,990	15%	28.8%
Colon and rectum	142,820	24,530	50%	66.1%
Respiratory tract				
Lung and bronchus	228,190	159,480	12%	17.5%
Urinary tract				
Kidney and renal pelvis	65,150	13,680	52%	73.3%
Urinary bladder	72,570	15,210	73%	79.8%
Reproductive system				
Breast	234,580	40,030	75%	90.3%
Ovary	22,240	14,030	37%	44.1%
Testis	7920	370	79%	96.5%
Prostate gland	238,590	29,720	67%	99.7%
Nervous system	23,130	14,080	22%	34.8%
Skin (melanoma only)	82,770	9480	80%	93.1%

Table 11   Cancer Incidences and Survival Rates in the United States

Data courtesy of the American Cancer Society and National Cancer Institute.

of malignant tumors must accomplish one of these two objectives to produce remission:

1. **The surgical removal or destruction of individual tumors.** Tumors containing malignant cells can be surgically removed or destroyed by radiation, heat, or freezing. These techniques are highly effective if the treatment is undertaken before metastasis has occurred. For this reason, early detection is important in improving survival rates for all forms of cancer.

2. **The killing of metastasized cells throughout the body.** This is much more difficult and potentially dangerous, because healthy tissues are likely to be damaged at the same time. At present, the most widely approved treatments are chemotherapy and radiation.

Chemotherapy may involve the administration of drugs that will either kill the cancerous tissues or prevent mitotic divisions. Traditional chemotherapy drugs damage actively dividing cells and also affect stem cells in normal tissues, and the side effects are usually unpleasant. For example, because some chemotherapy slows the regeneration and maintenance of epithelia of the skin and digestive tract, patients often lose their hair and experience nausea and vomiting. Several drugs are often administered simultaneously or in sequence, because, over time, cancer cells can develop a resistance to a single drug. Chemotherapy is used in the treatment of many kinds of metastasized cancer.

Massive doses of total body irradiation are sometimes used to treat advanced cases of *lymphoma*, a cancer of the immune system. In this rather drastic procedure, enough radiation is administered to kill all the blood-forming cells in the body. After treatment, new blood cells must be provided by a bone marrow transplant. In later sections dealing with the lymphatic system, we will discuss marrow transplants, lymphomas, and other cancers of the blood.

An understanding of molecular mechanisms and cell biology is leading to new approaches that may revolutionize cancer treatment. One approach focuses on the fact that cancer cells are ignored by the immune system. In **immunotherapy**, chemicals are administered that help the immune system recognize and attack cancer cells. More elaborate experimental procedures involve the creation of customized antibodies by the gene-splicing techniques discussed on page 33. The resulting antibodies are specifically designed to attack the tumor cells in each particular patient. This technique shows promise, but it remains difficult, costly, and very labor-intensive.

A second approach is targeted "designer" cancer drugs. One type of cancer, chronic myelogenous leukemia, involves the activity of an abnormal enzyme. A drug has been developed that inactivates this enzyme but has no effect on normal enzymes, and thus does not affect normal cells. In early trials of the drug imatinib (*Gleevec*), complete remission occurred in up to 95 percent of CML patients treated. Gleevec has also been effective against other types of cancer. These "Targeted Therapy" medicines usually have fewer side effects, but have been developed for just a few types of cancer.

Cancer and Survival

Advances in chemotherapy, radiation procedures, and molecular biology have produced significant improvements in the survival rates of several types of cancer patients. However, the improved survival rates indicated in **Table 11** reflect advances not only in therapy, but also in early detection. Much of the credit goes to

Tissue Level

increased public awareness and concern about cancer. In general, the odds of survival increase markedly if the cancer is detected early, especially before it undergoes metastasis. The American Cancer Society has identified seven "warning signs" that mean it's time to consult a physician. These signs are presented in Table 12. For persons at higher risk of cancer, various screening tests such as mammography for breast cancer and colonoscopy for colon cancer have improved early detection.

▶ Tissue Structure and Disease

Pathologists focus their activities on diagnosis rather than treatment. In their analyses, pathologists integrate anatomical and histological observations to determine the nature and severity of a disease. Because disease processes affect the histological organization of tissues and organs, **biopsies**, or tissue samples, often play a key role in their diagnoses.

▶Figure 17 diagrams the histological changes induced in the respiratory epithelium by one relatively common irritating stimulus, cigarette smoke. The normal respiratory epithelium is shown in ▶Figure 17a. The first abnormality to be observed in a smoker is **dysplasia** (dis-PLĀ-zē-uh), a change in the shape, size, and organization of tissue cells. The normal trachea (windpipe) and its branches are lined by a pseudostratified ciliated columnar epithelium. The cilia move a mucous layer that traps foreign particles and moistens incoming air. The drying and chemical effects of smoking first paralyze the cilia, halting the movement of mucus (▶Figure 17b). As mucus builds up, the individual coughs to dislodge it (the well-known "smoker's cough"). Dysplasia is generally a response to chronic irritation or inflammation, and the changes are reversible. However, dysplasia increases the risk of cancer formation in that tissue; in a tissue not subject to abnormal stresses, dysplasia may be the first indication of a developing cancer.

Epithelia and connective tissues may undergo more radical changes in structure, caused by the division and differentiation of stem cells. **Metaplasia** (me-tuh-PLĀ-zē-uh) is a structural change that dramatically alters the character of the tissue. In our example, over time heavy smoking causes the epithelial cells to lose their cilia altogether.

As metaplasia progresses, the epithelial cells produced by stem cell divisions no longer differentiate into ciliated columnar cells. Instead, they form a stratified squamous epithelium that provides greater resistance to drying and chemical irritation (▶Figure 17c). This epithelium protects the underlying tissues more effectively, but it eliminates the moisturization and cleaning properties of the epithelium. Cigarette smoke will now have an even greater effect on more delicate portions of the

Table 12	Seven Warning Signs of Cancer

Change in bowel or bladder habits

A sore that does not heal

Unusual bleeding or discharge

Thickening or lump in breast or elsewhere

Indigestion or difficulty in swallowing

Obvious change in a wart or mole

Nagging cough or hoarseness

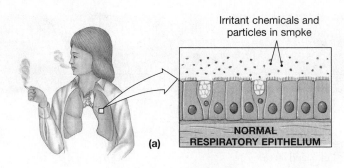

Irritant chemicals and particles in smoke

(a) NORMAL RESPIRATORY EPITHELIUM

Reversible

(b) The cilia of respiratory epithelial cells are damaged and paralyzed by exposure to cigarette smoke. These changes cause the local buildup of mucus and reduce the effectiveness of the epithelium in protecting deeper, more delicate portions of the respiratory tract.

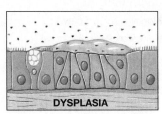

DYSPLASIA

Reversible

(c) In metaplasia, a tissue changes its structure. In this case the stressed respiratory surface converts to a stratified epithelium that protects underlying connective tissues but does nothing for other areas of the respiratory tract.

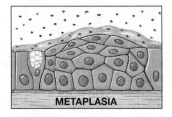

METAPLASIA

Irreversible

(d) In anaplasia, the tissue cells lose their resemblance to normal cells, begin dividing rapidly, and form a tumor.

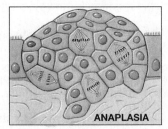

ANAPLASIA

▶Figure 17 **Changes in a Tissue under Stress.**

respiratory tract. Fortunately, metaplasia is reversible, and unless a malignant tumor has formed, the epithelium will gradually return to normal if the individual quits smoking.

In **anaplasia** (a-nuh-PLĀ-zē-uh), tissue organization breaks down. Tissue cells change size and shape, typically becoming unusually large or small (▶Figure 17d) and losing any resemblance to mature tissue cells. Anaplasia is characteristic of most if not all cancers; it occurs in smokers who develop one form of lung cancer. In anaplasia, the cells divide more frequently but not all divisions proceed in the normal way. Many of the tumor cells have abnormal chromosomes. Unlike dysplasia and metaplasia, anaplasia is irreversible.

COVERING, SUPPORT, AND MOVEMENT

The integumentary, skeletal, and muscular systems provide physical protection, structural support, and mobility, and together these organ systems account for more than half of the weight of the human body. These systems have diverse functions, but their high degree of structural and functional interdependence makes them a logical group. The skin covers and protects the entire body, but it is most closely associated with superficial skeletal muscles. Interactions between skeletal muscles and bones are so extensive that the two are often considered to be part of a single *musculo-skeletal system.*

There are direct anatomic connections among these systems. There are also physiological connections. For example, muscular contractions can only occur when the extracellular concentration of calcium is maintained within relatively narrow limits. Those limits can only be maintained with the help of a hormone derived from the skin that affects calcium mobilization from the skeleton. Because they are structurally and functionally interdependent, clinical problems that originate within one of these systems can have secondary effects on the others. We will encounter several examples as we proceed through the sections within this unit.

MAJOR SECTIONS INCLUDED WITHIN THIS UNIT:

The Integumentary System

The Skeletal System

The Muscular System

END-OF-UNIT CLINICAL PROBLEMS

The Integumentary System

The structures of the integumentary system include the skin and its accessory organs, such as hair, nails, and several types of exocrine glands. The integumentary system has a variety of functions, including protection of the underlying tissues, maintenance of body temperature, excretion of salts and water in sweat, cutaneous sensation, and the production of vitamin D_3.

The skin is the most visible organ of the body. As a result, abnormalities are easily recognized. A bruise, for example, typically creates a swollen and discolored area where blood has leaked into tissues surrounding damaged blood vessels. Changes in color, tone, texture, and the overall condition of the skin commonly accompany illness or disease. These changes can assist in diagnosis. For instance, extensive bruising without obvious cause may indicate a blood-clotting disorder, and yellowish skin, sclera, and mucous membranes may signify *jaundice*, a sign that often points to some type of liver disorder. The general condition of the skin can also be significant: In addition to color changes, changes in skin flexibility, elasticity, dryness, or sensitivity commonly follow malfunctions of other organ systems.

Examination of the Skin

In examining a patient, dermatologists (skin specialists) use a combination of investigative interviews ("What has been in contact with your skin lately?" or "How does it feel?") and physical examination to arrive at a diagnosis. The condition of the skin is carefully observed. Notes are made about the presence of **lesions**, which are changes in skin structure caused by trauma or disease processes. Lesions are also called **skin signs**, because they are measurable, visible abnormalities of the skin surface. **Figure 18** diagrams the most common skin signs and related disorders.

The distribution and appearance of lesions can be an important clue to the source of the problem. For instance, in *shingles* (*herpes zoster infection*), blisters on the skin occur in the area(s) of skin (dermatomes) innervated by the infected dorsal root sensory nerve ganglia. A ring of slightly raised, scaly (papular) lesions is typical of fungal infections that may affect the trunk, scalp, or nails. Examples of skin infections and allergic reactions are given in **Table 13**, with descriptions of the related skin signs. We consider skin lesions caused by trauma later, in the section titled "Trauma to the Skin" (p. 48).

Table 13 summarizes signs on the surface of the skin, but signs involving the accessory organs of the skin can also be important. For instance,

- Nails have a characteristic shape that can change due to an underlying disorder. An example is *clubbing* of the fingernails, a sign of *emphysema* or *congestive heart failure*. In these conditions, the fingertips broaden and the nails become distinctively curved, possibly the effect of hypoxia at the nail root.

- The condition of the hair can be an indicator of a person's overall health. For example, depigmentation and coarseness of hair occur in the protein deficiency disease *kwashiorkor*.

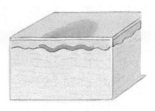

A flat **macule** is a localized change in skin color. Example: freckles

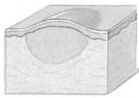

Accumulation of fluid in the papillary dermis may produce a **wheal**, a localized elevation of the overlying epidermis. Example: hives (urticaria)

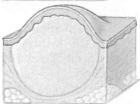

A **papule** is a solid elevated area containing epidermal and papillary dermal components. Example: mosquito or other insect bite

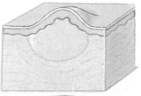

Nodules are large papules that may extend into the subcutaneous layer. Example: cyst

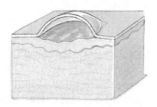

A **vesicle**, or blister, is a papule with a fluid core. A large vesicle may be called a bulla. Example: second-degree burn

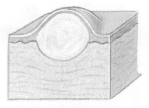

A **pustule** is a papule-sized lesion filled with pus. Example: acne pimple

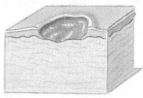

An **erosion**, or ulcer, may occur following the rupture of a vesicle or pustule. Eroded sites have lost part or all of the normal epidermis. Example: decubitus ulcer

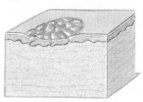

A **crust** is an accumulation of dried sebum, blood, or interstitial fluid over the surface of the epidermis. Examples: seborrheic dermatitis, scabs, impetigo

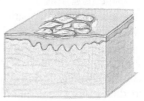

Scales form as a result of abnormal keratinization. They are thin plates of cornified cells. Example: psoriasis

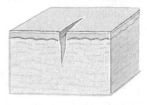

A **fissure** is a split in the integument that extends through the epidermis and into the dermis. Example: athlete's foot

▶ **Figure 18 Skin Signs.**

Table 13	Skin Signs of Various Disorders	
Cause	**Examples**	**Resulting Skin Lesions**
Viral infections	Chicken pox (Herpes varicella)	Lesions begin as macules and papules, but develop into vesicles
	Measles (rubeola)	A maculopapular rash that begins at the face and neck and spreads to the trunk and limbs
	Erythema infectiosum (fifth disease)	A maculopapular rash that begins on the cheeks (slapped-cheek appearance) and spreads to the limbs
	Herpes simplex	Raised vesicles that heal with a crust
Bacterial infections	Impetigo	Vesiculopustular lesions with exudate and yellow crusting
Fungal infections	Ringworm	An annulus (ring) of scaly papular lesions with central clearing
Parasitic infections	Scabies	Linear burrows with a small, very pruritic (itchy) papule at one end
	Lice (pediculosis)	Dermatitis: excoriation (scratches) due to pruritus (itching)
Allergies to medication	Penicillin	Wheals (urticaria/hives)
Food allergies	Eggs, certain fruits	Wheals
Environmental allergies	Poison ivy	Vesicles

Table 14	Examples of Vascular Lesions	
Lesion	**Features**	**Some Possible Causes**
Ecchymosis	Reddish purple, blue, or yellow bruising related to trauma	Trauma; blood-clotting disorder; some vitamin deficiencies; thrombocytopenia; increased tendency to bruise can be due to skin aging or sun damage
Hematoma	Pooling and possible clotting of blood, forming a mass; associated with pain and swelling	Trauma, broken blood vessel
Petechiae	Small red-to-purple pinpoint dots appearing in clusters	Leukemia; septicemia (toxins in blood); thrombocytopenia
Erythema	Red, flushed color of skin due to dilation of blood vessels in the skin	*Extensive*: drug reactions; *Localized*: first-degree burns, sunburn, contact dermatitis

Diagnosing Skin Conditions

Table 14 lists several major types of vascular lesions involving the skin. A single skin lesion can have multiple causes. This is one of the challenges facing dermatologists; the signs may be apparent, but the underlying causes may not. Making matters more difficult, many skin disorders produce the same signs and symptoms. For example, **pruritus** (proo-RĪ-tus), an irritating itching sensation, is an extremely common symptom associated with a variety of skin conditions. Questions about the patient's medical history, medications, possible sources of infection, and environmental exposure, as well as other signs, can be the key to making an accurate diagnosis.

Pain is another common symptom of many skin disorders. Although pain is unwelcome, cutaneous sensation is an important function of the integumentary system. Its importance is dramatically demonstrated in *Hansen's disease*, or *leprosy*. Hansen's disease is caused by a bacterium that has an affinity for cool regions of the body. The bacterium destroys cutaneous nerve endings that are sensitive to touch, pain, heat, and cold. Damage to the distal tissues occurs and accumulates unnoticed, because the person is no longer aware of painful stimuli. We will consider Hansen's disease further in a later section. Long-standing, poorly controlled diabetes mellitus can cause similar nerve damage, particularly to the feet. The injuries and non-healing infections that result may lead to amputations.

▶ **Figure 19** is an overview of skin disorders. Diagnostic tests that may prove useful in distinguishing among them include the following:

- **The scraping and microscopic examination of affected tissue**, a process often performed to check for fungal infections.
- **The culturing of fluid** removed from a lesion to identify infectious pathogens.
- **Biopsy** of affected tissue to view its cell structure.
- **Skin tests**, through which various types of skin or systemic disorders can be detected.

In a skin test, a localized area of the skin is exposed to an inactivated pathogen, a portion of a pathogen, or a substance capable of producing an immunologic reaction in sensitive individuals.

Exposure occurs by injection or surface application. For example, in a skin test for *tuberculosis*, a small quantity of tuberculosis antigens is injected *intradermally* (*intra*, within). If the individual has anti-tuberculin antibodies from past or current tuberculosis infection, a red papule or *erythema* (*erythros*, red; a change in skin color) and swelling will occur at the site of the injection 24–72 hours later. *Patch* or *scratch testing* is used to check sensitivity to *allergens*, environmental agents that can cause allergic reactions. In a patch test, the allergen is applied to the surface of the skin. In a scratch test the epidermis is slightly broken before applying the allergen. If erythema, swelling, or itching develops, the individual is sensitive to that allergen. **Table 15** summarizes information about common infections of the integumentary system.

Disorders of Keratin Production

Not all skin signs are the result of infection, trauma, or allergy; some are the normal response to environmental stresses. One common response is the excessive production of keratin, a process called **hyperkeratosis** (hī-per-ker-a-tŌ-sis). The most obvious effects—calluses and corns—are easily observed. Calluses are thickened patches that appear on already thick-skinned areas, such as the palms of the hands or the soles or heels of the foot, in response to chronic abrasion and distortion. Corns are more localized areas of excessive keratin production that form in areas of thin skin on or between the toes.

In **psoriasis** (so-RĪ-a-sis), stem cells in the stratum germinativum are unusually active, causing hyperkeratosis in specific areas, often the scalp, elbows, palms, soles, groin, or nails. Normally, an individual stem cell divides once every 20 days, but in psoriasis it may divide every day and a half. Keratinization is abnormal and typically incomplete by the time the outer layers are shed. The affected areas have red bases covered with vast numbers of small, silvery scales that continuously flake off. Psoriasis develops in 20–30 percent of individuals with an inherited tendency for the condition. Roughly 5 percent of the U.S. population has psoriasis to some degree, frequently aggravated by stress and anxiety. Most cases are painless and controllable, but not curable.

Xerosis (ze-RŌ-sis), or dry skin, is a common complaint of the elderly and people who live in arid climates. In xerosis, plasma membranes in the outer layers of skin gradually deteriorate and the stratum corneum becomes more a collection of scales than a single sheet. The scaly surface is much more permeable than an intact layer of keratin, and the rate of insensible perspiration increases. In persons with severe xerosis, the rate of insensible perspiration may increase by up to 75 times.

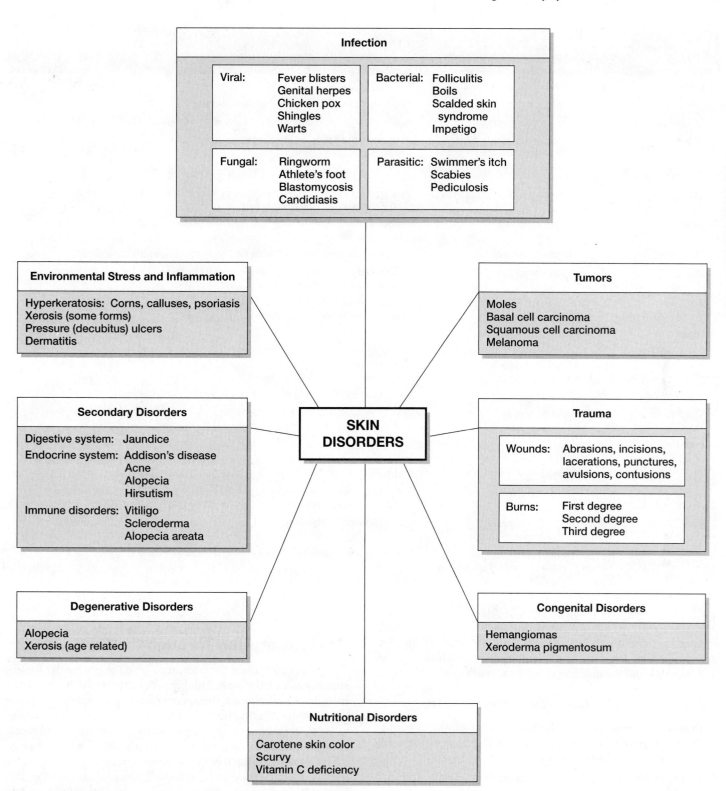

Infection

Viral: Fever blisters
Genital herpes
Chicken pox
Shingles
Warts

Bacterial: Folliculitis
Boils
Scalded skin
syndrome
Impetigo

Fungal: Ringworm
Athlete's foot
Blastomycosis
Candidiasis

Parasitic: Swimmer's itch
Scabies
Pediculosis

Environmental Stress and Inflammation

Hyperkeratosis: Corns, calluses, psoriasis
Xerosis (some forms)
Pressure (decubitus) ulcers
Dermatitis

Tumors

Moles
Basal cell carcinoma
Squamous cell carcinoma
Melanoma

Secondary Disorders

Digestive system: Jaundice
Endocrine system: Addison's disease
Acne
Alopecia
Hirsutism
Immune disorders: Vitiligo
Scleroderma
Alopecia areata

SKIN DISORDERS

Trauma

Wounds: Abrasions, incisions,
lacerations, punctures,
avulsions, contusions

Burns: First degree
Second degree
Third degree

Degenerative Disorders

Alopecia
Xerosis (age related)

Congenital Disorders

Hemangiomas
Xeroderma pigmentosum

Nutritional Disorders

Carotene skin color
Scurvy
Vitamin C deficiency

▶ **Figure 19** **Disorders of the Integumentary System.**

Pressure on the skin, another form of stress, can produce pressure ulcers, also known as *decubitus ulcers* or *bedsores*. *Decubitus* means "to lie down"; an *ulcer* is a localized loss of epithelium. Pressure ulcers form where dermal blood vessels are compressed against deeper structures such as bones or joints, so that local circulation is reduced enough to damage dependent tissues. Debilitated persons who cannot move to relieve such pressure are most vulnerable to these ulcers.

Dermatitis

Inflammation is a complex process that helps defend against pathogens and injury. Because skin contains an abundance of sensory receptors, inflammation can be very painful.

Dermatitis (der-muh-TĪ-tis) is an inflammation of the skin that involves primarily the papillary layer. The inflammation typically begins in a part of the skin exposed to infection or

Integumentary System

Table 15	Common Infectious Diseases of the Integumentary System	
Disease	**Organism (Name)**	**Description**
Bacteria		
Folliculitis	*Staphylococcus aureus*	Infections of hair follicles may form pimples and boils
Scalded skin syndrome	*Staphylococcus aureus*	In infants; large areas of epidermis blister, peel, and leave oozing red areas
Impetigo	*Staphylococci, Streptococci,* or both	Pustules form on skin, dry, and become yellow crusts
Viruses		
Oral herpes	Herpes simplex 1	Vesicles (blisters), also called cold sores or fever blisters, form usually on lips (vesicles heal, but may episodically reoccur)
Genital herpes	Herpes simplex 2	Lesions similar to those in oral herpes form on external genitalia; vesicles disappear and reappear
Chicken pox (*varicella*)	Herpes varicella-zoster	First infection: small, red macules form flaccid vesicles, which dry and become crusts
Shingles (*zoster*)	Herpes varicella-zoster	Reactivated infection; distinctive vesicular lesions form a pattern along sensory nerves; severe, prolonged pain may follow attacks
Warts	Human papillomaviruses	Rough papules form in the epidermis; genital warts may be sexually transmitted and may promote cervical cancer
Fungi		
Ringworm (*tinea*)	*Epidermophyton, Microsporum,* and *Trichophyton*	Dry, scaly lesions form on the skin in different parts of the body, including the scalp (*tinea capitis*), body (*tinea corporis*), groin (*tinea cruris*), foot (*tinea pedis*, which may be moist), and nails (*tinea unguium*)
Blastomycosis	*Blastomyces dermatitidis*	Pustules and abscesses form in the skin; may affect other organs
Candidiasis	*Candida albicans*	Normal inhabitant of the human body surface; red, frequently moist lesions form in skin infections (such as diaper rash); nails may also become infected
Parasites		
Swimmer's itch	*Schistosoma* worms (flukes)	Freshwater larval stages of schistosome worms (flukes) burrow into skin and cause itching
Scabies	*Sarcoptes scabiei* (itch mite)	Itch mite burrows and lays eggs in skin, often in areas between fingers and at the wrists; entrance marked by tiny, scaly swellings that become red and intensely itchy
Pediculosis	*Pediculus humanus* (human body louse)	Lice infestations on body and scalp; bites produce redness, dermatitis, and itching
	Phthirus pubis (pubic louse)	"Crabs"; lice infestation of the pubic area; bites also cause itching

irritated by chemicals, radiation, or mechanical stimuli. Dermatitis may produce an annoying itch. Sometimes, the condition can be quite painful, and the inflammation can spread rapidly across the entire integument.

Dermatitis has many forms, some of them common:

- **Contact dermatitis** generally occurs in response to strong chemical irritants. It produces an itchy rash that may spread to other areas; poison ivy is an example.

- **Eczema** (EK-se-muh), also called atopic dermatitis, can be triggered by temperature changes, fungi, chemical irritants, greases, detergents, or stress. Hereditary factors, environmental factors, or both can promote its development.

- **Diaper rash** is a localized dermatitis caused by a combination of moisture, irritating chemicals from fecal or urinary wastes, and flourishing microorganisms, frequently the fungus/yeast *Candida*.

- **Urticaria** (ur-ti-KAR-ē-uh), also called wheals or *hives*, is an extensive allergic response to a food, a drug, an insect bite, infection, stress, or some other stimulus.

Drug Absorption Through the Skin

The integument covers the body, making it the first barrier to substances entering the body. This barrier can sometimes be breached in order to administer therapeutic drugs transdermally. As the name implies, this method of introducing drugs requires the movement of the substance across the epidermal plasma membranes. Recall that plasma membranes are composed of a lipid bilayer, which creates a barrier that prevents water from rapidly entering or leaving the cell. However, lipid-soluble compounds can cross this barrier. If you dissolve a drug in an oil or some other lipid-soluble solvent, the drug may diffuse across the plasma membranes with the solvent. The movement is slow, primarily due to the density of the stratum corneum, but once a drug reaches the underlying connective tissues, it will be absorbed into the circulation.

A useful technique for long-term drug administration is the placement of a sticky, drug-containing patch over an area of thin skin (▶ Figure 20). To overcome the slow rate of diffusion, the patch must contain an extremely high concentration of the drug. This procedure is called *transdermal administration*. A single patch

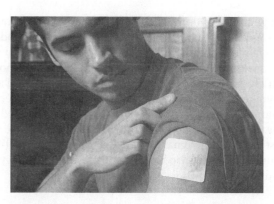

▶Figure 20 **A Skin Patch for Transdermal Drug Administration.**

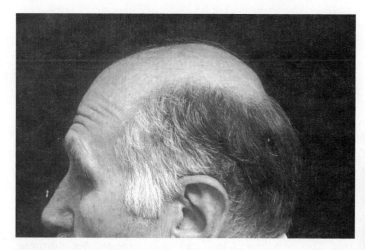

▶Figure 21 Male Pattern Baldness.

may work for several days, making daily pills unnecessary. Among the thirty-five drugs now approved to use transdermally are the following:

- Scopolamine, which by affecting the nervous system, can control the nausea associated with motion sickness.
- Nitroglycerin, which is used to improve blood flow within heart muscle to prevent a heart attack.
- Estrogens, which are administered to women to reduce symptoms of menopause or combined with progesterone for birth control.
- Various medications and drugs used to control high blood pressure.
- Nicotine, an addictive compound in tobacco, for use in suppressing the craving for a cigarette. The transdermal dosage of nicotine in the patch can gradually be reduced in small, controlled steps.

Dimethyl sulfoxide (*DMSO*) is a solvent that rapidly crosses the skin and other epithelium. It can carry some medicines with it. In the United States it is FDA approved to place in the urinary bladder to treat interstitial cystitis, and as a cryoprotectant when mixed with bone marrow or umbilical cord blood before freezing. Veterinarians use it topically (applied externally) to reduce inflammation in injured muscles and joints of large domesticated animals. In Great Britain and other countries it is used topically with anti-inflammatory and antiviral medicines.

Low-frequency ultrasound pulses and short bursts of high-voltage electricity can create temporary micropores in skin and are under study to transport several more medicines.

Baldness and Hirsutism

Hairs are dead, keratinized structures, so no amount of oiling, shampooing, or dousing with kelp extracts, vitamins, or nutrients will influence the follicle buried in the dermis. Skin conditions that affect follicles can contribute to hair loss. Temporary baldness can also result from exposure to radiation or to many of the toxic drugs used in cancer therapy. Two factors interact to cause most cases of baldness. A bald individual has a genetic susceptibility triggered by sufficiently large quantities of male sex hormones. Many women carry the genetic background for baldness, but unless major hormonal abnormalities develop, as in certain endocrine tumors, hair loss does not occur. (In some women, however, scalp hair does thin after menopause.)

Male pattern baldness affects the top of the head and forehead first, only later reducing the density of hair along the sides (▶**Figure 21**). Thus, hair follicles can be removed from the sides and implanted on the top or front of the head, temporarily delaying a receding hairline. This procedure is expensive, and not every hair transplant is successful. Finasteride is a prescription drug that blocks the production of one form of testosterone. In low doses, it slows the progression of male pattern baldness. *Minoxidil*, a drug originally marketed for the control of high blood pressure, appears to stimulate inactive hair follicles when rubbed onto the scalp. It is now available without a prescription as *Rogaine* and is most effective in preventing the progression of early hair loss. With both drugs, once stopped, hair loss resumes.

Alopecia areata (al-o-PĒ-shē-uh ar-ē-AH-ta) is a localized or generalized hair loss that can affect either sex. The cause is not known, and the severity and persistence of hair loss varies from case to case. The condition is associated with several disorders of the immune system and 20 percent of cases have a family history of alopecia. Topical steroid creams are often used but no completely effective treatment is known. Hair regrowth occurs in most cases.

Hirsutism (HER-soot-izm; *hirsutus*, bristly) is excess hair growth that can affect both sexes but is usually only considered a problem on women when it occurs in patterns generally characteristic of men. Because considerable overlap exists between the two genders in terms of normal hair distribution, and because racial, genetic, and cultural differences are significant, the precise definition is more often a matter of personal taste than of objective analysis. Age and sex hormones may play a role because hairiness increases late in pregnancy, and menopause produces a change in body hair patterns.

Severe hirsutism is associated with abnormal amounts of androgen (male sex hormone) production in either the ovaries or the adrenal glands, and treatment of the underlying endocrine abnormality is indicated. Shaving, plucking, and bleaching are common practices to minimize hirsutism. Cosmetically unwanted follicles can be permanently "turned off" by *electrolysis*, which destroys the follicle with a jolt of electricity, and lasers that damage or destroy pigmented follicles. Patients may also be treated with drugs that reduce or prevent androgen stimulation of the follicles. A topical cream for decreasing facial hair is also available.

Integumentary System

Folliculitis and Acne

Typical hair follicles have a sebaceous gland that secretes an oily sebum that lubricates the hair shaft and the surrounding skin. Although sebum has bactericidal (bacteria-killing) properties, under some conditions bacteria invade sebaceous glands. The presence of bacteria in sebaceous glands or follicles can produce **folliculitis** (fo-lik-ū-LĪ-tis), a local inflammation. If the duct of a sebaceous gland becomes blocked, a collection of pus in the dermis, or *boil*, develops. The usual treatment for a boil is to cut it open, or "lance" it, so that drainage and healing can occur.

Most individuals at puberty, especially those with a genetic tendency toward **acne** have larger-than-average sebaceous glands. When the ducts of these glands become blocked, their secretions accumulate and the bacterium *P. acnes* colonizes the area. This produces a localized inflammation. Sex hormone production, which accelerates at puberty, stimulates the sebaceous glands. Anxiety, stress, physical exertion, certain foods, or drugs may further encourage their secretory output.

The visible signs of acne are called **comedos** (ko-MĒ-dōz). Closed comedos ("whiteheads") contain accumulated, stagnant secretions. Open comedos ("blackheads") are open to the surface and contain more solid material. Although neither condition indicates the presence of dirt in the pores, washing may help keep superficial oiliness down.

Acne generally fades after sex hormone concentrations stabilize. Topical antibiotics, vitamin A derivatives such as *Retin-A*, or peeling agents may help reduce inflammation and minimize scarring. In cases of severe acne, one effective treatment involves antibiotic drugs. However, oral antibiotic therapy has risks, including the development of antibiotic-resistant bacteria, so this therapy is not used unless other treatment methods have failed. Truly dramatic improvements in severe cases have been obtained with the prescription drug *Accutane*. This compound is structurally similar to vitamin A, and it reduces oil gland activity on a long-term basis. A number of side effects, such as dry skin, lips, and eyes, as well as depression, have been reported; these disappear when the treatment ends. The use of Accutane during the first month of pregnancy carries a 25-times-normal risk of inducing birth defects, so women must avoid pregnancy while using this drug.

Trauma to the Skin

Trauma is a physical injury caused by some mechanical force. Skin trauma is common and several terms may be used to describe the resulting injuries. Such injuries generally affect all components of the integument, and each type of wound presents a different problem to clinicians attempting to limit damage and promote healing.

An **open wound** is a break in the epithelium. The major categories of open wounds are illustrated in ❱ **Figure 22**. **Abrasions** are the result of scraping (❱ **Figure 22a**). Bleeding may be slight, but a considerable area may be open to invasion by microorganisms. **Incisions** are linear cuts produced by sharp objects (❱ **Figure 22b**). Bleeding can be severe if deep vessels are damaged. The bleeding may help flush the wound, and closing the incision with bandages or stitches can limit the area open to infection while healing is under way. A **laceration** is a jagged, irregular tear in the surface of the skin produced by solid impact or by an object (❱ **Figure 22c**). Tissue damage may be extensive, and

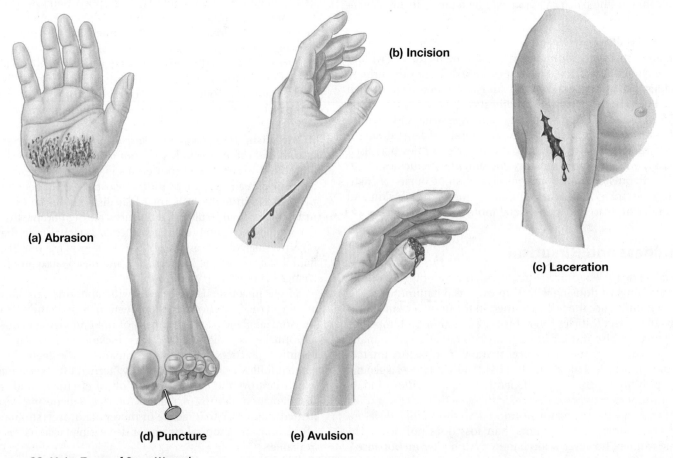

(a) Abrasion

(b) Incision

(c) Laceration

(d) Puncture

(e) Avulsion

❱ **Figure 22** Major Types of Open Wounds.

repositioning the opposing sides of the injury may be difficult. Despite the bleeding that generally occurs, lacerations are prone to infection. **Punctures** result when slender, pointed objects pierce the epithelium (◗ Figure 22d). Little bleeding results, and any microbes delivered under the epithelium in the process are likely to find conditions to their liking. In an **avulsion**, chunks of tissue are torn away by brute force in, for example, an auto accident or an explosion (◗ Figure 22e). Bleeding may be considerable, and even more serious internal damage may be present.

Closed wounds can affect any internal tissue, but because the epithelium is intact, the likelihood of infection is reduced. A **contusion** is a bruise causing bleeding in the dermis. "Black and blue" marks are familiar examples of contusions; most are not dangerous, but contusions of the head, such as "black eyes," may be the result of potentially life-threatening intracranial bleeding. Closed wounds caused by trauma severe enough to affect internal organs and organ systems are almost always serious threats to life.

Skin can regenerate effectively even after considerable damage. Burns are a type of trauma that can affect the epidermis, dermis, and deeper tissues. *First-degree burns* cause epidermal redness. *Second-degree burns* cause some dermal damage, and are associated with vesicle formation. *Third-degree burns* destroy all layers of the skin, including accessory glands and nerve endings.

Treatment of Skin Conditions

Topical **anti-inflammatory drugs**, such as the steroid *hydrocortisone*, can be used to reduce the redness and itching that accompany a variety of skin conditions. Some systemic (injected or swallowed) drugs may also be helpful; *aspirin* and *ibuprofen* are systemic drugs with anti-inflammatory properties. Topical or systemic antibacterial, antiviral, or antifungal compounds can treat a variety of skin infections. Isolated growths, such as skin tumors or warts, can be surgically removed. Alternatively, abnormal cells can be destroyed by electrical currents (**electrocautery**) or by freezing (**cryosurgery**). Ultraviolet radiation may help conditions such as acne or psoriasis, whereas the use of a sunscreen or a sunblock is important in preventing problems caused by UV exposure from the sun.

Aging Skin and Wrinkles

As we age our skin thins, the superficial cells aren't shed as frequently, and the surface becomes more prone to dryness. The dermis produces less collagen, and the elastin is less elastic. All this adds up to saggy, wrinkly, weaker skin. Usually this process starts in our twenties, with some genetic variation, but is accelerated by environmental factors, primarily sun exposure. Sun damage also causes age spots and freckles. Let sun-protected areas of your own skin serve as a control—compare the appearance and texture of the skin on your forearms, neck, or face with that on the usually sun-protected areas such as the buttocks.

To slow down premature skin aging, the American Academy of Dermatology (www.skincarephysicians.com) recommends 365-days-per-year sunscreen on exposed skin; avoiding deliberate tans, "tanning beds," and tanning salons; and covering up from 10 a.m. to 4 p.m., especially in summer. Sunlight contains both UVA and UVB rays. UVA rays get through window glass and into the deep layers of skin and are associated with wrinkles and the risk of melanoma skin cancers. UVB rays are blocked by window glass but cause sunburns and increased risk of all skin cancers. The SPF (skin protection factor) rating of sunscreens relates to sunburn (UVB) protection only. Most sunscreens are now "broad spectrum" and provide UVA protection as well.

Aging skin and its associated wrinkles are the basis of an estimated $16 billion per year "Cosmeceutical" industry in the United States. Prescription strength retinols such as tretinoin (derived from vitamin A) are FDA approved and regulated. These, combined with moisturizers, alpha-hydroxy acid exfoliators, and topical ascorbic acid (vitamin C) have some scientific evidence of effectiveness. Non–prescription strength retinols and extracts of fruits, herbs, teas, and vegetables are regulated for safety, but their effectiveness is unproven. Personal preference and not cost should probably guide which moisturizer, cleanser, exfoliator, and topical "nutraceutical" one may choose to use, since fragrance, packaging, and marketing probably cost more than the ingredients.

Muscles in the face and neck may contribute to deeper wrinkles. Reducing repetitive facial movements and sleeping on your back may help. Botox injections (p. 67) are effective for several months at a time but can also reduce facial expression.

Synthetic Skin

Skin grafts are used if large areas of skin have been completely destroyed. Pieces of undamaged skin from other areas of the body are used. Usually a partial thickness patch with the epidermis and part of the dermis is shaved off the donor site, cut in a mesh pattern to enlarge it, and tacked onto the burn area. An epithelium is reestablished at the donor site by epithelial cell migration from the edges and from follicles in the remaining dermis. After healing, accessory structures such as hair, sebaceous glands, and sweat glands are missing in the grafted area, and flexibility is reduced. After grafting, the complete reorganization and repair of the dermis and epidermis at the site of the injury takes approximately five years.

If the damaged area is large, there may not be enough normal skin available for grafting. Tissue-engineered skin-substitute products that consist of epidermal, dermal, and/or synthetic components either singly or in combination have been developed. *Epidermal culturing* can produce a new epithelial layer to cover a burn site. An epidermal sample from a burn patient can be cultured in a controlled environment that contains epidermal growth factors, fibroblast growth factors, and other stimulatory chemicals. This artificially produced epidermis can now be transplanted to cover the site of an injury. The larger the area that must be covered, the longer the culturing process continues. With enough time, the cells obtained in the original sample can provide enough epidermis to cover the entire body surface of a typical adult. The absence of a dermis makes this grafted epidermis less flexible and more fragile than normal—a particular problem around joints and areas of friction.

A second new procedure provides a model for dermal repairs that takes the place of normal tissue. A special synthetic skin, or "acellular wound dressing," is used. Several forms are available; most involve a biosynthetic dermis made of collagen-glycosaminoglycan scaffolding (usually from cows) and/or a silicon polymer covered by a layer of dead or live human epidermal cells. This combination is placed over the burn site either alone or covered by a cultured epidermal layer. Over time, fibroblasts and epidermal cells of the recipient migrate among the grafted collagen fibers, and gradually replace the model framework with their own. As with regular skin grafts, accessory structures are absent. Frequently several surgical procedures and additional skin grafts are required.

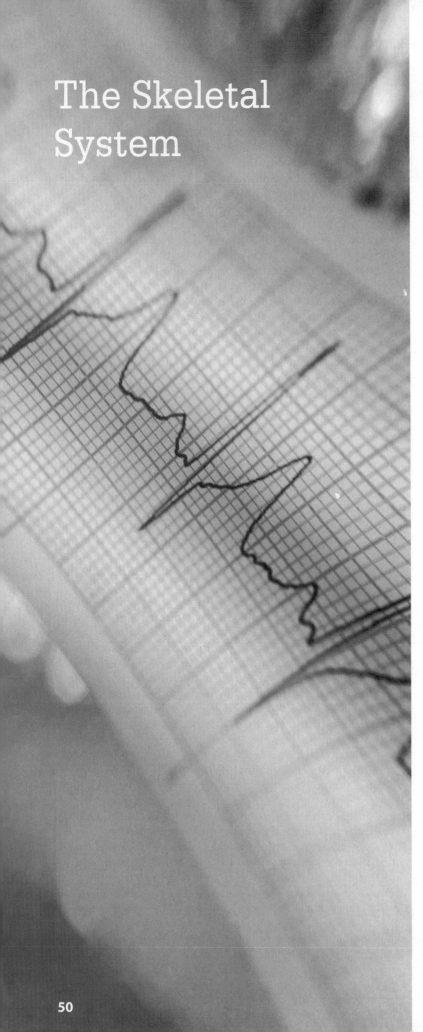

The Skeletal System

An Introduction to the Skeletal System and Its Disorders

The skeletal framework of the body is composed of at least 206 bones and the associated tendons, ligaments, and cartilages. The skeletal system has a variety of important functions, including the support of soft tissues, blood cell production, mineral and lipid storage, and, through its relationships with the muscular system, the support and movement of the body as a whole. Skeletal system disorders can thus affect many other systems. The skeletal system is in turn influenced by the activities of other systems. For example, weakness or paralysis of skeletal muscles will lead to a weakening of the associated bones and may cause changes in their relative positions.

The Dynamic Skeleton

Although the bones you study in the lab may seem to be rigid and permanent structures, the living skeleton is dynamic and undergoes continuous remodeling. The remodeling process involves (1) bone deposition by osteoblasts and (2) *osteolysis*, or dissolution of bone matrix, by osteoclasts. As indicated in ▶ **Figure 23**, the net result of the remodeling varies with the following five factors:

1. **The age of the individual.** During development, bone deposition occurs faster than bone resorption; as the amount of bone increases, the skeleton grows. At maturity, bone deposition and resorption are in balance. As the aging process continues, the rate of bone deposition declines and the bones become less dense. This gradual weakening, called *osteopenia*, begins at age 30–40 and may ultimately progress to osteoporosis (p. 56).

2. **The applied physical stresses.** Heavily stressed bones become thicker and stronger, and lightly stressed bones become thinner and weaker. Skeletal weakness can therefore result from muscular disorders, such as *myasthenia gravis* (p. 67) and the *muscular dystrophies* (p. 66), and from conditions that affect motor neurons of the central nervous system, such as spinal cord injuries (p. 83), strokes, *demyelination disorders* (p. 78), and *multiple sclerosis* (p. 83).

3. **Circulating hormone levels.** Changing levels of growth hormone (GH), androgens and estrogens, thyroid hormones, parathyroid hormone, and calcitonin increase or decrease the rate of mineral deposition in bone. As a result, many disorders of the endocrine system affect the skeletal system. For example,

   • Conditions affecting the skin, liver, or kidneys can interfere with calcitriol (a derivative of vitamin D) production. Vitamin D enhances calcium absorption along the intestinal tract.

   • Thyroid and parathyroid disorders can alter levels of thyroid hormones, parathyroid hormone, and calcitonin.

   • Pituitary gland disorders and liver disorders can affect the production of both GH and somatomedin.

   • Reproductive system disorders can alter circulating levels of androgens and estrogens.

   We will describe many of these conditions in the section dealing with the endocrine system.

4. **Rates of calcium and phosphate absorption and excretion.** For bone mass to remain constant, the rate of calcium and phosphate excretion, primarily at the kidneys, must be balanced by the rate of calcium and phosphate absorption at

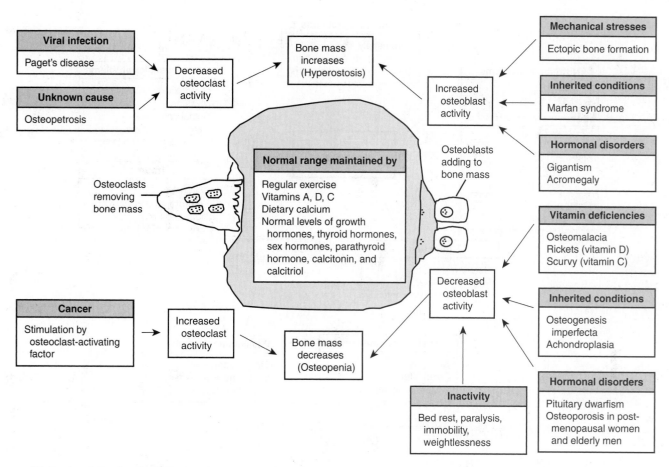

> **Figure 23 Factors Affecting Bone Mass.**

Skeletal System

the digestive tract. Kidney failure, dietary calcium and vitamin D deficiencies, and other problems that reduce calcium and phosphate absorption at the digestive tract will thus directly affect the skeletal system.

5. **Genetic or environmental factors.**   Genetic or environmental factors can affect the structure of bone or the remodeling process. A number of abnormalities of skeletal development, such as *Marfan syndrome* and *achondroplasia* (p. 55), are inherited. When bone fails to form embryonically in certain areas, underlying tissues can be exposed and associated functions can be altered. This type of abnormality occurs in a *cleft palate* and in *spina bifida*. Environmental stresses can alter the shape and contours of developing bones. For example, some cultures used boards to form an infant's skull to a shape considered fashionable, and nerve damage from spina bifida may cause the bones of an infant's feet to assume the equinovarus ("club foot") position prenatally. Environmental forces can also result in the formation of bone in unusual locations. These *heterotopic bones* (p. 54) may develop in a variety of connective tissues exposed to chronic friction, pressure, or mechanical stress. For example, cowboys in the 19th century sometimes developed heterotopic bones in the dermis of the thigh, from friction with the saddle.

Classification of Skeletal Disorders

> **Figure 24** diagrams the major classes of skeletal disorders that affect the structure and function of bones (> **Figure 24a**) and joints (> **Figure 24b**). Some of these disorders, such as *osteosarcoma*

and *osteomyelitis*, are the result of conditions that affect primarily the skeletal system; others, such as *acromegaly* and *rickets*, result from problems originating in other systems.

Traumatic injuries, such as fractures or dislocations, and infections also damage cartilages, tendons, and ligaments. A somewhat different array of conditions affects the soft tissues of the bone marrow. Red bone marrow contains the stem cells for red blood cells, white blood cells, and platelets. Blood diseases characterized by blood cell overproduction (*polycythemia*, *leukemia*, pp. 122, 129) or underproduction (several *anemias*, p. 128) result in bone marrow abnormalities.

Symptoms of Bone and Joint Disorders

A common symptom of a skeletal system disorder is pain. Because bone pain and joint pain are common symptoms associated with many bone and joint disorders, the presence of pain does not provide much help in identifying a specific disorder. A person may be able to tolerate chronic, aching bone or joint pain and therefore not seek medical assistance until more definitive symptoms appear. For example, a symptom that may reveal an underlying problem is a *pathologic fracture*. Pathologic fractures are the result of weakening of the skeleton by a disease process, such as *osteosarcoma* (a bone cancer). Such fractures at a tumor site can be caused by physical stresses that are easily tolerated by normal bones. Unless the usual course of the disease is understood, the minor trauma may be thought to have caused the fracture, when in fact it just exposed the underlying condition.

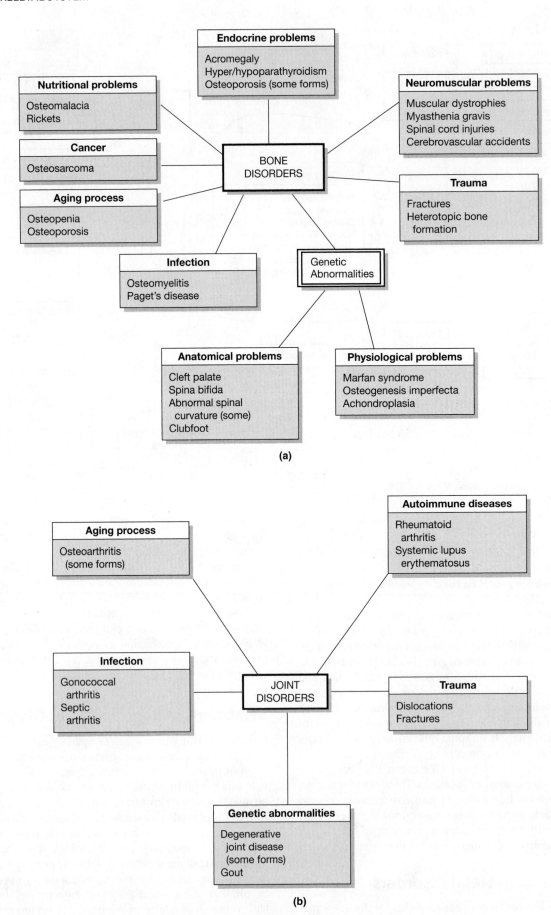

▶ **Figure 24** **An Overview of Disorders of the Skeletal System. (a)** Bone disorders. **(b)** Joint disorders.

Examination of the Skeletal System

The bones of the skeleton cannot be seen without relatively sophisticated equipment. However, a number of physical signs can assist in the diagnosis of a bone or joint disorder. Important factors noted in the physical examination include the following:

- **A limitation of movement or stiffness.** Many joint disorders, such as the various forms of arthritis, restrict movement or produce stiffness at one or more joints. Having a paired joint on the other limb can often provide a normal joint for comparison to the problematic one.

- **The distribution of joint involvement and inflammation.** In a *monoarthritic* condition, only one joint is affected. In a *polyarthritic* condition, several joints are affected simultaneously.

- **Sounds associated with joint movement.** Bony **crepitus** (KREP-i-tus) is a crackling or grating sound generated during the movement of an abnormal joint. The sound can result from the movement and collision of bone fragments following a joint fracture or from friction at an arthritic joint. If the periosteum is affected, movements generating crepitus are very painful.

- **The presence of abnormal bone deposits.** The callus is a thickened area of bone that develops at a fracture site as part of the repair process. Abnormal bone deposits can also develop around the joints in the fingers. These deposits are called *nodules*. When palpated, most nodules are solid and painless. Nodules, which can restrict movement, commonly form at the interphalangeal joints of the fingers in osteoarthritis.

- **Abnormal posture.** Bone disorders that affect the vertebral column can result in abnormal posture. The abnormality is most apparent when the condition alters the normal spinal curvature. Examples include *kyphosis*, *lordosis*, and *scoliosis* (p. 57). A condition involving an intervertebral joint, such as a herniated disc, will also produce abnormal posture and movement, as can muscle and nerve problems.

Table 16 summarizes descriptions of the most important diagnostic procedures and laboratory tests that can be used to obtain information about the status of the skeletal system.

Table 16	Examples of Tests Used in the Diagnosis of Bone and Joint Disorders	
Diagnostic Procedure	**Method and Result**	**Representative Uses**
X-ray of bone and joint	Standard x-ray; film sheet with radiodense tissues in white on a black background	Detects fractures, tumors, dislocations, reduction in bone density, and bone infections (osteomyelitis)
Bone scans	Injected radiolabeled phosphate accumulates in bones, and radiation emitted is converted into an image	Especially useful in diagnosis of metastatic bone cancer; detects fractures, early infections, and some degenerative bone diseases
Arthrocentesis	Insertion of a needle into joint for aspiration of synovial fluid	See section on analysis of synovial fluid (below)
Arthroscopy	Insertion of fiber-optic tubing into a joint cavity; displays interior of joint	Detects abnormalities of the menisci, ligaments, and articular surfaces; useful in differential diagnosis of joint disorders
MRI	Standard MRI produces computer-generated images	Detects bone and soft tissue abnormalities; noninvasive
DEXA	Dual Energy X-ray Absorptiometry: measures changes in bone density as small as 1 percent. Uses very small amount of radiation	Quantitates and monitors loss of bone density in osteoporosis and osteopenia
Laboratory Test	**Normal Values in Blood Plasma or Serum**	**Significance of Abnormal Values**
Alkaline phosphatase	Adults: 30–85 mIU/mL Children: 60–300 mIU/mL (higher values occur during adolescent bone growth)	Elevated levels occur in adults due to abnormal osteoblast activity; elevated levels occur in bone cancer, Paget's disease, and multiple myeloma
Calcium	Adults: 8.5–10.5 mg/dL Children 8.5–11.5 mg/dL	Elevated levels may occur in bone cancers, multiple fractures, hyperparathyroidism, and in prolonged immobilization
Phosphorus	Adults: 2.3–4.7 mg/dL Children: 4.0–7.0 mg/dL	Typically elevated when calcium levels are low; elevated levels occur in acromegaly, parathyroid disorders, and bone tumors
Uric acid	Adult males: 3.5–8.0 mg/dL Adult females: 2.8–6.8 mg/dL	Elevated levels occur with gout, which develops when uric acid crystals (products of purine metabolism) build up in a joint
Rheumatoid factor	Adults: A negative result is normal; 1:80 titer or higher is one criterion for rheumatoid arthritis	About 75 percent of people diagnosed with rheumatoid arthritis have a positive test for this factor; liver and collagen diseases, systemic lupus erythematosus, and aging also may give positive results
Synovial fluid analysis	WBC: <200/mm^3 RBC: none Glucose: <10 mg/dL below serum glucose levels Protein: 1.0–3.0 g/dL No uric acid crystals	Elevated white blood cell count suggests bacterial infection; mild elevation indicates inflammatory process; decreased glucose levels and decreased viscosity of fluid indicate inflammation of joint; uric acid crystals indicate gout

Skeletal System

Skeletal System

Endocrine Disorders

In **acromegaly** (*akron*, extremity + *megale*, great) an excessive amount of growth hormone, usually from a pituitary tumor, is released after puberty, when most of the epiphyseal cartilages have already closed. Cartilages and small bones respond to the hormone, however, resulting in abnormal growth at the hands, feet, lower jaw, skull, and clavicles (▶Figure 26a). Excessive growth, resulting in *gigantism*, occurs if there is hypersecretion of GH in childhood prior to epiphyseal closure. Inadequate production of growth hormone before puberty, by contrast, produces *pituitary dwarfism*. Persons with this condition are very short, but unlike achondroplastic dwarves (discussed below), their proportions are normal.

▶ Osseous Tissue and Skeletal Structure

Heterotopic Bone Formation

Heterotopic (*hetero*, different + *topos*, place), or ectopic (*ektos*, outside), bones are bones that develop in unusual places. Such bones dramatically demonstrate the adaptability of connective tissues. Mesenchymal stem cells can develop into bone, cartilage, or even fat and muscle. Physical or chemical events can stimulate the development of osteoblasts in normal connective tissues. For example, sesamoid bones develop within tendons near points of friction and pressure. Bone can also form within a large blood clot at the site of an injury or within portions of the dermis subjected to chronic abuse. Other triggers include foreign chemicals and problems that affect calcium excretion and storage.

Almost any connective tissues can be affected. Ossification within a tendon or around joints can painfully interfere with

movement. Bone can also form within the kidneys, between skeletal muscles, in the pericardium, in the walls of arteries, and around the eyes.

The excessive formation of bone is termed **hyperostosis** (hī-per-os-TŌ-sis). In the group of bone diseases characterized by *osteopetrosis* (os-tē-ō pe-TRŌ-sis; *petros*, stone), the total mass of the skeleton increases as a result of a decrease in osteoclast activity. Remodeling stops, and the shape, density, and strength of the bones change. Osteopetrosis in children produces a variety of skeletal deformities and may be fatal. Some cases of this relatively rare condition are genetic, whereas for other cases the cause is unknown.

Congenital Disorders of the Skeleton

Fibrodysplasia ossificans progressiva (*FOP*) is a rare single gene mutation disorder that involves the deposition of bone around skeletal muscles. The mutation affects a protein that helps control the replacement of cartilage with bone during skeletal maturation. A muscle injury can trigger rapid ossification in the injured area. The muscles of the skull, back, neck, and upper limbs may be gradually replaced by bone. The extent of the conversion can be seen in ▶Figure 25. ▶Figure 25a shows the skeleton of a healthy adult male; ▶Figure 25b shows the skeleton of a 39-year-old man with advanced FOP. Several of the vertebrae have fused into a solid mass, and major muscles of the back, shoulders, and hips have undergone extensive ossification. Treatment can be problematic, because any surgical excision may trigger more ossification.

There are more than 200 recognized inheritable disorders of connective tissues. Individual cases frequently result from a

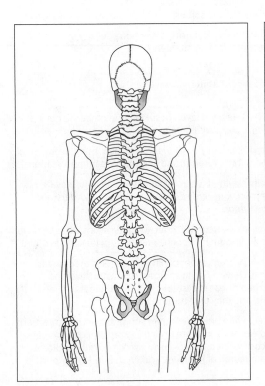

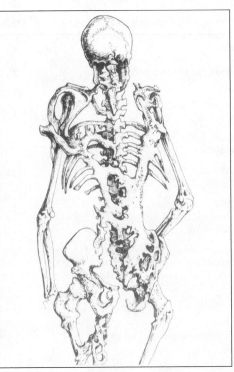

(a) (b)

▶Figure 25 **Heterotopic Bone Formation. (a)** The skeleton of a healthy adult male, posterior view. **(b)** The skeleton of an adult male with advanced myositis ossificans.

spontaneous mutation in either the oocyte or sperm of unaffected parents. Individuals with such conditions may transmit the disorder to their children. Osteogenesis imperfecta, Marfan syndrome, and achondroplasia are examples of inherited conditions characterized by abnormal bone formation.

Osteogenesis imperfecta, appearing in roughly 1 individual in 20,000, affects the organization of collagen fibers. Osteoblast function is impaired, growth is abnormal, and in severe forms the bones are very fragile, leading to repeated fractures and progressive skeletal deformation. Fibroblast activity is also affected, and the ligaments and tendons can be very "loose," permitting excessive movement at the joints. There are four recognized types, and large variations occur in symptom severity. The lifetime fracture count can range from just a few to hundreds, with the worst form often fatal soon after birth.

Marfan syndrome is also linked to defective connective-tissue structure. Most cases have a mutation in the gene that codes for the protein fibrillin 1. This protein is involved in strength and flexibility of connective tissue and also affects growth factors.

Extremely long and slender limbs, the most obvious physical indication of this disorder, result from excessive cartilage formation at the epiphyseal cartilages (Figure 26b). The skeletal effects are striking, but arterial wall weaknesses are more dangerous.

Achondroplasia (ā-kon-drō-PLĀ-sē-uh) also results from abnormal epiphyseal activity. The affected child's epiphyseal cartilages form into bone unusually slowly, with little longitudinal growth, and the adult has short, stocky limbs (Figure 26c). Although other skeletal abnormalities occur, the trunk is normal in size, and sexual and mental development remain unaffected. An adult with achondroplasia is known as an *achondroplastic dwarf*. The condition results from mutation of the *FGFR3* gene on chromosome 4 that affects a fibroblast growth factor, and contribution of the abnormal gene by one parent produces this condition; this form of inheritance is called *autosomal dominance*. Eighty percent of cases are the result of spontaneous mutations. If both parents have achondroplasia, the chances are that 25 percent of their children will

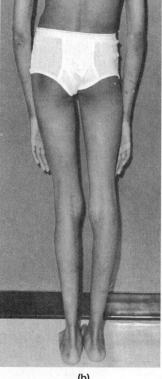

(a)   (b)   (c)

 Figure 26 Disorders of Bone Formation. (a) Acromegaly. **(b)** Marfan syndrome. **(c)** Achondroplasia.

be unaffected, 50 percent will be affected to a similar degree as their parents, and 25 percent will inherit two abnormal genes, leading to severe abnormalities and early death.

Bone Growth and Repair

Despite the body's considerable capacity for bone repair, not every fracture heals as expected. Steel plates, rods, and screws are used in severe fractures to help hold the ends in place while calluses form. This immobilizes the bone fragments and facilitates the repair process. A *delayed union* is a repair that proceeds more slowly than anticipated. *Non-union*, or no repair, can occur as a result of inadequate blood supply, infection, and continued movement at the fracture site that prevents complete callus formation. Non-union can cause significant disability.

Several techniques can induce bone repair. Surgical bone grafting is the most common treatment for delayed or non-union. Each year in the United States, roughly 500,000 people receive *bone grafts*; worldwide the total number is approximately 2.2 million. Bone grafts are transplants of bone, done to stimulate bone repair. The bone fragments, which serve as an "osteoconductive" scaffold, are commonly taken from another part of the individual's body (an *autograft*), such as the iliac crest. However, because the inserted bone is ultimately destroyed and replaced, sterilized bone fragments from donors—cadavers (an *allograft*) can also be used to establish a framework for the repair process. Thorough sterilization to prevent blood-borne diseases, including AIDS, is required. One advantage of autografts is the presence of bone-forming cells in the graft and around the injury site. An experimental method of inducing bone repair with these osteoprogenitor cells involves mixing the patient's bone marrow cells into a soft matrix of bone collagen and ceramic to form a composite graft. This combination is used like a putty at the fracture site. Mesenchymal stem cells in the marrow divide, producing chondrocytes that create a cartilaginous patch, which is later converted to bone by periosteal cells. In some cases, simply injecting marrow cells into the fracture site has been successful, even without the insertion of an autograft.

Synthetic "bio-ceramics" and polymers, often "seeded" with natural or recombinate cell growth factors, which stimulate the conversion of osteoprogenitor cells into active osteoblasts, are being studied as well. This approach is called "osteoinductive stimulation."

Bone Mineralization and Metabolism Disorders

In **osteomalacia** (os-tē-ō-ma-LĀ-shē-uh; *malakia*, softness) the size of the skeletal elements does not change, but their mineral content decreases, softening the bones. The osteoblasts work hard, but the matrix doesn't accumulate enough calcium salts. This condition, also called *rickets* when it occurs in children, occurs in adults or children who are deficient in vitamin D, which is required for normal calcium absorption from the intestines. Several renal diseases, an inadequate diet, and insufficient sun exposure can all contribute to the disease. Treatment involves addressing any underlying disorder and vitamin D and calcium supplementation.

With aging, bone mass declines, potentially progressing through osteopenia to **osteoporosis** (os-tē-ō-por-Ō-sis; *porosus*, porous), where the reductions in bone mass and density are sufficient to compromise normal function. Maximal bone density is reached in our twenties and decreases as we age. Factors such as weight-bearing physical activity, proper nutrition providing vitamin D and dietary calcium, and sex hormone levels affect peak bone mass. Men have higher peak bone mass than women. Inadequate calcium intake in teenagers reduces peak bone density and increases the risk of osteoporosis. The distinction between the "normal" osteopenia of aging and the clinical condition of osteoporosis is a matter of degree.

Current estimates indicate that 29 percent of women between the ages of 45 and 79 can be considered osteoporotic. The increase in incidence after menopause has been linked to a decrease in the production of estrogens (female sex hormones). The incidence of osteoporosis in men of the same age is estimated at 18 percent.

The excessive fragility of osteoporotic bones commonly leads to breakage, and subsequent healing is impaired. Over a lifetime, up to 50 percent of American women and 25 percent of American men may suffer an osteoporosis-related bone fracture. Vertebrae may collapse, distorting the vertebral articulations, putting pressure on spinal nerves, and causing prolonged pain. A fall may cause a hip fracture of the femoral neck. Up to 20 percent of older persons who suffer a hip fracture die within a year of injury. Supplemental estrogens, dietary changes to elevate calcium levels in blood, exercise that stresses bones and stimulates osteoblast activity, and the administration of calcitonin by nasal spray appear to slow, but not prevent, the development of osteoporosis. The inhibition of osteoclast activity by drugs called *bisphosphonates*, such as *Fosamax*, can reduce the risk of spine and hip fractures in elderly women and improve bone density. For long-term use, exercise, dietary calcium and vitamin D, and bisphosphonates are currently preferred. From 1996 to 2004 the U.S. age-adjusted hip fracture rate decreased 25 percent (from 1060 per 100,000 population to 850 per 100,000).

Infectious diseases that affect the skeletal system become more common as individuals age. In part, this fact reflects the higher incidence of fractures, combined with slower healing and the reduction of immune defenses.

Osteomyelitis (os-tē-ō-mī-e-LĪ-tis; *myelos*, marrow) is a painful and destructive bone infection generally caused by bacteria. Prolonged antibiotic treatment is often needed. Heredity and environmental factors, including the possibility of a viral infection, appear to be responsible for **Paget's disease**, also known as **osteitis deformans** (os-tē-Ī-tis de-FOR-manz). Three different genes have been associated with Paget's disease. It may affect up to 5 percent of the population over 70, especially those of Western European heritage. The number of new cases has been declining. In Paget's disease normal bone remodeling is distorted. Localized osteoclastic activity accelerates, producing areas of acute osteoporosis, and osteoblast bone formation produces abnormal matrix proteins with excessive, weak new bone. The result is a gradual deformation of the skeleton. Bisphosphonate treatment may slow the progression of the disease, as it does osteoporosis, by reducing osteoclast activity.

▶ The Axial Skeleton

Sinus Problems and Septal Defects

When an irritant is introduced into the nasal passages, our bodies work to remove the irritant. Large particles or strong chemical agents can cause us to sneeze, expelling a sizable amount of air quickly and forcefully. This sweeps some offending

Skeletal System

▶ **Figure 27 Sinus Medications.** Pharmacies often contain a dazzling array of products designed to treat sinusitis in its various forms.

particles or chemicals out with the air. Smaller particles or milder irritants trigger the production of mucus by the epithelium of the paranasal sinuses. The mucus stream flushes the nasal surfaces clean, often removing irritants such as pepper, pollen grains, or dust. In *allergic rhinitis* an overactive immune response to an allergen such as pollen or house dust mites causes excessive mucosal swelling, mucous production, and sneezing.

A sinus infection, however, is another matter entirely. A viral or bacterial infection produces an inflammation of the mucous membrane of the nasal cavity. As swelling occurs, the communicating passageways narrow. Drainage of mucus slows, the sinuses fill, and the individual experiences headaches and a feeling of pressure within the facial bones. This condition of sinus inflammation and congestion is called **sinusitis**. The maxillary sinuses are commonly involved. Because gravity does little to assist mucus drainage from these sinuses, the effectiveness of the flushing action is reduced, and pressure on the sinus walls typically increases.

The relief of pain associated with sinusitis is the basis of a large over-the-counter (OTC) drug market in the United States (▶ **Figure 27**). Every major pharmaceutical company has at least one product designed specifically to relieve sinus pressure. The active ingredients in these preparations are compounds that dry the epithelial linings, reduce pain, and restrict further swelling. Antihistamines and vasoconstrictors that reduce mucus production combined with pain relievers such as acetaminophen or ibuprofen are often included. The ingredients and dosages differ

very little from one oral "cold and sinus" product to the next. However, adult doses and concentrations are higher than in pediatric compounds, and there is potential risk of inadvertent misuse in children. Children under 4 years old have more risk of side effects from these drugs, and the FDA has recommended "Do not use under 4 years" labeling of these OTC medicines. Careful attention to dosage guidelines with regard to age and weight (for children) and maximum daily dosage is required. Marketing and packaging play a major role in determining which OTC remedy dominates the market at any given time.

Temporary sinus problems may follow exposure of the mucous epithelium to chemical irritants or invading microorganisms. Chronic sinusitis may accompany chronic allergies or occur as the result of a **deviated** (nasal) **septum**. In this condition, the nasal septum has a bend in it, generally at the junction between the bony and cartilaginous regions. Septal deviation often blocks the drainage of one or more sinuses, producing chronic cycles of infection and inflammation. A deviated septum can result from injuries to the nose. Many boxers and hockey players suffer from a deviated septum because their sports subject them to numerous blows to the soft tissues of the nose. It can usually be corrected or improved by surgery.

Kyphosis, Lordosis, and Scoliosis

The vertebral column has to move, balance, and support the trunk and head with multiple bones and joints involved. Conditions or events that damage the bones, muscles, and/or nerves can result in distorted shapes and impaired function. In **kyphosis** (kī-FŌ-sis), the normal thoracic curvature becomes exaggerated posteriorly, producing a "round-back" appearance (▶ **Figure 28a**). This condition can be caused by (1) osteoporosis with compression fractures affecting the anterior portions of vertebral bodies, (2) chronic contractions in muscles that insert on the vertebrae, or (3) abnormal vertebral growth.

In **lordosis** (lor-DŌ-sis), or "swayback," both the abdomen and buttocks protrude abnormally (▶ **Figure 28b**). The cause is an anterior exaggeration of the lumbar curvature. This may result from abdominal obesity or weakness in the muscles of the abdominal wall. It is common in adults with achondroplasia.

Scoliosis (skō-lē-Ō-sis) is an abnormal lateral and rotational curvature of the spine (▶ **Figure 28c**). This condition may result from

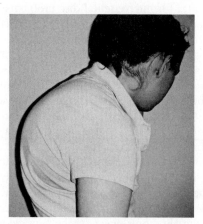

(a) Kyphosis

(b) Lordosis of pregnancy

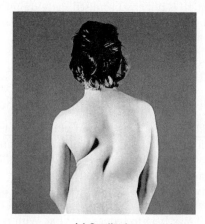

(c) Scoliosis

▶ **Figure 28 Abnormal Curvatures of the Spine.**

Skeletal System

developmental problems, such as incomplete vertebral formation, damage to vertebral bodies, or from neuromuscular paralysis from problems such as cerebral palsy, muscular dystrophy, spina bifida, or polio. Historically, severe scoliosis, or "hunchback" was often caused by tuberculosis infection of the vertebrae. In the antibiotic era, most cases of scoliosis are unexplained, or "idiopathic," and distort a previously normal spine. Idiopathic scoliosis generally appears in girls during adolescence, when periods of growth are most rapid. Treatment may consist of a combination of monitoring for progression and braces. Braces may slow progression of curvature until growth finishes, and limit final curvature. Severe cases with curves of over 40 degrees can be treated through surgical straightening with implanted metal rods or cables. Unfortunately, less invasive electrical muscle stimulation, exercise programs, and manipulation have not been found to be effective treatments for severe scoliosis.

▶ Spina Bifida

During the third week of embryonic development, the vertebral arches form around the developing spinal cord. In the condition called *spina bifida* (SPĪ-nuh BI-fi-duh; *bifidus*, cut into two parts), the most common neural tube defect (NTD), a portion of the spinal cord develops abnormally such that the adjacent vertebral arches do not form. Because the vertebral arch is incomplete, the membranes (or *meninges*) that line the dorsal body cavity bulge outward. This is the most common developmental abnormality of the nervous system, occurring at a rate of up to 1 case per 2000 births. Both heredity and maternal diet, particularly the amount of folic acid present before and during early pregnancy, have been linked to NTDs. Women who may become pregnant are advised to take 400 micrograms of folic acid daily, and to assist in this, food in the United States containing wheat, rice, and corn has been fortified with folic acid since 1998. Probably as a result, the incidence of NTDs in the United States dropped significantly.

The spinal region affected and the severity of the condition vary widely. It is most common in the lower thoracic, lumbar, or sacral region, typically involving three to six vertebrae. Variable degrees of paralysis occur distal to the affected vertebrae. Surgical repairs can close the gap in the menigeal and vertebral wall. Orthopedic and urogenital complications can be significant. Mild cases involving the sacral and lumbar regions may pass unnoticed, because neural function is not compromised significantly and "baby fat" may mask the fact that some of the spinous processes are missing.

▶ The Appendicular Skeleton
Problems with the Ankle and Foot

The ankle and foot are subjected to a variety of stresses during normal daily activities. In a *sprain*, a ligament connecting two bones is stretched to the point at which some of the collagen fibers are torn. The ligament remains functional, and the structure of the joint is not affected. The most common cause of a **sprained ankle** is a forceful inversion of the foot that stretches the lateral ligament. An ice pack may help to reduce swelling. With rest and support, the ankle should heal in about three weeks. A strain is a similar injury to either a muscle or its tendon.

In more serious sprains, the entire ligament can be torn apart, or the connection between the ligament and the lateral malleolus can be so strong that the bone breaks instead of the ligament. In general, a broken bone heals more quickly and effectively than does a completely torn ligament. A dislocation may accompany such injuries.

In a **dancer's fracture**, the proximal portion of the fifth metatarsal is broken. Most such cases occur while the body weight is being supported by the longitudinal arch of the foot (as in ballet dancing). A sudden shift in weight from the medial portion of the arch to the lateral, less elastic border breaks the fifth metatarsal close to its distal articulation.

Someone with *flat feet* loses or never develops the longitudinal arch. Rarely, this is caused by fused foot bones and may require surgery. Children have very mobile articulations and elastic ligaments, so they commonly have "flexible," flat feet. Their feet look flat only while they are standing, and the arch appears when they stand on their toes or sit down. In most cases, the condition disappears as growth continues. "Fallen arches" develop as tendons and ligaments stretch and become less elastic. Up to 40 percent of adults may have "flat feet," but no action is necessary unless pain develops. Individuals with abnormal arch development are most likely to suffer metatarsal injuries. Obese individuals and those who must constantly stand or walk on the job are also likely candidates. *Claw feet and toes* are produced by muscular abnormalities. The median longitudinal arch becomes exaggerated as the plantar flexors overpower the dorsiflexors, but the metatarsophalangeal joints are dorsiflexed. Causes include muscle degeneration and nerve paralysis. The condition tends to get progressively worse with age.

Congenital equinovarus, or *clubfoot*, occurs in about 1 in 1000 births. In some cases it is an inherited developmental abnormality. The condition develops when abnormal muscle development distorts the bones and joints of one or both feet; the tibia and ankle are also affected. Many infants born with spina bifida have clubfoot as well. Early treatment with serial casts to stretch tight muscles and ligaments into the proper position is usually successful.

▶ Articulations
Diagnosing and Treating Intervertebral Disc Problems

The most common sites for intervertebral disc problems are between vertebrae C_5 and C_6, L_4 and L_5, and L_5 and S_1. A clinician may be able to determine the location of the injured disc by noting the distribution of abnormal sensations and related spinal reflexes. For example, someone with a herniated disc at L_4-L_5 will experience pain in the hip, the groin, the posterior and lateral surfaces of the thigh, the lateral surface of the calf, and the superior surface of the foot. A herniation at L_5-S_1 produces pain in the buttocks, the posterior thigh, the posterior calf, and the sole of the foot. Most lumbar disc problems can be treated successfully with some combination of limited activity, muscle relaxants, painkillers, and physical therapy. Training to make proper body movements, especially when lifting and bending, may reduce the risk of relapse when combined with muscle-strengthening exercises.

Surgery to relieve the symptoms may rarely be required in some cases involving lumbar disc herniation. Various methods of treatment involve removing the offending disc, inserting material to restore

Skeletal System

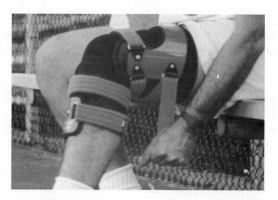

▶ Figure 29 A Knee Brace.

the position of the vertebrae, and possibly fusing the vertebral bodies together to prevent relative movement between them. Accessing the disc may require that the laminae of the nearest vertebral arch be removed. For this reason, the procedure is known as *laminectomy* (la-mi-NEK-to-mē). In cases in which the herniated portion of the disc does not extend far into the vertebral foramen, portions of the disc can be removed by a small tool that is guided to the site by radiological imaging. This minimally invasive procedure is faster and less invasive than a laminectomy.

Knee Injuries

Athletes place tremendous stresses on their knees. Ordinarily, the medial and lateral menisci move as the position of the femur changes. Placing a lot of weight on the knee while it is partially flexed can trap a meniscus between the tibia and femur, resulting in a break or tear in the cartilage. In the most common injury, the lateral surface of the leg is driven medially, tearing the medial meniscus. In addition to being quite painful, the torn cartilage may restrict movement at the joint. It can also lead to chronic problems and the development of a "trick knee"—a knee that feels unstable. Sometimes the meniscus can be heard and felt popping in and out of position when the knee is extended. To prevent such injuries, or to reduce subsequent damage, a knee brace can be worn (▶ Figure 29).

Other knee injuries involve tearing one or more stabilizing ligaments or damaging the patella. Torn ligaments can be difficult to correct surgically, and healing is slow. Rupture of the anterior cruciate ligament (ACL) is a common sports injury that affects women two to eight times more often than men. Twisting on an extended weight-bearing knee is frequently the cause. Nonsurgical treatment with exercise and braces is possible, but requires a change in activity patterns. Reconstructive surgery using part of the patellar tendon or an allograft from a cadaver tendon may allow a return to active sports.

Arthroscopy and Joint Injuries

An **arthroscope** uses fiber optics to permit the exploration of a joint without major surgery. Optical fibers are thin threads of glass or plastic that conduct light. The fibers can be bent around corners, so they can be introduced into a knee or other joint and moved around, enabling the physician to see and diagnose problems inside the joint. Arthroscopic surgical treatment of the joint is possible at the same time. This procedure, called **arthroscopic surgery**, has greatly simplified the treatment of knee and other joint injuries. ▶ Figure 30a is an arthroscopic view of the interior of an injured knee, showing a damaged meniscus. Small pieces of cartilage can be removed and the meniscus surgically trimmed. A total **meniscectomy**, the removal of the affected cartilage, is generally avoided, because it leaves the joint prone to develop degenerative joint disease. New tissue-culturing techniques may someday permit the replacement of the meniscus or even the articular cartilage.

Arthroscopy is an invasive procedure with some risks. Magnetic resonance imaging (MRI) is a safe, noninvasive, and cost-effective method of viewing and examining soft tissues around the joint. It improves the diagnostic accuracy of knee injuries, and reduces the need for diagnostic arthroscopies. It can also help guide the arthroscopic surgeon; ▶Figures 30b and ▶30c are MRI views of the knee joint. Notice the image clarity and the soft-tissue details visible in these scans.

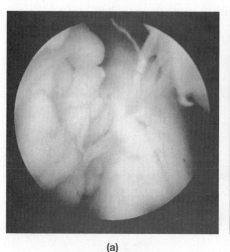

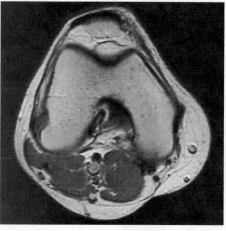

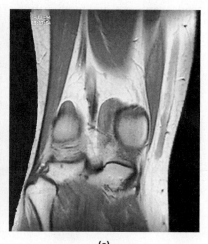

(a)   (b)   (c)

▶ Figure 30 **Arthroscopy and MRI Scans of the Knee. (a)** An arthroscopic view of a damaged knee, showing the torn edge of an injured meniscus. **(b)** A transverse and **(c)** a frontal MRI scan of the knee joint.

Knee Replacement Surgery

If a knee joint is severely damaged by arthritis with unremitting pain and impaired mobility despite medication and other physical treatments, a total knee replacement surgery may be considered. First done in 1968, more than 600,000 surgeries are performed annually in the United States. The new joint surfaces consist of metal and plastic components, and replacement of the damaged natural joint provides significant pain relief and improved mobility. However, the replacement joint is not as strong or durable (most last 10 to 20 years) as the normal, uninjured knee. Older persons who do not do activities that overstress the new joint, are not severely obese, and do not have other medical problems that preclude major surgery are the best candidates for surgery.

Rheumatism, Arthritis, and Synovial Function

Rheumatism (ROO-ma-tizm) is a general term that indicates pain and stiffness affecting the skeletal system, the muscular system, or both. There are several major forms of rheumatism. **Arthritis** (ar-THRĪ-tis) includes all the rheumatic diseases that affect synovial joints. Arthritis always involves damage to the articular cartilages, but the specific cause varies. For example, arthritis can result from bacterial or viral infection, injury to the joint, metabolic problems, or autoimmune disorders.

Proper synovial function requires healthy articular cartilages. When damage occurs to an articular cartilage, the matrix begins to break down; the exposed hyaline cartilage changes from a slick, smooth gliding surface to a rough feltwork of bristly collagen fibers. This feltwork drastically increases friction, damaging the cartilage further and producing pain. Eventually, the central area of the articular cartilage may completely disappear, exposing the underlying bone.

Fibroblasts are attracted to areas of friction, and they begin tying the opposing bones together with a network of collagen fibers. This network may later be converted to bone, locking the articulating elements into position. Such a bony fusion, called **ankylosis** (an-ke-LŌ-sis), eliminates the friction, but only at the drastic cost of making movement impossible.

The diseases of arthritis are usually considered as either degenerative or inflammatory. **Degenerative diseases** begin at the articular cartilages, and modification of the underlying bone and inflammation of the joint occur secondarily. **Inflammatory diseases** start with the inflammation of synovial tissues, and damage later spreads to the articular surfaces. We will consider one example of each type.

Osteoarthritis (os-tē-o-ar-THRĪ-tis), also known as **degenerative arthritis** or **degenerative joint disease** (DJD), generally affects older individuals. In the U.S. population over half of persons older than 65 years of age show x-ray signs of this disease. The condition seems to result from cumulative wear and tear on the joint surfaces and commonly affects weight-bearing joints including the knees, hips, lower back, and neck. The distal finger joints are also frequently involved.

Perhaps 30 percent of individuals with DJD have a gene that codes for an abnormal, weaker form of collagen that differs from the normal protein in only 1 of its 1000 amino acids. This collagen may lead to premature cartilage breakdown and arthritis.

Rheumatoid arthritis is a serious inflammatory condition that affects over 2 million people in the United States, and close to 75 percent of them are women. The cause is uncertain; genetic and environmental factors have all been proposed. The patient's own immune system attacks the synovial membrane of joints, which become swollen and inflamed, a condition known as **synovitis** (sī-nō-VĪ-tis). The cartilaginous matrix begins to break down, and the process accelerates as dying cartilage cells release lysosomal enzymes. General anti-inflammatory medicines, such as aspirin and ibuprofen, or corticosteroids may help control inflammation and pain. Other "disease-modifying antirheumatic drugs" (DMARDs) may slow or stop joint damage. DMARDs include anticancer drugs such as methotrexate and newer biologic agents that target the immune system. Trials of combination therapy using several drugs may bring further improvement. Ankylosis, where a joint fuses, was relatively common in the past when complete rest was routinely prescribed for rheumatoid arthritis patients and is rarely seen today.

Nutritional supplements such as *glucosamine* and *chondroitin sulfate* appear to have some anti-inflammatory effect (though smaller than that of aspirin), and a number of people report less pain while taking these supplements. Surgical procedures can realign or redesign the affected joint. In extreme cases involving the hip, knee, elbow, or shoulder, the defective joint can be replaced by an artificial one. Prosthetic (artificial) joints, such as those shown in ❯ **Figure 31**, are weaker than natural ones, but elderly people seldom stress them to their limits.

(a) Shoulder

(b) Hip

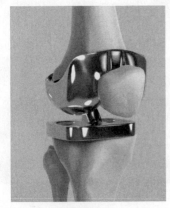

(c) Knee

❯ **Figure 31 Artificial Joints.**

Hip Fractures, Aging, and Arthritis

Most hip fractures (~90 percent) are associated with osteoporosis, and they usually develop after a fall in individuals over age 60. The associated immobility leads to a 5–20 percent increase in deaths and a 15–25 percent decrease in the likelihood of independent living in the year following the fracture. Minimally displaced fractures of the femoral neck may be treated with the surgical placement of pins or screws, giving support that may promote an early return to normal activities. When the injury is severe, the vascular supply to the joint may be damaged. As a result, two problems can develop:

1. **Avascular necrosis.**   The mineral deposits in the bone of the pelvis and femur are turned over very rapidly, and osteocytes have high energy demands. In *avascular necrosis*, a reduction in blood flow to the femoral head first injures and then kills the osteocytes. When bone maintenance stops in the affected region, the matrix begins to break down.

2. **Degeneration of articular cartilages.**   The chondrocytes in the articular cartilages absorb nutrients from the synovial fluid, which circulates around the joint cavity as the bones change position. A fracture of the femoral neck is generally followed by joint immobility and poor circulation to the synovial membrane. The combination results in a gradual deterioration of the articular cartilages of the femur and acetabulum.

In younger patients at risk for these complications, surgical repair with realignment of the bone, held in place by plates and screws, may preserve the articular surfaces. For older patients, replacement of the femoral head with a prosthetic head is often preferable.

If both the femoral head and the acetabulum are damaged, or if severe arthritis is present, both parts of the joint can be replaced. In a "total hip" replacement, the damaged portion of the femur is removed, and an artificial femoral head and neck are attached by a spike that extends into the marrow cavity of the shaft. The femoral head may be cemented in place, or carefully fitted so cement is not needed. Special cement is used to attach a new articular surface to the acetabulum. Joint replacement eliminates most pain and restores joint motion. They are generally implanted in older people, who seldom subject them to severe stress. Current models deteriorate in 10 to 20 years, so younger patients may need multiple replacements. Close to 300,000 total hip replacement surgeries are done in the United States each year.

Skeletal System

The Muscular System

An Introduction to the Muscular System and Its Disorders

The muscular system includes more than 700 skeletal muscles that are directly or indirectly attached to the skeleton by tendons or aponeuroses. The muscular system produces movement across joints as the contractions of skeletal muscles pull on the attached bones. Muscular activity does not always result in movement, however; it can also help stabilize skeletal elements and prevent movement. Skeletal muscles are also important in guarding entrances and exits of internal passageways, such as those of the digestive, respiratory, urinary, or reproductive systems, and in generating heat to maintain a stable body temperature.

The Diagnosis of Muscular Disorders

Signs of Muscular Disorders

Skeletal muscles normally contract only under the command of the nervous system. For this reason, clinical observation of muscular activity may provide indirect information about the nervous system, as well as direct information about the muscular system. The assessment of a patient's facial expressions, posture, speech, and gait can be an important part of a physical examination. Classical signs of muscle disorders include the following:

- **Gower's sign** is a distinctive method of standing up from a sitting or lying position on the floor when proximal muscles are weak. This sign is characteristic of young children with *muscular dystrophy* (p. 66). They move from a sitting position to a standing position by pushing the trunk off the floor with the hands and then moving the hands to the knees. The hands are then used as braces to force the body into the standing position. The extra support is necessary, because the pelvic muscles are too weak to swing the weight of the trunk over the legs.

- **Ptosis** is a drooping of the upper eyelid. It may be seen in *myasthenia gravis* (p. 67), *botulism* (p. 66), and *myotonic dystrophy* (p. 66), or it may follow damage to cranial nerve III, which innervates the *levator palpebrae superioris muscle* of the eyelid.

- **A muscle mass**—an abnormal dense region within a muscle—is sometimes seen or felt in a skeletal muscle. A muscle mass results from torn muscle or tendon tissue, a hematoma, or bone deposition, as in fibrodysplasia ossificans progressiva.

- Abnormal contractions may indicate problems with the muscle tissue or its innervation. *Muscle spasticity* exists when a muscle has excessive tone. A *muscle spasm* is a sudden, strong, and painful involuntary contraction.

- **Muscle flaccidity** exists when the relaxed skeletal muscle appears soft and loose and its contractions are very weak or absent. Flaccidity is often related to nerve disorders.

- **Muscle atrophy** is skeletal muscle deterioration, or wasting, due to disuse, immobility, or interference with the normal innervation.

- Abnormal patterns of muscle movement, such as *tics, choreiform movements*, or *tremors*, and muscular paralysis are generally caused by nervous system disorders. We will describe these movements further in sections dealing with abnormal nervous system function.

Symptoms of Muscular Disorders

Two common symptoms of muscular disorders are pain and weakness in the affected skeletal muscles. Possible causes of muscle weakness are diagrammed in ▶Figure 32.

Possible causes of muscle pain include the following:

- **Muscle trauma.** Examples of traumatic injuries to a skeletal muscle include a laceration, a deep bruise or crushing injury, a muscle strain or tear, and a damaged tendon.

- **Muscle infection.** Skeletal muscles can be infected by viruses, as in some forms of myositis (muscle inflammation). These types of infections generally produce pain that is restricted to the affected muscles. Diffuse muscle pain can develop in the course of other infectious diseases, such as influenza.

- **Adverse medication effects.** Myalgia is defined as muscle pain without signs of myopathy, which is muscle damage, and is detected by elevated muscle enzyme levels. Although several factors may cause myalgia, it is a known side effect of

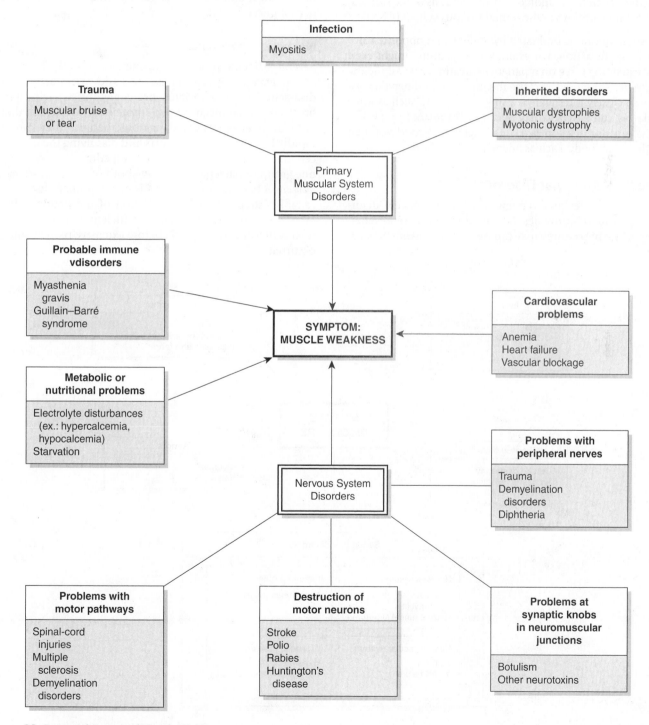

▶Figure 32 **Potential Causes of Muscle Weakness.**

several drugs. For example, the statin drugs, which are widely used to lower cholesterol levels and reduce the risk of heart attacks, are associated with myalgia in 5 of every 100 patients treated; detectable myopathy (muscle wasting) occurs in 1 of every 1000 patients, and dangerous myopathy in 1 of every 10,000 patients. Most of the affected patients may tolerate lower doses or a change in medication.

- **Related problems with the skeletal system.** Muscle pain can result from skeletal problems such as arthritis (p. 60) or a sprained ligament near the point of muscle origin or insertion.
- **Problems with the nervous system.** Muscle pain can be related to the inflammation of sensory neurons or the stimulation of pain pathways in the central nervous system (CNS).

Muscle strength can be evaluated by applying an opposite force against a specific action. For example, an examiner might exert a gentle extending force on a patient's forearm while asking the patient to flex the elbow. Because the muscular and nervous systems are so closely interrelated, a single symptom, such as muscle weakness, can have a variety of causes (❱ Figure 32). Muscle weakness can also develop as a consequence of a condition that affects the entire body, such as anemia or starvation.

Diagnosis of Muscular Disorders

❱Figure 33 provides an overview of muscular system disorders, and **Table 17** provides information about representative diagnostic procedures that can be used to categorize such disorders.

❱ Disorders of Muscle Tissue

Disorders considered in this section can affect skeletal muscle tissue throughout the body.

Disruption of Normal Muscle Organization

A variety of disorders are characterized by a disruption in the structural organization of skeletal muscles. We will consider only a few representative examples: necrotizing fasciitis, characterized by a breakdown in the connective tissues of skeletal muscles; and fibromyalgia and chronic fatigue syndrome, two muscle disorders of uncertain origin.

Necrotizing Fasciitis

Several bacteria produce enzymes such as *hyaluronidase* or *cysteine protease*. Hyaluronidase breaks down hyaluronic acid and disassembles the associated proteoglycans. Cysteine protease breaks down connective-tissue proteins. The bacteria that produce these enzymes are dangerous, because they can spread rapidly by liquefying the matrix and dissolving the intercellular cement that holds epithelial cells together. The *streptococci* are one group of bacteria that secrete both of these enzymes. *Streptococcus A* bacteria are involved in many human diseases, most notably "strep throat," an infection of the pharynx. In most cases, the immune response is sufficient to contain and ultimately defeat these bacteria before extensive tissue damage has occurred.

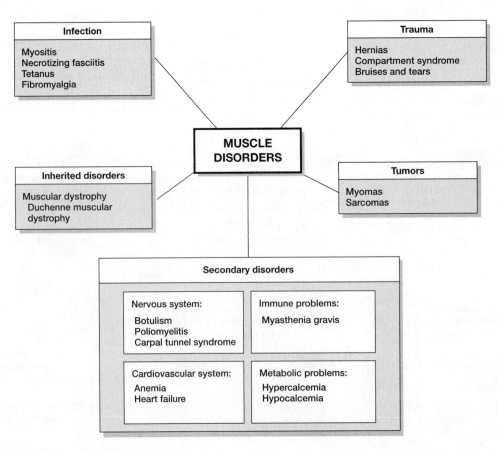

❱Figure 33 **Disorders of the Muscular System.**

Table 17	Examples of Tests Used in the Diagnosis of Muscle Disorders	
Diagnostic Procedure	**Method and Result**	**Representative Uses**
Muscle biopsy	Removal of a small amount of affected muscle tissue	Identifies histologic muscle disease
Electromyography (EMG)	Insertion of a probe that transmits measurements of electrical activity in contracting muscles	Abnormal EMG readings occur in disorders such as myasthenia gravis, amyotrophic lateral sclerosis (ALS), and muscular dystrophy
MRI	Standard MRI	Detects muscle, ligament, or tendon damage and associated soft-tissue abnormalities
Laboratory Test	**Normal Values in Blood Plasma or Serum**	**Significance of Abnormal Values**
Aldolase	Adults: less than 8 U/L 22–59 mU/L (SI units)	Elevated levels occur in muscular dystrophy but not in myasthenia gravis or multiple sclerosis
Aspartate aminotransferase (AST or SGOT)	Adults: 7–40 U/L Children: 15–55 U/L	Elevated levels occur in some muscle diseases and can occur after exercise
Creatine phosphokinase (CPK or CK)	Adults: 30–180 IU/L	Elevated levels occur in muscular dystrophy and myositis or after extreme exercise
Lactate dehydrogenase (LDH)	Adults: 100–190 U/L	Elevated levels occur in some muscle diseases and in lactic acidosis
Electrolytes		
Potassium	Adults: 3.5–5.0 mEq/L	Decreased levels of potassium can cause muscle weakness
Calcium	Adults: 8.5–10.5 mg/dL	Decreased calcium levels can cause muscle tremors and tetany; increased levels can cause muscle flaccidity

However, tabloid newspapers have a field day with stories of "killer bugs" and "flesh-eating bacteria." The details are horrific: Minor cuts become major open wounds, and interior connective tissues dissolve. This condition is called **necrotizing fasciitis**. Untreated, it is fatal. Even with rapid diagnosis, aggressive surgical removal of affected tissue, and antibiotics, the mortality rate is extremely high. The pathogen responsible is a strain of *Streptococcus A* that overpowers immune defenses and swiftly invades and destroys soft tissues. Moreover, the pathogens erode their way along the fascial wrapping that covers skeletal muscles and other organs. In most cases, myositis also occurs, followed by degeneration of the muscle tissue. Some form of highly aggressive infectious soft-tissue invasion occurs 75–150 times annually in the United States. The rapid development of extreme pain in an otherwise minor wound is one warning symptom.

Fibromyalgia, Chronic Fatigue Syndrome, and Myofascial Pain Syndrome

Fibromyalgia (-*algia*, pain) is a disorder that has formally been recognized only since the mid-1980s. Although first described in the early 1800s, the condition is still somewhat controversial, because the symptoms cannot be definitively linked to any anatomical or physiological abnormalities. However, physicians now recognize a distinctive pattern of symptoms with the diagnostic criteria of widespread musculoskeletal pain for 3 months or more, and tenderness in 11 or more of 18 specific tender points. Sleep disorders, depression, and irritable bowel syndrome also occur alongside fibromyalgia.

Fibromyalgia may be the most common musculoskeletal disorder affecting women under 40 years of age; from 3 to 6 million individuals in the United States may have this condition. The four most common tender points are (1) the medial surface of the knee, (2) the area distal to the lateral epicondyle of the humerus, (3) the area near the external occipital crest of the skull, and (4) the junction between the second rib and its costal cartilage. An additional clinical criterion is that the pains and stiffness cannot be explained by other mechanisms.

Most of the symptoms mentioned could be attributed to other problems. For example, chronic depression can, by itself, lead to fatigue and poor-quality sleep. As a result, the pattern of tender points is the diagnostic key to fibromyalgia. This symptom distinguishes fibromyalgia from **chronic fatigue syndrome** (CFS). The current symptoms accepted as a definition of CFS include (1) a sudden onset, generally following a viral infection; (2) disabling fatigue, lasting 6 months or longer; (3) muscle weakness and pain; (4) sleep disturbance; (5) fever; and (6) enlargement of cervical lymph nodes. Roughly twice as many women as men are diagnosed with CFS. In many ways, both disorders resemble chronic pain syndromes, and the same treatment strategies may be effective.

Myofascial pain syndrome is another poorly understood chronic musculoskeletal pain disorder that is regionally localized. Locating trigger points both active and latent are part of the diagnosis and physical therapy involving stretching and local pain relief with either injection of local anesthetic or topical sprays may reduce pain.

Myofascial pain is more common in women and is associated with depression, sleep disorders, fatigue, and behavioral disturbances, as with all chronic pain conditions. It may occur in association with trauma, overuse, or conversely, fixed posture from repetitive office desk or computer work. Unless myofascial pain evolves into a fibromyalgia type clinical syndrome, most persons improve. Until better understanding of the disease process is achieved,

Muscular System

treatment is supportive and nonspecific including localized physical therapy and medication to improve sleep and/or depression.

Attempts to link either fibromyalgia or CFS to a viral infection, adrenal gland dysfunction, or to some physical or psychological trauma have not been successful, and the causes remain unknown. Most laboratory tests are normal. For both conditions, treatment is at present limited to relieving symptoms when possible. For example, anti-inflammatory medications may help relieve pain, anti-depressant medications may improve sleep patterns and reduce depression, and exercise programs may help maintain a normal range of motion. Reassurance that fibromyalgia is not progressive, crippling, or life-threatening may help sufferers.

The Muscular Dystrophies

The **muscular dystrophies** (DIS-trō-fēz) are inherited diseases caused by a mutation in one of the many genes that affect muscle function. Progressive muscle weakness and deterioration occur with variable severity depending on the type of mutation. One of the most common and best understood conditions is **Duchenne muscular dystrophy (DMD)**, which may be inherited from the mother or, in 30 percent of cases, from spontaneous mutation. Symptoms of this form of muscular dystrophy start in childhood, commonly between the ages of 3 and 7. The condition is caused by mutations of the DMD gene, which codes for the protein dystrophin. This gene is located on the X chromosome and generally affects males (it can occur in females only if a very rare form of chromosome duplication occurs). DMD occurs at an incidence of roughly 30 per 100,000 male births. In its most severe form, with only 0–5 percent of the normal amount of dystrophin in muscle fibers, affected children develop progressive muscular weakness, and require wheelchairs by age 8 to 12. Most individuals die before age 30, due to respiratory paralysis or cardiac problems. Skeletal muscles are primarily affected, although X-linked dilated cardiomyopathy is caused by mutation of the same gene.

Dystrophin is a large protein that attaches thin filaments of the sarcomeres to an anchoring protein on the sarcolemma, providing mechanical strength to the muscle fiber and connecting the myofibrils to the sarcolemma and the extracellular matrix. In children with DMD, calcium channels remain open for an extended period, and calcium levels in the sarcoplasm rise to the point at which key proteins denature. Inflammation occurs and the muscle fiber then degenerates resulting in elevated plasma levels of the muscle enzyme creatine phosphokinase. Steroid treatment slows progression for up to three years at the price of significant side effects. Rats with a form of DMD treated by various forms of gene and stem cell therapy have sometimes improved, giving hope that an effective treatment may be possible.

Women carrying the defective gene are asymptomatic, but each of their sons will have a 50 percent chance of developing DMD. Prenatal genetic testing allows for carrier detection and fetal diagnosis of DMD by 8 weeks gestation.

Myotonic Dystrophy

Myotonic dystrophy is a form of muscular dystrophy that occurs in the United States with an incidence of 13.5 per 100,000 population worldwide. It involves both muscle weakness and reduced ability to voluntarily relax muscles, with resulting "stiffness." Symptoms may develop in infancy, but more commonly they develop after puberty. As with other forms of muscular dystrophy,

adults developing myotonic dystrophy experience a gradual reduction in muscle strength and control. Problems with other systems, especially the cardiovascular and digestive systems, typically develop. There is no effective treatment.

The inheritance of myotonic dystrophy is unusual, because children of an individual with this condition commonly develop symptoms that are more severe than those of the parent, a situation called *anticipation*. The increased severity of the condition appears to be related to the increasing length of the abnormal area of the affected gene on chromosome 19. For some reason, the nucleotide sequence of that part of the gene gets repeated several times, and the number can increase from generation to generation. Evidence indicates that the extra nucleotides interfere in some way with the transcription of an adjacent gene involved with the control of muscle tone.

Problems with the Control of Muscle Activity

Another group of disorders interferes with normal neuromuscular communication by affecting either the nerve's ability to issue commands or the muscle's ability to respond. Anything that interferes with neural function or with excitation–contraction coupling will cause muscular paralysis. Three examples are worth noting:

1. **Botulism** results from the consumption of canned or smoked foods contaminated with a bacterial toxin. The toxin prevents the release of ACh at the synaptic terminals, leading to a potentially fatal muscular paralysis.

2. The progressive muscular paralysis of **myasthenia gravis** results from the loss of ACh receptors at the junctional folds. The primary cause is a misguided attack on the ACh receptors by the immune system. Genetic factors play a role in predisposing individuals to this condition.

3. The motor paralysis caused by **polio** is the result of viral damage to motor neurons.

Botulism

Botulinus (bot-ū-LĪ-nus) **toxin** prevents the release of acetylcholine (ACh) at synaptic terminals, thereby producing a severe and potentially fatal paralysis of the skeletal muscles. A case of botulinus poisoning is called **botulism**.[1] The toxin is produced by the bacterium *Clostridium botulinum*, which does not need oxygen to grow and reproduce. Because this bacterium can live quite well in a sealed can or jar, most cases of botulism are linked to improper canning or storing procedures, followed by failure to cook the food adequately before it is eaten. Home-canned tuna or beets, smoked fish, and cold soups are the foods most commonly linked to botulism. Boiling for a half-hour destroys both the toxin and the bacteria.

Symptoms generally begin 12–36 hours after a contaminated meal is eaten. The initial symptoms are typically disturbances in vision, such as double vision or a painful sensitivity to bright light. These symptoms are followed by other sensory and motor problems, including blurred speech and an inability to stand or walk.

---

[1] The disorder was described more than 200 years ago by German physicians who treated patients who were poisoned by dining on contaminated sausages. *Botulus* is the Latin word for "sausage."

Roughly half of botulism patients experience intense nausea and vomiting. These symptoms persist for days to weeks, followed by a gradual recovery; some patients are still in recovery after a year.

The major risk of botulinus poisoning is respiratory paralysis and death by suffocation. Treatment is supportive: bed rest, observation, and, if necessary, the use of a mechanical respirator. In severe cases, an antitoxin and drugs that promote the release of ACh, such as *guanidine hydrochloride*, may be administered. The overall mortality rate in the United States is about 7 percent.

A commercial form of botulinum toxin, "Botox," is injected into selected muscles to treat problems such as spasms of the eyelids, neck, urinary bladder, limbs, and unequal extraocular muscle strength in strabismus (crossed eyes). Cosmetic use of Botox to reduce facial wrinkles is the most common use. Overactive axillary apocrine sweat glands and excessive nasal mucus discharge both respond to Botox because the sympathetic nerves innervating these glands release ACh. Regardless of the intended use, the effect is temporary and thus may need to be repeated.

Myasthenia Gravis

Myasthenia gravis (mī-as-THĒ-nē-uh GRA-vis) is characterized by a general muscular weakness that tends to be most pronounced in the muscles of the arms, head, and chest. The first symptom is generally a weakness of the eye muscles and drooping eyelids. Facial muscles are commonly weak as well, and the individual develops a peculiar smile known as the "myasthenic snarl." As the disease progresses, weakness of the pharynx leads to problems with chewing and swallowing, and holding the head upright becomes difficult.

The muscles of the upper chest and upper limbs are next to be affected. All the voluntary muscles of the body may ultimately be involved. Severe myasthenia gravis produces respiratory paralysis, with a mortality rate of 5–10 percent. However, the disease does not always progress to such a life-threatening stage. For example, roughly 20 percent of people with the disease experience no symptoms other than eye problems.

The condition results from a decrease in the number of ACh receptors on the motor end plate. Before the remaining receptors can be stimulated enough to trigger a strong contraction, the ACh molecules are destroyed by cholinesterase. As a result, muscular weakness develops.

The primary cause of myasthenia gravis appears to be a malfunction of the immune system. Roughly 70 percent of individuals with myasthenia gravis have an abnormal thymus, an organ involved with the maintenance of normal immune function. In myasthenia gravis, the immune response attacks the ACh receptors of the motor end plate as if they were foreign proteins. For unknown reasons, 1.5 times as many women as men are affected. The typical age at onset is 20–30 for women, versus over 60 for men. Estimates of the incidence of this disease in the United States range from 2 to 10 cases per 100,000 population.

One approach to therapy involves the administration of drugs, such as *neostigmine*, that are termed **cholinesterase inhibitors**. As their name implies, these compounds are enzyme inhibitors; they tie up the active sites on the enzyme where cholinesterase normally binds ACh. With cholinesterase activity reduced, the concentration of ACh at the synapse can rise enough to stimulate the surviving receptors and produce muscle contraction. Corticosteroid therapy is typically beneficial, as is surgical removal of the thymus, which may yield prolonged remission.

Polio

Because skeletal muscles depend on their motor neurons for stimulation, disorders that affect the nervous system can have an indirect effect on the muscular system. **Polio** is caused by the *poliovirus*, a virus that does not produce clinical symptoms in roughly 95 percent of infected individuals. The virus produces variable symptoms in the remaining 5 percent. Some individuals develop a nonspecific illness resembling the flu. Other individuals develop a brief *meningitis* (p. 81), an inflammation of the protective membranes that surround the CNS. In still another group of people, the virus attacks somatic motor neurons in the CNS. When those neurons are destroyed, the lack of motor innervation causes paralysis and progressive atrophy of the dependent muscles.

In this third form of the disease, the individual develops a fever 7–14 days after infection. The fever subsides, but recurs roughly a week later, accompanied by muscle pain, cramping, and paralysis of some or all muscles in one or more limbs. Respiratory paralysis may also occur, and the mortality rate of this form of polio is 2–5 percent in children and 15–30 percent in adults. If the individual survives, some degree of recovery generally occurs over a period of up to six months.

For unknown reasons, the survivors of paralytic polio may develop progressive muscular weakness in muscles not previously affected 20–30 years after the initial infection. This *postpolio syndrome* is characterized by fatigue, muscle pain and weakness, and, in some cases, muscle atrophy. There is no treatment for the condition, although rest seems to help.

Polio has been almost completely eliminated from the world through an aggressive vaccination program. In 1988 there were an estimated 350,000 cases, whereas in 2012 there were only 223 cases reported. The World Health Organization in 2013 stated that sustained, endemic polio exists only in Nigeria, Pakistan, and Afghanistan. However, polio-free countries can suffer imported polio from these three countries as well as rare polio cases related to mutated oral vaccines. A Global Eradication Initiative has as its goals: (a) increasing vaccination levels in the three countries where polio remains endemic, (b) keeping vaccination levels high in polio-free countries, and (c) performing focused mass vaccination campaigns when a case of polio occurs. The goal is total eradication of the disease, just as smallpox was eliminated in 1977.

For decades, the oral Sabin vaccine has been the most widely used because it is easy to administer and provides better immune stimulation than the injectable vaccine. The oral vaccine contains live but weakened (attenuated) viruses, whereas the injected vaccine contains completely inactivated viruses. However, the oral vaccine does carry a 1 in 1 million risk that the immunized person will develop polio. In 2000, polio had become so rare worldwide that the CDC recommended only the injected vaccine for infants in the United States. Unfortunately, many parents today decline to immunize their children against the poliovirus, because they assume that the disease has been "conquered." Failure to immunize is a mistake, because (1) there is still *no cure* for polio, (2) the virus remains in the environment in many areas of the world, and (3) up to 38 percent of children aged 1–4 years have not been immunized. A major epidemic could possibly develop very quickly if the virus was "imported" into the United States.

Delayed Onset Muscle Soreness

You have probably experienced muscle soreness the day after a period of physical exertion (**▶ Figure 34**). Considerable

▶Figure 34 **Delayed Onset Muscle Soreness.** A rigorous workout can cause lingering pain whose origins are uncertain.

controversy exists over the source and significance of this pain, which is known as *delayed onset muscle soreness* (DOMS) and has several interesting characteristics:

- DOMS is distinct from the soreness you experience immediately after you stop exercising. The initial short-term soreness is probably related to the biochemical events associated with muscle fatigue.
- DOMS generally begins several hours after the exercise period ends and may last 3 or 4 days.
- The amount of DOMS is highest when the activity involves eccentric contractions (where a muscle elongates despite producing tension). Activities dominated by concentric or isometric contractions produce less soreness.
- Levels of CPK and myoglobin are elevated in the blood, indicating damage to muscle plasma membranes. The nature of

the activity (eccentric, concentric, or isometric) has no effect on these levels, nor can the levels be used to predict the degree of soreness experienced.

Three mechanisms have been proposed to explain DOMS:

1. Small tears may exist in the muscle tissue, leaving muscle fibers with damaged membranes. The sarcolemma of each damaged muscle fiber permits the loss of enzymes, myoglobin, and other chemicals that may stimulate nearby pain receptors.
2. The pain may result from muscle spasms in the affected skeletal muscles. In some studies, stretching the muscle involved after exercise can reduce the degree of soreness.
3. The pain may result from tears in the connective tissue framework and tendons of the skeletal muscle.

Some evidence supports each of these mechanisms, but it is unlikely that any one tells the entire story. For example, muscle fiber damage is certainly supported by biochemical findings, but if that were the only factor, the type of activity and level of circulating enzymes would be correlated with the level of pain experienced, and such is not the case.

Power, Endurance, and Energy Reserves

▶Figure 35 compares the power/endurance curves for anaerobic and aerobic activities. The first half minute of peak activity is totally supported by the mobilization of ATP and CP (creatine phosphate) reserves (▶Figure 35a). Thereafter roughly two-thirds of the energy requirements of skeletal muscles operating at peak activity levels are met via glycolysis, with associated lactic acid generation.

At modest activity levels, a skeletal muscle can rely on aerobic respiration to provide ATP. At peak levels of activity, the muscle relies primarily on anaerobic metabolism. The level of activity at which the muscle must begin relying on anaerobic metabolism to meet its energy demands is called the **anaerobic threshold**.

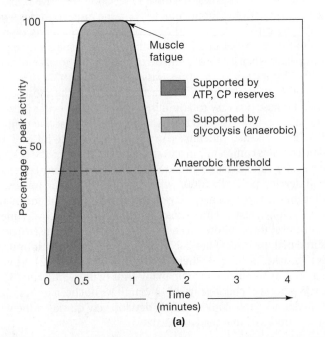

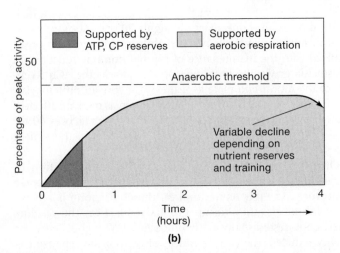

▶Figure 35 **Muscular Performance and Endurance. (a)** At peak levels of activity, skeletal muscles rely primarily on glycolysis for ATP production, with associated lactic acid production. Initial burst activity is supported by ATP and CP reserves. Muscles operating at peak levels fatigue rapidly. **(b)** Muscular activity can continue for extended periods when ATP demands are kept below the anaerobic threshold.

Muscular System

Muscular System

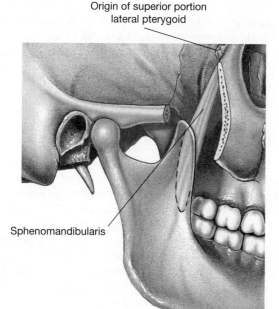

▶ **Figure 36 The Sphenomandibularis Muscle.**

If energy demands are kept below the anaerobic threshold, muscular activity can be continued until nutrient sources are exhausted. In a trained athlete, muscle fatigue may not occur for several hours (▶ Figure 35b).

What's New in Anatomy?

Like most people, you might assume that every anatomical structure in the human body was described centuries ago, and that nothing is "new" in the field of anatomy. Many people were surprised in 1996, however, when anatomical researchers at the University of Maryland documented the existence of a previously unknown skeletal muscle. This "new" muscle, named the *sphenomandibularis muscle*, extends from the lateral surface of the sphenoid to the mandible (▶ Figure 36). The sphenomandibularis muscle assists the muscles of mastication.

The research began with a computer analysis of the Virtual Human database, a digitized photographic atlas of cross-sectional anatomy. That initial work was then supported by careful cadaver dissections. Although the discovery of the sphenomandibularis muscle remains controversial (it may have been described previously as a portion of the temporalis muscle), it is a good example of how modern technologies are providing new perspectives on the human body.

▶ Muscular System Disorders

Hernias

When the abdominal muscles contract forcefully, pressure in the abdominopelvic cavity can increase dramatically. That pressure is applied to internal organs. If the individual exhales at the same time, the pressure is relieved because the diaphragm can move upward as the lungs collapse. But during vigorous isometric

exercises or when lifting a weight while holding one's breath, pressure in the abdominopelvic cavity can rise to 106 kg/cm^2, roughly 100 times the normal pressure. A pressure that high can cause a variety of problems, including hernias. A **hernia** develops when a visceral organ or part of an organ protrudes abnormally through an opening in a surrounding muscular wall or partition. There are many types of hernias; here we will consider only *inguinal* (groin) *hernias* and *diaphragmatic hernias*.

Late in the development of male fetuses, the testes descend into the scrotum by passing through the abdominal wall at the **inguinal canals**. In adult males, the sperm ducts and associated blood vessels penetrate the abdominal musculature at the inguinal canals as the *spermatic cords*, on their way to the abdominal reproductive organs. In an inguinal hernia, the inguinal canal enlarges and the abdominal contents, such as a portion of the greater omentum, small intestine, or (more rarely) urinary bladder, enter the inguinal canal (▶ Figure 37). If the herniated structures become trapped or twisted, surgery may be required to prevent serious complications. Inguinal hernias are not always caused by unusually high abdominal pressures: Injuries to the abdomen or inherited weakness or distensibility of the canal can have the same effect.

The esophagus and major blood vessels pass through openings in the diaphragm, the muscle that separates the thoracic and abdominopelvic cavities. In a **diaphragmatic hernia**, abdominal organs slide into the thoracic cavity. If entry is through the *esophageal hiatus*, the passageway used by the esophagus, a *hiatal hernia* (hī-Ā-tal; *hiatus*, a gap or opening) exists. The severity of the condition depends on the location and size of the herniated organ or organs. Hiatal hernias are very common, and most go unnoticed, although they may increase the severity of gastric acid entry into the esophagus (gastroesophogeal reflux disease, or GERD, commonly known as heartburn). Radiologists see them in about 30 percent of individuals whose upper gastrointestinal tracts are examined with barium-contrast techniques.

When clinical complications other than GERD develop, they generally do so because abdominal organs that have pushed into the thoracic cavity are exerting pressure on structures or organs there. Like inguinal hernias, a diaphragmatic hernia can

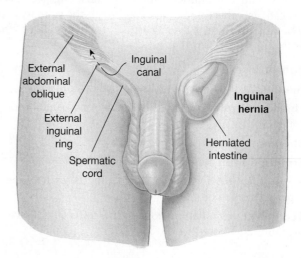

▶ **Figure 37 An Inguinal Hernia.**

Muscular System

result from congenital factors or from an injury that weakens or tears the diaphragm. If abdominal organs occupy the thoracic cavity during fetal development, the lungs may be poorly developed at birth.

Sports Injuries

Sports injuries affect amateurs and professionals alike. A five-year study of college football players indicated that 73.5 percent experienced mild injuries, 21.5 percent moderate injuries, and 11.6 percent severe injuries during their playing careers. Contact sports are not the only activities that show a significant injury rate: A study of 1650 joggers running at least 27 miles a week reported 1819 injuries in a single year.

Muscles and bones respond to increased use by enlarging and strengthening. Poorly conditioned individuals are therefore more likely than people in good condition to subject their bones and muscles to intolerable stresses. Training is also important in minimizing the use of antagonistic muscle groups and in keeping joint movements within the intended ranges of motion. Planned warm-up exercises before athletic events stimulate circulation, improve muscular performance and control, and help prevent injuries to muscles, joints, and ligaments. Stretching exercises after an initial warm-up will stimulate blood flow to muscles and help keep ligaments and joint capsules supple. Such conditioning extends the range of motion and may prevent sprains and strains when sudden loads are applied.

Dietary planning can also be important in preventing injuries to muscles during endurance events, such as marathon running. Emphasis has commonly been placed on the importance of carbohydrates, leading to the practice of "carbohydrate loading" before a marathon. But while operating within aerobic limits, muscles also utilize amino acids extensively, so an adequate diet must include both carbohydrates and proteins.

Improved playing conditions, equipment, and regulations also play a role in reducing the incidence of sports injuries. Jogging shoes, ankle or knee braces, helmets, mouth guards, and body padding are examples of equipment that can be effective. The substantial penalties now earned for personal fouls in contact sports have reduced the numbers of neck and knee injuries.

Several traumatic injuries common to those engaged in active sports can also affect nonathletes, although the primary causes may be very different. A partial listing of activity-related conditions includes the following:

- **Bone bruise:** bleeding within the periosteum of a bone
- **Bursitis:** an inflammation of the bursae at joints
- **Muscle cramps:** prolonged, involuntary, and painful muscular contractions
- **Sprains:** tears or breaks in ligaments
- **Strains:** tears in muscles or tendons
- **Stress fractures:** cracks or breaks in bones subjected to repeated stresses or trauma
- **Tendinitis:** an inflammation of the connective tissue surrounding a tendon

Finally, many sports injuries would be prevented if people who engage in regular exercise would use common sense and recognize their personal limitations. It can be argued that some athletic events, such as the ultramarathon, place such excessive stresses on the cardiovascular, muscular, respiratory, and urinary systems that these events cannot be recommended, even for athletes in peak condition.

Intramuscular Injections

Drugs are commonly injected through hollow needles into tissues rather than directly into the bloodstream (accessing blood vessels may be technically more complicated). An **intramuscular (IM) injection** introduces a fairly large amount of a drug, which will then enter the circulation gradually. The drug is introduced into the mass of a large skeletal muscle. Uptake is generally faster and accompanied by less tissue irritation than when drugs are administered *intradermally* or *subcutaneously* (injected into the dermis or subcutaneous layer, respectively). Depending on the size of the muscle, up to 5 mL of fluid may be injected at one time, and multiple injections are possible. A decision on the injection technique and the injection site is based on the type of drug and its concentration.

For IM injections, the most common complications involve accidental injection into a blood vessel or piercing of a nerve. The sudden entry of massive quantities of a drug into the bloodstream can have fatal consequences, and damage to a nerve can cause motor paralysis or sensory loss. Thus, the site of injection must be selected with care. Bulky muscles that contain few large vessels or nerves are ideal sites. The gluteus medius muscle or the posterior, lateral, superior part of the gluteus maximus muscle is commonly selected. The deltoid muscle of the arm, about 2.5 cm (1 in.) distal to the acromion, is another effective site (▶ **Figure 38**). Probably most satisfactory from a technical point of view is the vastus lateralis muscle of the thigh; an injection into this thick muscle will not encounter vessels or nerves, but may cause pain later when the muscle is used in walking. This is the preferred injection site in infants before they start walking, as their gluteal and deltoid muscles are relatively small. The site is also used in elderly patients or others with atrophied gluteal and deltoid muscles.

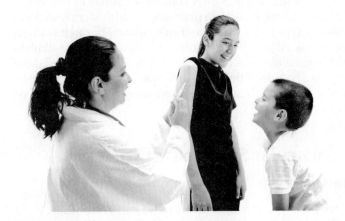

▶ **Figure 38 Intramuscular Injection in the Arm.**

1. Ann, who is 35 years old, complains of progressively worsening, fluctuating pain at her knees and the joints of her hands for several weeks. The pain is at its worst in the early morning. The symptoms are symmetrical—the same joints on both sides of her body are affected. She also reports low-grade fever and fatigue at times. Physical examination reveals limited movement of her wrists and inflammation with spongy swelling around the knee joints. Slight ulnar deviation of the fingers at the painful metacarpophalangeal joints of her hands is noted. There are no skin lesions present. Many different disorders can cause joint pain. Study the following information to make a preliminary diagnosis:

   • Synovial fluid analysis: cloudy, reduced viscosity, absence of bacteria, no uric acid crystals

   • X-ray studies of the hand: detectable deterioration of articular cartilages in metacarpophalangeal joints

   • Serum rheumatoid factor: 1:60 titer, elevated erythrocyte sedimentation rate

2. The mother of a 5-year-old boy reports that her son cannot rise from a sitting to a standing position without support from his arms "walking up his thighs." He also falls frequently. Physical examination reveals proximal muscle weakness, especially in the lower limbs. The physician orders bone x-rays, a serum aldolase test, and nerve conduction tests. Two of the test results are abnormal, and the physician makes a preliminary diagnosis of muscular dystrophy. This diagnosis is later confirmed by muscle biopsy. Which of the three results was (were) abnormal?

3. Taking a shortcut across campus Jeremy jumps down from a ledge and breaks his ankle. After six weeks the cast that encircled his leg from his tibial tuberosity to his metatarsophalangeal joints is removed. What cutaneous and muscular changes could be anticipated?

NOTES:

Answers to these problems can be found on page 241.

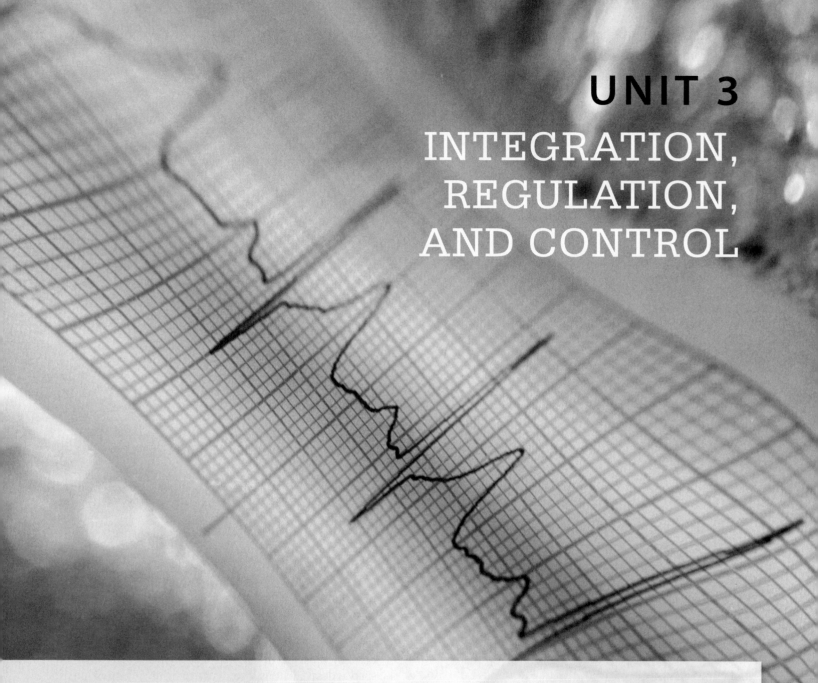

UNIT 3
INTEGRATION, REGULATION, AND CONTROL

Homeostasis is continually threatened. Locomotion must be precisely coordinated, injuries avoided, pathogenic organisms overcome, and tissues repaired. The electrolyte and water concentrations of body fluids must be adjusted as metabolic and excretory demands disturb the existing balance, and energy reserves mobilized or restored as dietary or metabolic conditions fluctuate over time. The homeostatic maintenance of a normal tissue environment affects every living cell in our bodies. But in most cases individual cells are unable to do anything about the environmental characteristics that affect them.

Instead, several organ systems detect variations from normal and respond to restore homeostasis. The pattern, speed, consistency, and nature of the response can vary widely.

Two organ systems, the nervous system and the endocrine system, are responsible for this kind of coordination and control. When either of these complex systems malfunctions, the effects are widespread and potentially devastating. Because the range of disorders and the range of diagnostic procedures and potential therapies are so broad, we will only consider representative disorders and general diagnostic and therapeutic patterns in this unit.

MAJOR SECTIONS INCLUDED WITHIN THIS UNIT:

The Nervous System
The Endocrine System
END-OF-UNIT CLINICAL PROBLEMS

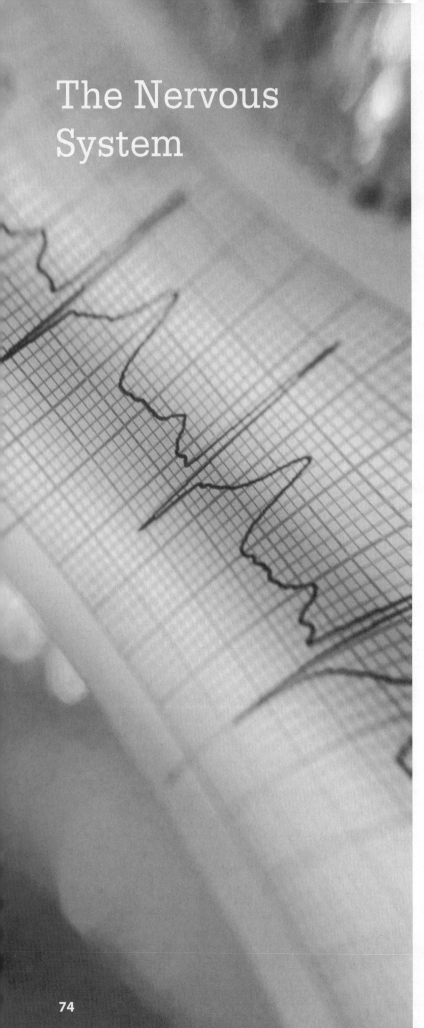

The Nervous System

▶ An Introduction to the Nervous System and Its Disorders

The nervous system is a highly complex and interconnected network of neurons and supporting neuroglia. Neural tissue is extremely delicate, and the characteristics of the extracellular environment must be kept within narrow homeostatic limits for it to function well. When homeostatic regulatory mechanisms break down under the stress of genetic or environmental factors, such as infection or trauma, or the cumulative effects of aging, then symptoms of neurological disorders appear.

Literally hundreds of disorders affect the nervous system. A *neurological examination* attempts to trace the source of a problem through an evaluation of the sensory, motor, behavioral, and cognitive functions of the nervous system. ▶ **Figure 39** introduces several major categories of nervous system disorders. We will discuss many of these examples in the sections that follow. **Table 18** summarizes representative infectious diseases of the nervous system.

The Symptoms of Neurological Disorders

The nervous system has varied and complex functions, and the symptoms of neurological disorders are equally diverse. However, a few symptoms accompany a wide variety of disorders:

- **Headache.** Headache seems to be a universal experience, with 70 percent of people reporting at least one headache each year. Most of these are *muscle contraction (tension-type) headaches*, which are thought to be due to head and neck muscle tension, or *migraine headaches*, which are now thought to have neurovascular origins (see pp. 75–76). Only rarely are these conditions related to life-threatening problems.

- **Muscle weakness.** Muscle weakness can have an underlying neurological basis, as we noted in the section "Signs and Symptoms of Muscular System Disorders." (See ▶ **Figure 33**, p. 64.) The examiner must determine the primary cause of the symptom for most effective treatment. Myopathies (muscle disease) must be differentiated from neurological diseases such as demyelinating disorders, neuromuscular junction dysfunction, and peripheral nerve damage.

- **Paresthesias.** Loss of feeling, numbness, or tingling sensations may develop after damage to (1) a sensory nerve (cranial or spinal nerve) or (2) sensory pathways in the central nervous system (CNS). The effects can be temporary or permanent. For example, a *pressure neuropathy* which may have loss of muscle function as well as sensory changes, may last a few minutes, whereas the paresthesia that develops distal to an area of severe spinal cord damage will probably be permanent (pp. 83–84).

The Neurological Examination

During a physical examination, information about the nervous system is obtained indirectly by assessing sensory, motor, and intellectual functions. Examples of factors noted in the physical examination include:

- **State of consciousness.** There are many levels of consciousness, ranging from unconscious and incapable of being

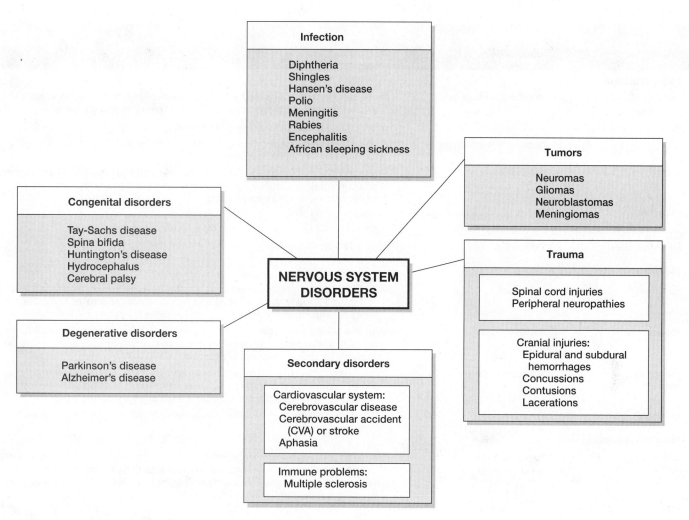

Figure 39 Nervous System Disorders.

Nervous System

aroused, to fully alert and attentive, to hyperexcitable. We introduce the terms assigned to the various levels of consciousness in a later section (p. 98).

- **Reflex activity.** The state of peripheral sensory and motor innervation can be checked by testing specific reflexes. For example, the *knee-jerk reflex* will not be normal if damage has occurred in associated segments of the lumbar spinal cord, their spinal nerve roots, or the peripheral nerves involved in the reflex. Musculoskeletal abnormalities also affect reflex responses, but such abnormalities can usually be differentiated through further examination.

- **Abnormal speech patterns.** Normal speech involves intellectual processing, motor coordination at the speech centers of the brain, precise respiratory control, the regulation of tension in the vocal cords, and the adjustment of the musculature of the palate and face. Problems with the selection, production, or use of words commonly follow damage to the cerebral hemispheres, as in a stroke (pp. 81, 90).

- **Abnormal motor patterns.** An individual's posture, balance, and mode of walking, or gait, are useful indicators of the level of motor coordination and strength. Clinicians also ask about abnormal involuntary movements that may

indicate a *seizure*—a temporary disorder of cerebral function (pp. 90–91).

A number of diagnostic procedures and laboratory tests can be used to obtain additional information about the status of the nervous system. **Table 19** summarizes information about these procedures.

Headaches

Almost everyone has experienced a **headache** at one time or another. The National Institute of Neurological Disorders and Stroke (NINDS, a part of the National Institutes of Health [NIH]) lists four headache types: vascular (or neurovascular; primarily migraine) muscle contraction (tension), traction, and inflammatory. However, the International Headache Society has two categories of headaches: Primary (with 17 sub-types) that have no obvious anatomical cause, and Secondary (with 14 sub-types) that can be traced to structural problems, such as trauma or infection.

Diagnosis and treatment pose a number of problems, primarily because headaches can be produced by a wide variety of conditions (one source notes 316 established causes). The most

Table 18   Examples of Infectious Diseases of the Nervous System

Disease	Organism(s)	Description
Bacterial Diseases		
Hansen's disease (leprosy)	*Mycobacterium leprae*	Progresses slowly in coolest areas of body; invades nerves and produces sensory loss and motor paralysis; tissues may degenerate after unperceived trauma
Bacterial meningitis		Inflammation of the spinal or cranial meninges
	Haemophilus influenzae	Previously common meningitis; usually infects children (age 2 months–5 years); vaccine available and 99% effective
	Neisseria meningitidis	Meningococcal meningitis; usually infects children and adults ages 5–40 years; young adults living closely together (military recruits and college students in dorms) at higher risk; vaccines somewhat effective
	Streptococcus pneumoniae	Streptococcal meningitis; usually infects children and adults over age 40; high mortality rate (40%); effective vaccine available
Brain abscesses	Various bacteria	Focal infection increases in size and compresses the brain
Viral Diseases		
Poliomyelitis	Polioviruses	Polio has different forms; only one attacks motor neurons, leading to paralysis of limbs and muscle atrophy. Vaccine is available. Disease almost eliminated in most of the world by ongoing immunization
Rabies	Rabies virus	Virus invades the central nervous system through peripheral nerves. Untreated cases are fatal; treatment involves rabies antitoxin and vaccination; pre-exposure vaccine available for those at high risk of exposure
Encephalitis	Various encephalitis	Inflammation of the brain; fever and headache; no vaccine is available for most virus forms. Transmission occurs by mosquitoes. Eastern equine encephalitis is most lethal (50%–75% mortality rate). West Nile virus is more reported
Meningitis	Various viruses	Generally less severe than bacterial, usually resolves with symptomatic treatment
Parasitic Diseases		
African sleeping sickness	*Trypanosoma brucei*	Caused by a flagellated protozoan; infection occurs through bite of tsetse fly; sickness infects blood, lymph nodes, and then nervous system. Symptoms include headache, tiredness, weakness, and paralysis, before coma and death; no vaccine is available but drug treatment is effective

Nervous System

common causes of headache appear to be either neurovascular or muscular problems.

For most people, headaches are primary, tension-type with moderate pain all over the head. There is a sensation of pressure or tightness, the pain is poorly localized, and it is often associated with muscle tension, such as tight neck muscles. Tension-type headaches do not have the associated neurovascular features that define migraine headaches: aura, throbbing, often unilateral severe pain, light sensitivity, and nausea or vomiting.

Other headaches develop secondarily due to underlying problems including:

- **CNS disorders**, such as infections (meningitis, encephalitis, rabies) or brain tumors
- **Trauma**, such as a blow to the head
- **Cardiovascular disorders**, such as a stroke (p. 81)
- **Metabolic disturbances**, such as low blood sugar

Some symptoms, if they occur with a headache, raise the possibility of serious underlying problems that are associated with secondary headaches. Examples of such symptoms include (1) sudden, severe headache or sudden headache associated with a stiff neck; (2) headaches associated with fever, convulsions, or accompanied by confusion or loss of consciousness; (3) headaches following a blow to the head, or associated with severe pain in the eye or ear; (4) persistent headache in a person who was previously headache free; and (5) recurring headache in children.

Evidence from functional MRI scans indicates that migraine sufferers have increased neural excitability in the cerebral cortex. Many migraine episodes begin with an "aura," painless sensations such as flashing lights or blurred vision. An aura probably reflects cerebrocortical excitation followed by neural suppression. Associated neurochemical release affects a portion of the midbrain known as the *dorsal raphe*. Electrical stimulation of the dorsal raphe can produce changes in cerebral blood flow with cerebrovascular dilation and constriction resulting in the classic throbbing migraine headache. Several drugs with antimigraine action inhibit neurons there. The most effective drugs stimulate a class of serotonin receptors that are abundant in the dorsal raphe. For some migraine sufferers, the headaches have been linked to inherited abnormalities in genes that control cellular activity in migraine critical parts of the brain.

The triggers for **muscle contraction (tension-type) headaches** probably involve a combination of factors, but sustained contractions of the neck and facial muscles are commonly implicated. Tension headaches can last for days or can occur daily over longer periods. Some tension headaches may accompany depression or anxiety. Cause and effect in this case may be a "which came first, the chicken or egg?" type of question.

Table 19	Representative Diagnostic and Laboratory Tests for Nervous System Disorders	
Diagnostic Procedure	**Method and Result**	**Representative Uses**
Lumbar puncture (LP) (spinal tap)	Needle aspiration of CSF from the subarachnoid space in the L_3–L_4 or L_4–L_5 lumbar area of the vertebral column	See Analysis of CSF (below) for diagnostic uses
Skull x-ray	Standard x-ray	Detects fractures and possible sinus involvement or other bony abnormalities
Electroencephalography (EEG)	Electrodes placed on the scalp detect electrical activity of the brain, produces graphic record	Detects abnormalities in frequency and amplitude of brain waves due to epilepsy, less precise than MRI or CT scans for anatomical disorders
Computerized tomography (CT) scan of the brain	Standard CT; contrast media are commonly used	Detects tumors; cerebrovascular abnormalities, such as acute bleeding from strokes or aneurysms; trauma scars; ischemic strokes; areas of edema, or atrophy (tissue loss)
Cerebral angiography and digital subtraction angiography (DSA)	Dye is injected into an artery of the neck, and the movement of the dye is observed via serial x-rays; DSA transfers location of dye information onto a computer for image enhancement	Detects abnormalities in the cerebral vessels, such as aneurysms, ruptures, blockages
Positron emission tomography (PET) scan	Radiolabeled compounds injected into bloodstream accumulate at specific areas of the brain; may help diagnosis of Parkinson's disease and mental illness; the radiation emitted is monitored by a computer that generates a reconstructed image	Determines blood flow to specific regions of the brain; detects focal points of brain metabolic activity; diagnoses Alzheimer's disease
Magnetic resonance imaging (MRI)	Standard MRI; contrast media are commonly used to enhance visualization	Detects brain tumors, hemorrhaging, edema, spinal-cord injury, and other soft tissue structural abnormalities
	Functional MRI (fMRI). Maps brain activity while the subject is performing a task	Quantifies brain activity in specific areas, has largely replaced PET scans in detecting and studying normal and pathologic brain function
Laboratory Test	**Normal Values**	**Significance of Abnormal Values**
Pressure of CSF	50–180 mm H_2O	Pressure higher than 180 mm H_2O is considered abnormal, possibly indicating hemorrhaging, brain tumor, blood clots around the brain, or infection
Color of CSF	Clear and colorless	Cloudiness suggests hemorrhage, incidental bleeding from puncture technique, or infection
Analysis of CSF		
Glucose in CSF	40–70 mg/dL	Decreased levels occur with CNS bacterial or fungal infections
Protein in CSF	20–50 mg/dL	Elevated levels occur in some infectious processes, such as meningitis and encephalitis, also during inflammatory processes, including multiple sclerosis or brain tumor
Cells present in CSF	No RBCs present; WBC count should be less than 5 per mm³	RBCs appear with subarachnoid hemorrhage and increased number of neutrophils occurs in bacterial infections, such as bacterial meningitis; increase in lymphocyte numbers occurs in viral meningitis and multiple sclerosis; cancer cells from a brain tumor may be shed into CSF
Culture of CSF	Organism causing infectious process in brain or spinal cord can be cultured for identification and determination of sensitivity to antibiotics	Culture can determine causative agent in bacterial meningitis or brain abscess. Takes at least 2 days
Polymerase Chain Reaction (PCR)	Negative	Rapid same day test that identifies many viral meningitis cases
Latex Agglutination (LA)	Negative	Rapid same day test for bacterial meningitis antigens in CSF
IgG	6–44.2 mg/dL	Elevated in multiple sclerosis and some other inflammatory problems
IgG index	0.29–0.59	Ratio of IgG to albumin in CSF divided by IgG to albumin in blood; the ratio is elevated in multiple sclerosis

Nervous System

▶ Disorders of Neural Tissue

Axoplasmic Transport and Disease

With a soft flutter of wings, dark shapes drop from the sky onto the backs of grazing cattle. Each shape is a small bat whose scientific name, *Desmodus rotundus,* is less familiar than the popular term, *vampire bat.* Vampire bats inhabit tropical and semitropical areas of North, Central, and South America. They range from the Texas coast to Chile and southern Brazil. These rather aggressive animals are true vampires, subsisting on a diet of fresh blood. Over the next hour, every bat in the flight—which may number in the hundreds—will consume about 65 mL of blood through small slashes in the skin of their prey.

As unpleasant as this blood collection may sound, it is not the blood loss that is the primary cause for concern. The major problem is that these bats can be carriers of the rabies virus. **Rabies** is an acute disease of the central nervous system. The rabies virus can infect any mammal, wild or domestic. With few exceptions (worldwide there have been seven documented survivors), the result is death within three weeks of developing symptoms. For unknown reasons, bats can survive rabies infection for an indefinite period. That is why they are effective carriers of the disease. Because many bat species, including vampire bats, form dense colonies, a single infected individual can spread the disease throughout the entire colony.

Rabies is generally transmitted to people through the bite of a rabid animal. About 55,000 cases of human rabies occur each year worldwide, the majority of them the result of dog bites. Only one or two of those cases, however, occur in the United States. Most U.S. cases of rabies are caused instead by the bites of wild animals such as raccoons, foxes, skunks, or bats, because most dogs and cats in the United States are vaccinated against rabies. The incubation period in these animals is highly variable; most eventually become ill, although bats may be infected carriers for several years. Dogs and cats usually become visibly ill within weeks of infection; this accounts for the duration of animal quarantine periods, which are designed to prevent the spread of rabies to non-infected islands such as England, New Zealand, and Hawaii.

Although these bites generally involve peripheral sites, such as the hand or foot, the symptoms are caused by CNS damage. The virus present at the site of the injury is absorbed by the synaptic knobs of peripheral nerves in the region. It then gets a free ride to the CNS, courtesy of retrograde flow. The asymptomatic incubation period is usually from one to seven weeks. During the first few days of illness, the individual may experience headache, fever, muscle pain, nausea, and vomiting. The afflicted person then enters the central nervous system phase marked by extreme excitability, hallucinations, muscle spasms, and disorientation. Swallowing becomes difficult, accounting for an early name for rabies: *hydrophobia* (literally, fear of water). The accumulation of saliva makes the individual appear to be "foaming at the mouth." Coma and death soon follow. Anyone exposed to the patients' saliva should receive preventive treatment.

Post-exposure preventive treatment (PEP) was developed by Louis Pasteur in the late 1800s in France. The modern version of PEP must begin almost immediately after exposure and consists of an injection of immune globulin antibodies against the rabies virus, followed by a series of vaccinations against rabies. Without such treatment, rabies infection in humans, with seven documented exceptions, is always fatal; but even this post-exposure treatment may not be sufficient after a massive infection, which can lead to death in as little as four days. Individuals such as veterinarians or field biologists who are at high risk of exposure thus commonly take a pre-exposure series of vaccinations. These injections start the immune defenses and improve the effectiveness of the post-exposure treatment.

Rabies is perhaps the most dramatic example of a clinical condition directly related to axoplasmic transport. However, many toxins, including heavy metals and toxins released by the tetanus bacteria, some pathogenic bacteria, and other viruses, use this mechanism to enter the CNS.

Demyelination Disorders

Demyelination disorders are linked by a common symptom: the destruction of myelinated axons in the CNS and peripheral nervous system (PNS). The mechanism responsible for this loss differs in each disorder. We will consider only the major categories:

- **Heavy-metal poisoning.** Chronic exposure to heavy-metal ions, such as arsenic, lead, or mercury, can lead to damage of neuroglia and to demyelination. As demyelination occurs, the affected axons deteriorate, and the condition becomes irreversible. Historians note several examples of heavy-metal poisoning with widespread impact. For example, the contamination of drinking water with lead used in plumbing pipes has been cited as one factor in the decline of the Roman Empire. Lead exposure from paint and auto exhaust at levels associated with brain damage in young children led to the removal of lead from U.S. gasoline in 1986 and house paint in 1978. In 1986, lead was banned from plumbing materials as well. Batteries and modern electronic devices (computers, cell phones, and their batteries and accessories) contain heavy metals and are classified as hazardous household waste by the U.S. Environmental Protection Agency (EPA). Many of these metals are valuable, and recycling efforts are increasing.

  Well into the 19th century, mercury used in the preparation of felt presented a serious occupational hazard for those employed in the manufacture of stylish hats. Over time, mercury absorbed through the skin and across the lungs accumulated in the CNS, producing neurological damage that affected both physical and mental functioning. (This effect is the source of the expression "mad as a hatter.") In the 1950s Japanese fishermen and their families working in Minamata Bay, Japan, collected and consumed seafood contaminated with mercury discharged from a nearby chemical plant. Mercury poisoning affected hundreds of people and caused crippling birth defects. Less severe problems have affected children born to mothers in the midwestern United States who ate large amounts of fish during pregnancy. As a result, pregnant women are now advised to limit fish consumption. (For unknown reasons, the flesh of some species of fish contains relatively high levels of mercury.)

- **Diphtheria.** Diphtheria (dif-THĒ-rē-uh; *diphthera,* leather + *-ia,* disease) is a disease that results from a bacterial infection of the respiratory tract or the skin. The diphtheria toxin damages Schwann cells and destroys myelin sheaths in the peripheral nervous system. This patchy demyelination leads to sensory and motor problems. The toxin also affects cardiac muscle cells, and heart enlargement and heart failure may occur. The fatality rate for untreated cases ranges from 35 to 90 percent, depending on the site of infection and the subspecies of bacterium. Because an effective vaccine (which is frequently combined with the tetanus vaccine) exists, cases are relatively rare in countries with adequate health care. Russia experienced an epidemic in the 1990s when vaccines were not widely available there.

- **Multiple sclerosis.**   Multiple sclerosis (skler-Ō-sis; *sklerosis*, hardness), or **MS**, is a disease characterized by recurrent incidents of autoimmune inflammatory demyelination that affects axons in multiple areas of the central nervous system, primarily the optic nerve, brain, and spinal cord. Common symptoms include partial loss of vision and problems with speech, balance, and general motor coordination, including bowel and urinary bladder control. The time between incidents and the degree of recovery vary from case to case. Each incident leaves a greater degree of functional impairment. In about one-third of all cases, the disorder is steadily progressive, and of those patients with the relapsing/remitting form, 2 to 3 percent a year develop progressive disease. The average age at the first attack is 30–40 years;

the incidence among women is twice that among men. In some patients, various forms of immune suppression and modulation including corticosteroid or interferon injections have reduced relapse rates and slowed the progression of the disease. We will discuss MS further in a later section (p. 83).

How Drugs Disrupt Neural Function

Many toxins work by interfering with normal synaptic function. Modified forms of some toxins and similar compounds are useful as drugs. Drugs can affect neural activity at various sites on the neuron, including sites along axons, and especially at the synapse. In general, drugs that interfere with key steps in synaptic activity act by (1) interfering with neurotransmitter synthesis, (2) altering the rate of neurotransmitter release, (3) preventing neurotransmitter inactivation, or (4) preventing neurotransmitter binding to receptors. **Table 20** and ▶**Figure 40** provide information about specific chemical compounds that affect ACh activity.

Table 20	Drugs Affecting Acetylcholine (ACh) Activity at Synapses	
Drug	**Mechanism**	**Effects**
Hemicholinium	Blocks ACh synthesis	Produces symptoms of synaptic fatigue
***Botulinus* toxin**	Blocks ACh release directly	Paralyzes voluntary muscles; produced by a Clostridia bacteria; responsible for a deadly type of food poisoning, and blocks some autonomic nervous system effects. Small intramuscular injection used to treat muscle spasms in the neck or around the eyes; used cosmetically to reduce wrinkles on the face and reduce excessive sweating in injected areas. Effects last several months
Barbiturates	Decrease rate of ACh release	Produce muscle weakness; depress CNS activity; administered as sedatives, anesthetics, and seizure (epilepsy) control
Lidocaine (Xylocaine)	Reduces membrane permeability to sodium	Prevents stimulation of sensory neurons; used as a local anesthetic
Tetrodotoxin (TTX) Saxitoxin (STX) Ciguatoxin (CTX)	Blocks sodium ion channels Similar to TTX Similar to TTX	Eliminates production of action potentials; produced by some marine organisms during normal metabolic activity
Neostigmine	Anticholinesterase, prevents ACh inactivation by cholinesterase, does not enter the CNS	Produces sustained contraction of skeletal muscles; other effects on cardiac muscle, smooth muscle, and glands (esp. saliva production); used to treat myasthenia gravis and to counteract overdoses of tubocurarine
Donepezil	Cholinesterase inhibitor, reversible CNS effect, benefit thought from increase in cerebral levels of acetylcholine	Slows progression of early Alzheimer's disease by up to three years; similar medicines are rivastigmine, galantamine
Insecticides (organophosphates) (malathion, parathion, etc.) and nerve gases	Similar to neostigmine, but irreversible effect	Similar to neostigmine but not used as medicine
***d*-tubocurarine**	Prevents/blocks ACh binding to neuromuscular postsynaptic receptor sites	Paralyzes voluntary muscles; curare produced by South American plant
Nicotine	Binds to ACh receptor sites	Low doses facilitate voluntary muscles; high doses cause paralysis; an active ingredient in cigarette smoke; addictive for most people
Succinylcholine	Reduces sensitivity to ACh	Paralyzes voluntary muscles; used to produce temporary muscular relaxation during surgery
Atropine	Competes with ACh for binding sites on postsynaptic membrane	Reduces heart rate, smooth muscle activity; decreases salivation; dilates pupils; produces skeletal muscle weakness at high doses; produced by deadly nightshade plant

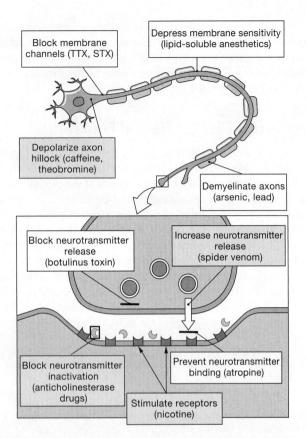

Depress membrane sensitivity
(lipid-soluble anesthetics)

Block membrane
channels (TTX, STX)

Depolarize axon
hillock (caffeine,
theobromine)

Demyelinate axons
(arsenic, lead)

Block neurotransmitter
release
(botulinus toxin)

Increase neurotransmitter
release
(spider venom)

Block neurotransmitter
inactivation
(anticholinesterase
drugs)

Prevent neurotransmitter
binding (atropine)

Stimulate receptors
(nicotine)

▶ **Figure 40** **The Mechanisms of Drug Action at Various Sites in Neurons.** The synapse depicted is a cholinergic synapse. Factors that facilitate neural function and make neurons more excitable are shown in white boxes; factors that inhibit or depress neural function are shown in gray boxes.

A comparable range of drugs is available to target other types of chemical synapses.

Botulinus toxin is responsible for the primary symptom of *botulism*, a widespread paralysis of skeletal muscles. Botulinus toxin blocks the release of ACh at the presynaptic membrane of cholinergic motor neurons. The same substance that causes this potentially fatal illness is now widely used to treat muscle spasms and tics, and even excessive sweating (ACh is the neurotransmitter used by the sympathetic nerves to activate sweat glands). Plastic surgeons and dermatologists use it for cosmetic purposes. Minute quantities of the toxin (available under the name "Botox") are injected under the skin to paralyze small facial muscles that cause wrinkles. (The treatment must be repeated every few months.) The venom of the black widow spider has the opposite effect, causing a massive release of ACh that produces intense, painful muscular cramps and spasms.

Other drugs primarily affect the postsynaptic membrane. **Anticholinesterase drugs**, or *cholinesterase inhibitors*, block the breakdown of ACh by acetylcholinesterase. The result is an exaggerated and prolonged stimulation of the postsynaptic membrane. At neuromuscular junctions, this abnormal stimulation produces an extended and extreme state of contraction. Military nerve gases block cholinesterase activity for weeks, although few persons exposed to them are likely to live long enough to regain normal synaptic function. Most animals including insects utilize ACh as a neurotransmitter, and

anticholinesterase (AChE) drugs, such as *malathion*, are in widespread use in insect-control projects.

Drugs such as **atropine** and **d-tubocurarine** prevent ACh from binding to the postsynaptic receptors. These drugs work on different types of ACh receptors. Atropine can be administered to counteract the effects of AChE poisoning on ACh receptors in smooth and cardiac muscles; *d*-tubocurarine is a derivative of *curare*, a plant extract used by certain South American tribes to paralyze their prey. Curare and related compounds induce paralysis by preventing the stimulation of the neuromuscular junction by ACh. Other compounds, including **nicotine**, an active ingredient in cigarette smoke, bind to the receptor sites and stimulate the postsynaptic membrane. No enzymes exist that remove these compounds, and their effects are relatively prolonged.

Neurotoxins in Seafood

Among the chemicals that exert effects at various sites along axons are several toxins found in seafood, including *tetrodotoxin* (TTX), *saxitoxin* (STX), and *ciguatoxin* (CTX), which block sodium ion channels. These toxins are unaffected by cooking, freezing, drying or stomach acid, and usually do not change the look, taste, or odor of the seafood. Neurophysiology research laboratories have long used TTX to study the electrical properties of neurons. After blocking sodium ion channels, researchers can monitor subtle changes in the transmembrane potential that occur at or above the threshold—changes that would otherwise be masked by an influx of sodium ions. At low doses, these toxins can produce abnormal sensations and interfere with muscle control. Higher doses can cause respiratory paralysis and death.

Tetrodotoxin (tet-RŌ-dō-tok-sin), or **TTX**, is found in the liver, gonads, and blood of certain Pacific puffer fish species, and a related compound is found in the skin glands of some salamanders. Tetrodotoxin selectively blocks voltage-regulated sodium ion channels, effectively preventing nerve cell activity. The usual result is death from respiratory muscle paralysis. Despite the risks, some Japanese consider the puffer fish a delicacy, served under the name *fugu*. Specially licensed chefs prepare the dish, carefully removing the potentially toxic organs. Nevertheless, a mild tingling and sense of intoxication are considered desirable, and several people die each year as a result of improper preparation of fugu.

Saxitoxin (sak-si-TOK-sin), or **STX**, can have a similarly lethal effect. Saxitoxin and related poisons are produced by several species of marine microorganisms. When these organisms undergo a population explosion, they may color the surface waters, producing a "red tide." Eating seafood that has become contaminated by feeding on the toxic microbes can result in **paralytic shellfish poisoning** (**PSP**; from clams, mussels, or rarely oysters) or **ciguatera** (sig-wa-TAR-a) caused by **ciguatoxin** (sēg-wa-TOK-sin; *cigua*, a sea snail) (**CTX**; from fish). Mild cases produce symptoms of paresthesias and a curious reversal of hot-versus-cold sensations. Severe cases, which are relatively rare, result in coma and death, if untreated, due to respiratory paralysis.

Tay–Sachs Disease

Tay–Sachs disease is a genetic abnormality that involves the metabolism of *gangliosides*, important components of neuron plasma membranes. It is a *lysosomal storage disease* caused by abnormal lysosome activity. Individuals with this condition lack the enzyme needed to break down one particular ganglioside, which accumulates in the lysosomes of CNS neurons and causes them to deteriorate. Affected infants seem normal at birth, but neurological problems begin to appear within six months. The progression of symptoms includes muscle weakness, blindness, seizures, and death, generally before age 4.

No effective treatment exists, but prospective parents can be tested to determine whether they carry the gene (found on chromosome 15) responsible for this condition. The disorder is most prevalent in Ashkenazi Jews of Eastern European origin. A prenatal test is available to detect Tay–Sachs disease in a fetus.

Conditions That Disrupt Neural Function

Two important types of conditions within the body can disrupt normal CNS and PNS functions: (1) changes in the extracellular environment and (2) an inability to meet the metabolic demands of active neurons.

Changes in the Extracellular Environment

Changes in the extracellular environment, such as changes in ionic composition, or body temperature, can alter the resting membrane potential or disrupt the metabolic operations that support the generation of action potentials:

- **Changes in hydrogen ion concentration (pH) can have dramatic effects on neural activity.** The normal extracellular pH is 7.35–7.45. If the pH rises above this range, neurons are facilitated: At a pH near 7.8, they begin to generate action potentials spontaneously, producing severe convulsions. If the pH drops below the range, neurons are inhibited: At a pH near 7.0, the rate of generation of action potentials is so low that the nervous system shuts down, and the individual becomes completely unresponsive.

- **Changes in the ionic composition of the extracellular fluids have a direct effect on neural function.** Fluctuations in Na^+ or K^+ concentrations, such as those caused by dehydration or kidney disease, may either facilitate or inhibit neural activity by partially depolarizing or hyperpolarizing the plasma membrane. For example, an elevated extracellular K^+ concentration, **hyperkalemia** (hī-per-ka-LĒ-mē-uh; *kalium*, potassium + *haima*, blood) interferes with repolarization and suppresses the generation of action potentials. The result is a general weakness progressing to paralysis of skeletal muscles and death due to cardiac arrest. Abnormally high or low extracellular Ca^{2+} concentrations affect synaptic function by reducing or enhancing calcium entry into the synaptic knob, thereby modifying the amount of neurotransmitter released. Altered Ca^{2+} levels also produce direct effects on the excitability of membranes.

- **Changes in body temperature have a direct effect on the activity of neurons.** When body temperature rises, neurons become more excitable; an individual with a high fever may experience hallucinations or convulsions. When body temperature falls, neurons become inhibited; an individual whose body temperature has fallen below normal limits will be lethargic and confused and may lose consciousness.

Abnormal Metabolic Activity

The brain accounts for just 2 percent of your body weight, but it requires 18 percent of your resting energy consumption. Active neurons need ATP to support (1) the synthesis, release, and recycling of neurotransmitter molecules; (2) the movement of materials to and from the cell body via axoplasmic transport; (3) the maintenance of the normal resting potential; and (4) the recovery from action potentials through the activity of the sodium–potassium exchange pump. Each time an action potential occurs, sodium ions enter and potassium ions leave the cell; over time, ATP must be expended to maintain normal cytoplasmic ion concentrations. When impulses are generated at high frequencies, the energy demands are enormous.

Neurons normally derive ATP solely through aerobic mechanisms. Because their cytoplasm does not contain glycogen reserves, neurons are totally dependent on a continuous and reliable supply of both oxygen and glucose from the blood. In cases of severe malnutrition, neural function deteriorates as the body's energy reserves are exhausted.

Neural function is also impaired if the local circulation and its supply of oxygen is restricted or, worse yet, shut off. If the circulation to a region is interrupted for just a few seconds, neurons in that region will be injured. The longer the interruption, the more severe the injury will be. In a **stroke**, the blood supply to the brain is interrupted, either by a circulatory blockage or by vascular rupture. The degree of functional impairment after a stroke is determined by (1) the location and size of the region deprived of circulation, and (2) the duration of the circulatory interruption. Interruption of brain blood flow will kill the affected neurons in 4–10 minutes.

Several inherited abnormalities in neural function are characterized by metabolic problems in neurons. For example, *Tay–Sachs disease* (pp. 80–81) is a genetic abnormality involving the metabolism of *gangliosides*, which are important components of neuron plasma membranes.

▶ Disorders of the Spinal Cord and Spinal Nerves

Meningitis

The warm, dark, nutrient-rich environment of the cranial or spinal meninges provides ideal conditions for a variety of bacteria, fungi, and viruses. Microorganisms that cause brain infections (encephalitis) (see **Table 18**, p. 76) may sometimes cause meningitis as well; the same bacteria may be associated with middle ear, sinus, throat, and lung infections. These pathogens may gain access to the meninges by traveling within blood vessels or by entering at sites of vertebral or cranial injury. Headache, chills, high fever, disorientation, and rapid heart and respiratory rates appear as higher centers are affected. Without treatment, delirium, coma, convulsions, and death may follow within hours to days.

The most common clinical assessment involves checking for a "stiff neck" by asking the patient to touch the chin to the chest. Meningitis affecting the cervical portion of the spinal cord results in a marked increase in the muscle tone of the extensor muscles of the neck. So many motor units become activated that flexion of the neck becomes painfully difficult, if not impossible.

Nervous System

The mortality rate for viral or bacterial forms of meningitis ranges from 1 to more than 50 percent, depending on the type of virus or bacteria, the age and health of the individual, and other factors. Bacterial meningitis can be combated with antibiotics, the maintenance of proper fluid and electrolyte balance, and other supportive care. Survivors may have permanent neurological damage, including loss of sight or hearing, seizures, or mental retardation. The incidence of the most common forms of childhood bacterial meningitis, caused by *Haemophilus influenzae* (a bacteria despite its name) and streptococcal pneumonia, have been dramatically reduced (99 percent and 75 percent, respectively) by immunization. There were 20,000 cases of *Haemophilus influenzae* meningitis each year in the USA before the vaccine was introduced in 1988; for the last 15 years there have been fewer than 70 cases per year.

Spinal Taps and Myelography

Tissue samples, or **biopsies**, are taken from many organs to assist in diagnosis. For example, when a liver or skin disorder is suspected, small plugs of tissue are removed and examined for signs of cell damage or abnormal cell growth, or are used to identify the microorganism causing an infection. Unlike many other tissues, however, neural tissue consists largely of cells rather than extracellular fluids or fibers. Samples of neural tissue are seldom removed for analysis, because the body usually does not replace extracted or damaged neurons. Instead, small volumes of cerebrospinal fluid are collected and analyzed. Cerebrospinal fluid is intimately associated with the neural tissue of the CNS. Accordingly, when pathogens, cell debris, or metabolic wastes are present in the CNS, they can be detected in the cerebrospinal fluid (see **Table 19**, p. 77).

The withdrawal of CSF, a procedure called a *spinal tap*, must be done with care to avoid injuring the spinal cord. The adult spinal cord extends inferiorly only as far as vertebra L_1 or L_2. Between vertebra L_2 and the sacrum, the spinal meninges remain intact, but they enclose only the relatively sturdy components of the cauda equina and a significant quantity of CSF. When the vertebral column is flexed, a needle can be inserted between the inferior lumbar vertebral spinous processes, usually at either the L_3–L_4 or L_4–L_5 space, (the source of the alternative term Lumbar Puncture [LP]) and into the subarachnoid space, with minimal risk to the cauda equina. Risks are low primarily because (1) the volume of CSF is much greater than the volume of neural tissue, and (2) the needle is inserted with a plug inside it, so it will usually push aside, rather than penetrate, any nerve roots encountered. (The primary purpose of the plug, called a *stylet*, is to prevent the introduction of superficial tissues into the CSF; the stylet is withdrawn from the needle after insertion.) A small amount (3–9 mL) of fluid is collected from the subarachnoid space (▶ **Figure 41a**). The fluid is not withdrawn by suction, which could pull nerve roots against the tip of the needle and injure them. Instead, the fluid drips out under its own pressure, in a way that resembles the collection of maple syrup from a maple tree. Spinal taps are performed to administer spinal anesthesia, when CNS infection is suspected, or when severe back pain, headaches, disc problems, and some types of strokes are diagnosed.

Myelography is the introduction of radiopaque dyes into the CSF of the subarachnoid space. Because the dyes are opaque to x-rays, the CSF appears white on an x-ray image (▶ **Figure 41b**). Any blockage, tumors, inflammation, or adhesions that distort or

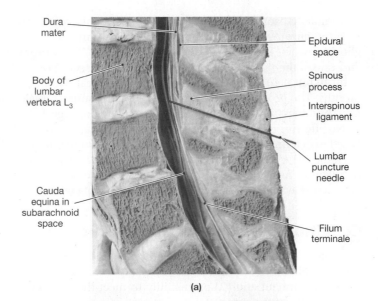

(a)

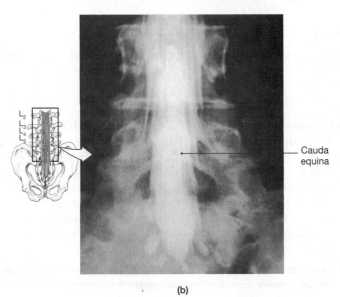

(b)

▶ **Figure 41 Spinal Taps and Myelography. (a)** The lumbar puncture needle is in the subarachnoid space, between the nerves of the cauda equina. The needle has been inserted between the third and fourth lumbar spinous processes, pointing at a superior angle. **(b)** A myelogram of the inferior lumbar region. An x-ray image of the spinal cord after a radiopaque dye has been introduced into the CSF.

divert CSF circulation appear in silhouette. The combination of a spinal tap and myelography can detect CNS blockage, infection, inflammation, or some cancers that secondarily involve the CNS, including leukemia (cancer of the white blood cells). Treatment may then involve injecting antibiotics, steroids, or anticancer drugs into the subarachnoid space. Noninvasive MRI scans have largely replaced myelography, except in cases where MRI is contraindicated, as in patients with cardiac pacemakers.

Damage to Spinal Tracts

Damage to spinal tracts can produce loss of sensation and/or motor control. The nature of the sensory and motor deficit depends on the site of damage and the specific tracts involved.

We will consider three examples of conditions that produce sensory and motor problems by their impact on spinal tracts: *multiple sclerosis, polio,* and *spinal trauma.*

Multiple Sclerosis

Multiple sclerosis (MS), introduced in the discussion of demyelination disorders (p. 79), is a disease that produces muscular paralysis and sensory losses through demyelination. Historically MS was diagnosed when neurological abnormalities were "separated in space and time"—meaning that patients suffered multiple episodic problems involving multiple areas of the nervous system. The initial symptoms, frequently involving some combination of paresthesias, clumsiness, and problems with vision and loss of bladder control (incontinence), are the result of myelin degeneration within the white matter of the lateral and posterior columns of the spinal cord, the optic nerves, or along tracts within white matter of the brain. Histological examination of the neural tissue shows areas of sclerosis (*skleros,* hardening, a scar) called **plaques**. MRI scans that detect these abnormalities have improved the accuracy of diagnosis, prognosis, and treatment. Spinal-cord involvement produces weakness, tingling sensations, and a loss of "position sense" for the limbs. For most patients, those with relapsing/remitting MS, improvement occurs. During subsequent attacks the residual effects build up. The cumulative sensory and motor losses may progressively lead to a generalized muscular paralysis with 50 percent of patients requiring help walking within 15 years after the initial attack.

Evidence suggests that this condition is linked to problems in the immune system, and to focal breakdown of the blood–brain barrier that usually reduces access of immune system components into the CNS. Probably caused by a combination of genetic and environmental factors, the result is autoimmune damage to myelin sheaths primarily in the white matter of the CNS. Individuals with MS have lymphocytes that do not respond normally to foreign proteins. Because several viral proteins have amino acid sequences similar to those of normal myelin, it has been proposed that MS results from a case of mistaken identity. For unknown reasons, MS appears to be associated with cold and temperate climates. Relative lack of vitamin D related to less sunlight may have some role. It has been suggested that individuals who develop MS may have an inherited susceptibility to a virus and that this susceptibility is exaggerated by environmental conditions. The yearly incidence in the United States averages around 50 cases for every 100,000 people in the population. Improvement has been noted in some MS patients treated with short courses of corticosteroids for relapses and long-term treatments with several medicines that suppress the misdirected immune response (immune modulators). These disease-modifying drugs have reduced the rate of relapse and progression of the disease.

Polio

The viral disease *polio* causes paralysis by destroying somatic motor neurons. This disorder, introduced on page 67, has been almost eliminated by an aggressive immunization program. Immunization continues because polio still occurs in some areas of the world. The disease could be brought into the United States at any time, leading to an epidemic among unimmunized persons.

Spinal Trauma

Physical damage to the spinal cord by trauma from a severe auto crash or other accident can cause permanent paralysis because the damaged tracts seldom undergo even partial repair. Extensive damage to the spinal cord axons causes loss of motor and sensory function below the level of injury. Damage superior to the seventh cervical vertebra affects motor control (partial or total paralysis) and sensation of the upper and lower limbs, a condition termed *quadriplegia* (also called *tetraplegia*). *Paraplegia,* the loss of motor control of the lower limbs, may follow damage to the thoracic spinal cord.

Less-severe injuries affecting the spinal cord or cauda equina produce variable symptoms of sensory loss or motor paralysis that reflect the specific nuclei, tracts, or spinal nerves involved. We will consider one example—the loss of peripheral sensation along the distribution of a spinal nerve—in a later section.

Spinal-Cord Injuries and Experimental Treatments

At the outset, severe injury to the spinal cord frequently produces a period of sensory and motor paralysis termed **spinal shock**. The skeletal muscles weaken or become paralyzed, neither somatic nor visceral reflexes function, and the brain no longer receives sensations of touch, pain, heat, or cold. The location and severity of the injury determine how long these symptoms persist and how much nerve function is preserved.

Violent jolts, such as those associated with blows or gunshot wounds near the spinal cord, can cause **spinal concussion** without visibly damaging the cord. Spinal concussion may produce spinal shock, but the symptoms are temporary, and recovery may be complete in a matter of hours. More serious injuries, such as vertebral fractures, generally involve physical damage to the spinal cord. In a **spinal contusion**, hemorrhages occur in the meninges, pressure rises in the cerebrospinal fluid, and the white matter of the spinal cord may degenerate at the site of injury. Partial recovery over a period of weeks may leave some functional losses. Recovery from a **spinal laceration** by vertebral fragments or other foreign bodies tends to be far slower and less complete. **Spinal compression** occurs when the spinal cord becomes squeezed or distorted within the vertebral canal. In a **spinal transection**, the cord is completely severed. At present, surgical procedures cannot repair a severed spinal cord.

Many spinal-cord injuries involve some combination of compression, laceration, contusion, and partial transection. Relieving pressure and stabilizing the affected area through axial traction and/or surgery (such as spinal decompression and fusion, the immobilization of adjacent vertebrae) may prevent further damage and allow the injured spinal cord to recover as much as possible. The initial trauma triggers many chemical and immunological problems at the cellular level that further damage the spinal cord, and this secondary cellular damage continues for several days after the injury. Treatment within 8 hours of injury with the corticosteroid *methylprednisilone* has been shown to be of some modest benefit. In complete or incomplete quadriplegia, any small improvement in motor strength in one or more muscles can provide important functional gains.

More understanding of the chemical and immune responses that contribute to secondary cell death and damage after injury may lead to other protective treatments to limit loss of nerve function. In addition, several avenues of research are now under way with the goal of restoring normal function after spinal injury. These experimental therapies can be broadly categorized as either biological or technological.

Biological Methods

A major biological line of investigation involves the introduction of stem cells and the biochemical control of nerve growth and regeneration.

Treated with embryonic stem cells at the injury site nine days after a crushing injury to the spine, laboratory rats recovered some limb mobility and strength. Oligodendrocytes, astrocytes, and functional neurons developed at the injury site. Neural stem cells have also been proposed for the treatment of strokes, Parkinson's disease, and Alzheimer's disease. The recent discovery that the adult brain contains populations of inactive stem cells has opened a new line of investigation: What are the factors that activate resident stem cells?

Neurons are influenced by a combination of growth promoters and growth inhibitors. Damaged myelin sheaths apparently release an inhibitory factor that slows the repair process. Researchers have made an antibody, *IN-1*, that inactivates the inhibitory factor released in the damaged spinal cords of rats. The treatment stimulates repairs, even in severed spinal cords.

A partial listing of compounds known to affect nerve growth and regeneration includes **nerve growth factor** (NGF), **brain-derived neurotrophic factor** (BDNF), **neurotrophin-3** (NT-3), **neurotrophin-4** (NT-4), **glial growth factor**, **glial maturation factor**, **ciliary neurotrophic factor**, and **growth-associated protein 43** (GAP-43). Many of these factors have now been synthesized by means of gene-splicing techniques, and sufficient quantities are available to permit their use in experiments. Initial results are promising, and these factors, in various combinations, are being evaluated for the treatment of CNS injuries and the chronic degeneration seen in Alzheimer's disease and Parkinson's disease.

Assistive Technology

Assistive technologies in the form of wheelchairs, modified cars and other tools, and accessible homes and buildings, help paralyzed people live more varied lives. Computers and other electronic devices with mechanical elements can be connected to the nervous system to make neural prostheses.

When the spinal cord is injured above spinal segment C_3, respiratory paralysis results and patients require continuous mechanical ventilation. If the phrenic nerve, which arises from segments C_3 to C_5 and innervates the diaphragm, is not injured, respiratory cycles can be controlled by electrical stimulation of the phrenic nerve. The electrodes are implanted either along the phrenic nerve or intramuscularly on the diaphragm.

Commercially available bicycles that use computers to stimulate specific muscles and muscle groups electrically in a predetermined sequence can increase leg muscle mass and improve overall fitness. The technique is called *functional electrical stimulation*, or FES. This approach involves surface electrodes attached to the skin over leg muscles where low-level electrical pulses cause coordinated contractions and pedaling motions. With this equipment and lightweight braces, quadriplegics have walked several hundred yards and paraplegics several thousand.

Another device helps quadriplegics control some hand movement by using shoulder muscle movements. The *Parastep* system, which uses a microcomputer controller and surface electrodes, is available now for exercise and limited ambulation. Using a network of wires woven into the fabric of close-fitting garments to create a set of electronic "hot pants," a paraplegic woman completed several miles of the 1985 Honolulu Marathon, and more recently another paraplegic woman walked down the aisle at her wedding. However, muscular stimulation seems most beneficial in reducing the atrophy of unused muscles and improving overall fitness and health, rather than providing full mobility in paralyzed persons.

Such technological solutions can provide only a degree of motor control without accompanying sensation. Everyone would prefer a biological procedure that would restore the functional integrity of the nervous system.

Damage and Repair of Peripheral Nerves

If a peripheral axon is damaged but not displaced, normal function may eventually return as the cut stump grows across the injury site, away from the soma and along its former path. For normal function to be restored, several things must happen: The severed ends must be relatively close together (1–2 mm, or 0.04–0.08 in.); they must remain in proper alignment; and there must be no physical obstacles between them, such as the collagen fibers of scar tissue. These conditions can be created in the laboratory, using experimental animals and individual axons or small fascicles. But in accidental injuries to peripheral nerves, the edges are likely to be jagged; intervening segments may be lost entirely, and elastic contraction in the surrounding connective tissues may pull the cut ends apart and misalign them.

Until recently, the surgical response would involve trimming the injured nerve ends, sewing them together neatly, and hoping for the best. Surgical microscopes and other techniques are helpful but this procedure is not very successful, in part because scalpels do not produce a smoothly cut surface and in part because the thousands of broken axons would never be perfectly aligned. Moreover, axons are not highly elastic, so if a large segment of the nerve was removed, crushed, or otherwise destroyed, there would be no way to bring the intact ends close enough to permit regeneration. In such instances, a **nerve graft** could be inserted, using a section from some other, less important peripheral nerve. The functional results were even less likely to be wholly satisfactory, because the growing axonal tips had to find their way across not one but two gaps, resulting in limited and random re-innervation. Nevertheless, any return of function was better than none! Sterilized, cell-free sections of peripheral nerves from a cadaver can be used as grafts when a lengthy gap needs to be bridged.

Research now focuses on the physical and biochemical control of nerve regeneration, using the growth factors introduced above. Another promising strategy is the use of a synthetic or biological sleeve with nerve growth factors to guide and stimulate nerve growth. The sleeve may be a tube with an outer layer of silicone around a vein or an inner layer of cowhide collagen. When filled with muscle tissue or cultured Schwann cells (both of which seem to contain nerve growth factors), or saturated with purified growth factors, these grafts approach the performance of nerve grafts in animal tests.

Peripheral Neuropathies

Peripheral neuropathies are characterized by regional losses of sensory and/or motor function (palsy). They have many causes including focal or diffuse nerve trauma, infection, metabolic disorders, and toxic exposure. *Mononeuropathies* have focal single nerve involvement and result from local damage. *Polyneuropathies* are diffuse, symmetric, and distal, having the most effect on nerves with long axons. Mononeuropathy symptoms relate to a single nerve or plexus, whereas polyneuropathy symptoms commonly start as stocking-glove numbness or pains and distal, bilateral motor problems. Diabetes mellitus (p. 113) is the most common cause of polyneuropathy. **Brachial neuropathies** result from injuries to the brachial plexus or its branches. **Crural neuropathies** involve the nerves of the lumbosacral plexus.

Mononeuropathies appear for several reasons. The *pressure neuropathies* are especially interesting; a familiar, but mild, example is the experience of having an arm or leg "fall asleep." Areas of the limb become numb, and afterward an uncomfortable "pins-and-needles" sensation, or **paresthesia**, accompanies the return to normal function.

These incidents are seldom clinically significant, but they provide graphic examples of the effects of more serious neuropathies that can last for days to months, or even be permanent. In **radial nerve palsy**, pressure on the back of the arm interrupts the motor function of the radial nerve, so the extensors of the wrist and fingers are paralyzed. This condition is also known as "Saturday night palsy," because falling asleep on a couch with your arm over the seat back (or beneath someone's head) can produce the right combination of pressures. Students may also be familiar with **ulnar palsy**, which can result from prolonged contact between an elbow and a desk. The ring and little fingers lose sensation, and the fingers cannot be adducted. *Carpal tunnel syndrome* is a neuropathy resulting from compression of the median nerve at the wrist where it passes deep to the flexor retinaculum in company with the flexor tendons. Repetitive flexion/extension cycles at the wrist can irritate these tendon sheaths; the swelling that results leads to nerve compression.

Persons who carry large wallets in their hip pockets may develop symptoms of **sciatic compression** after they drive or sit in one position for extended periods. As nerve function declines, the individuals notice lumbar or gluteal pain, numbness along the back of the leg, and weakness in the leg muscles. Similar symptoms result from the compression of nerve roots that form the sciatic nerve by a distorted lumbar intervertebral disc. This condition is termed **sciatica**, and one or both lower limbs may be affected, depending on the site of compression.

Shingles and Leprosy (Hansen's Disease)

"Shingles" develops in persons who were previously infected with the *herpes varicella-zoster* virus. The initial infection, usually in early childhood, produces symptoms known as chicken pox. After recovery, the virus remains dormant within neurons of the anterior gray horns of the spinal cord. It is not known what triggers the reactivation of this pathogen. The reactivated virus involves neurons within the dorsal roots of spinal nerves and sensory ganglia of cranial nerves. This disorder produces a distinctive, painful, blistered rash whose distribution corresponds to that of the affected sensory nerves

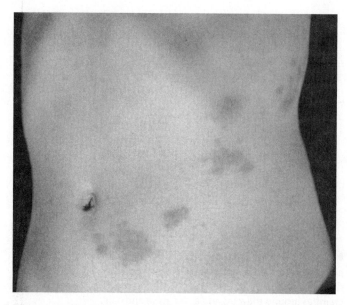

Figure 42   Shingles. The side of a person with shingles. The skin eruptions follow the distribution of dermatomal innervation.

(**Figure 42**). The virus is present in the blister fluid and can cause chicken pox in exposed, susceptible persons. A vaccine given to people over age 60 reduces their risk of getting shingles by up to 60 percent. Children given a form of the vaccine are protected from chicken pox; the hope is that they will avoid shingles as well.

Most people who contract shingles suffer just a single episode in their adult lives. However, the problem can recur, most often in people with weakened immune systems, including those with AIDS or some forms of cancer. Treatment for shingles involves large doses of the antiviral drug *acyclovir* (*Zovirax*) or related drugs.

Leprosy, also known as *Hansen's disease*, is an infectious disease caused by the bacterium *Mycobacterium leprae*. The bacteria invade peripheral nerves, growing faster in the cooler areas of the body, especially in the skin and distal limbs, producing sensory losses. The disease presents different clinical appearances that result from varied immune responses of affected persons. Only about 5 percent of those exposed to *Mycobacterium leprae* become infected. It is a disease that progresses slowly, and symptoms may not appear for up to 20 years after infection. Untreated, over time, decreased sensation and motor paralysis develop, and the combination can lead to recurring injuries and infections. The eyes, nose, hands, and feet may develop deformities as a result of infection and neglected injuries (**Figure 43**).

There are about 2000 cases of Hansen's disease in the United States, with about 100 new cases diagnosed annually, mostly in immigrants from tropical areas where there are more cases of the disease. After 20 years of World Health Organization sponsorship an estimated 14 million cases have been cured worldwide, and the number of new cases has decreased by 95 percent. Since the 1940s, the disease has been successfully treated with drugs, and early treatment prevents deformities. Multi-drug treatment with three drugs has been used since 1961 and cures leprosy in 6 to 24 months. The World Health Organization has provided free medicine to all patients since 1995. Treated individuals are not infectious, and the practice of confining "lepers" in isolated compounds has thankfully been discontinued.

Nervous System

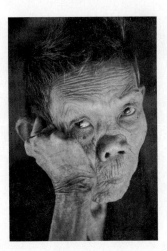

Figure 43 Leprosy (Hansen's Disease). The distal limbs may become deformed as untreated leprosy progresses. The untreated disease also affects facial features, typically starting with degenerative changes around the eyes and at the nose and ears.

Reflexes and Diagnostic Testing

A neurological examination evaluates the sensory, motor, behavioral, and cognitive functions of the nervous system. The techniques involved were considered on pages 74–75; they range from observation and asking questions to monitoring brain function through sophisticated scanning procedures.

Many somatic reflexes can be assessed through careful observation and the use of simple tools. The procedures are easy to perform, and the results can provide valuable information about the location of damage to the spinal cord or spinal nerves. By testing a series of spinal and cranial reflexes, the examiner can assess the function of sensory pathways and motor centers throughout the spinal cord and brain. Neurologists test many reflexes; only a few are so generally useful that physicians make them part of a standard physical examination. Representative examples are shown in **Figure 44**, and **Table 21** lists somatic reflexes that can be used in this way.

The *abdominal reflex* (**Figure 44a**) is a superficial reflex that is normally present in adults. In this reflex, a light stroking of the skin produces a reflexive twitch in the abdominal muscles that moves the navel toward the stimulus. The reflex is facilitated by descending commands; it disappears after descending tracts have been damaged. The *patellar reflex, biceps reflex* (**Figure 44b**), *triceps reflex* (**Figure 44c**), and *ankle-jerk reflex* (**Figure 44d**) are stretch reflexes controlled by specific segments of the spinal cord. Testing these reflexes provides information about the corresponding spinal segments. For example, a normal patellar reflex indicates that spinal nerves and spinal segments L_2–L_4 and the muscles they innervate are undamaged.

Figure 44 Somatic Reflexes.

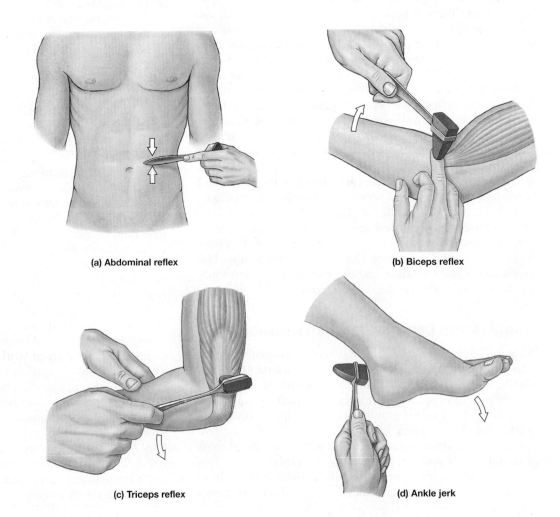

(a) Abdominal reflex

(b) Biceps reflex

(c) Triceps reflex

(d) Ankle jerk

Table 21	Reflexes Used in Diagnostic Testing				
Reflex	**Stimulus**	**Afferent Nerve(s)**	**Spinal Segment**	**Efferent Nerve(s)**	**Normal Response**
Superficial Reflexes					
Abdominal reflex	Light stroking of skin of abdomen	T_7–T_{12}, depending on region stroked	T_7–T_{12}, at level of arrival	Same as afferent	Contraction of abdominal muscles that pulls navel toward the stimulus
Cremasteric reflex	Stroking of skin of upper thigh	Femoral nerve	L_1	Genitofemoral nerve	Contraction of cremaster, elevation of scrotum
Plantar reflex	Longitudinal stroking of sole of foot	Tibial nerve	S_1, S_2	Tibial nerve	Flexion at toe joints
Anal reflex	Stroking of region around the anus	Pudendal nerve	S_4, S_5	Pudendal nerve	Constriction of external anal sphincter
Stretch Reflexes					
Biceps reflex	Tap to tendon of biceps brachii muscle near its insertion	Musculocutaneous nerve insertion	C_5, C_6	Musculocutaneous nerve	Flexion at elbow
Triceps reflex	Tap to tendon of triceps brachii muscle near its insertion	Radial nerve	C_6, C_7	Radial nerve	Extension at elbow
Brachioradialis reflex	Tap to forearm near styloid process of the radius	Radial nerve	C_5, C_6	Radial nerve	Flexion at elbow, supination, and extension at wrist
Patellar reflex	Tap to patellar tendon	Femoral nerve	L_2–L_4	Femoral nerve	Extension at knee
Ankle-jerk reflex	Tap to calcaneal tendon	Tibial nerve	S_1, S_2	Tibial nerve	Extension (plantar flexion) at ankle

Nervous System

Abnormal Reflex Activity

In **hyporeflexia**, normal reflexes are weak, but apparent, especially with reinforcement. In **areflexia** (ā-rē-FLEK-sē-uh; a-, without), normal reflexes fail to appear, even with reinforcement. Hyporeflexia or areflexia may indicate temporary or permanent damage to skeletal muscles, dorsal or ventral nerve roots, spinal nerves, the spinal cord, or the brain.

Hyperreflexia occurs when higher centers maintain a high degree of facilitation along the spinal cord. Under these conditions, reflexes are easily triggered and the responses may be grossly exaggerated.

This effect also results from compression of the spinal cord or diseases that target higher centers or descending tracts. One potential result of hyperreflexia is the appearance of alternating contractions in opposing muscles. When one muscle contracts, it stimulates the stretch receptors in the opposing muscle. The stretch reflex then triggers a contraction in that muscle, and this contraction stretches receptors in the original muscle. This self-perpetuating sequence, which can be repeated indefinitely, is called **clonus** (KLŌ-nus). In a hyperreflexive person, a tap on the patellar tendon will set up a cycle of kicks rather than just one or two.

A more extreme hyperreflexia develops if the motor neurons of the spinal cord lose contact with higher centers. In many cases, after a severe spinal injury, the individual first experiences a temporary period of areflexia known as spinal shock (p. 83). When the reflexes return, they may respond in an exaggerated fashion, even to mild stimuli. For example, the lightest touch on the skin may produce a massive withdrawal reflex with intense muscle spasms strong enough to break bones. In the **mass reflex**, the entire spinal cord becomes hyperactive for several minutes, issuing exaggerated skeletal muscle and visceral motor commands.

❱ Disorders of the Brain and Cranial Nerves

Cranial Trauma

Traumatic brain injury (TBI) may result from harsh contact between the head and another object or from a severe jolt. Head injuries account for over half the deaths attributed to trauma. Every year roughly 1.4 million cases of TBI occur in the United States. Approximately 50,000 people die, and another 80,000 have long-term disability. We presented the characteristics of spinal concussion, contusion, and laceration on page 83; comparable descriptions apply to brain injuries.

Concussions may accompany even minor head injuries. A concussion may involve transient confusion with abnormal mental status, temporary loss of consciousness, and some degree of amnesia (pp. 96–97). Physicians examine concussed individuals quite closely and may x-ray or CT scan the skull to check for fractures or cranial bleeding. Mild concussions produce a brief interference in brain function, possibly brief unconsciousness, and little memory loss. Severe concussions produce extended periods of unconsciousness and abnormal neurological functions. Severe concussions are typically associated with **contusions** (bruises), hemorrhages, or **lacerations** (tears) of the brain tissue; the possibilities for recovery vary with the areas affected. Extensive damage to the reticular formation can produce a permanent state of unconsciousness, and damage to the lower brain stem generally proves fatal.

Wearing helmets during activities such as bike, horse, skateboard, or motorcycle riding, contact sports like football and hockey, and when batting and base running in baseball provides protection for the brain. Seat belts give similar protection in the event of a motor vehicle accident. If a concussion does occur, restricted activities, including delay in return to the activity that led to the injury, is recommended.

Epidural and Subdural Hemorrhages

A severe head injury may damage meningeal vessels and cause bleeding into the cranial cavity. The most serious cases involve an arterial break. Because arterial blood pressure is relatively high, the volume of leaked blood expands rapidly, and problems can develop just minutes after the injury. If blood is forced between the dura mater and the cranium, the condition is known as an **epidural hemorrhage**. The elevated fluid pressure then distorts the underlying tissues of the brain. The individual loses consciousness from minutes to hours after the injury, and death follows in untreated cases. An epidural hemorrhage involving a damaged vein does not produce as much leaked blood or massive symptoms immediately, and the individual may not have neurological problems for hours, days, or even weeks after the original injury. As a result, the problem may not be noticed until the nervous tissue has been damaged by distortion, compression, and secondary hemorrhaging. Epidural arterial hemorrhages are rare, occurring in fewer than 1 percent of head injuries. However, the mortality rate is 100 percent in untreated cases, and over 50 percent even after the blood pool has been removed and the damaged vessels have been closed.

Subdural hemorrhages are roughly twice as common as epidural hemorrhages. The most common source of blood is a small vein or one of the dural sinuses. The term subdural hemorrhage may be misleading, because in many cases blood enters the meningeal layer of the dura mater, flowing deep to the epithelium that contacts the arachnoid membrane rather than between the dura mater and the arachnoid. Because the blood pressure in a venous subdural hemorrhage is lower than that in an arterial epidural hemorrhage, the volume of blood leaked and distortion produced gradually increases (over weeks to months or even longer). The effects on brain function can be quite variable and difficult to diagnose

Hydrocephalus

The adult brain is surrounded by the inflexible bones of the cranium. The enclosed cranial cavity contains two fluids—blood and cerebrospinal fluid (CSF)—and the soft tissues of the brain. The total volume cannot change, so when the volume of blood or CSF increases, the volume of the brain must decrease. If, for example, a dural or subarachnoid blood vessel ruptures, the fluid volume increases as blood collects in the cranial cavity. The rising intracranial pressure compresses the brain, leading to neural dysfunction that can end in unconsciousness and death.

Any increase in the rate of CSF production is normally matched by an increase in the rate of CSF removal at the arachnoid granulations. If this equilibrium is disturbed, clinical problems appear as the intracranial pressure increases. The volume of CSF will increase if the rate of formation accelerates or the rate of removal decreases. In either event, the increased fluid volume compresses and distorts the brain. Increased rates of formation can accompany head injuries, but the most common problems arise from masses (such as tumors or abscesses), scarring from meningeal infection (meningitis), or developmental abnormalities. These conditions restrict the normal circulation and reabsorption of CSF. Because CSF production continues, the ventricles gradually expand, compressing the surrounding neural tissues and causing brain function to deteriorate.

Infants are especially sensitive to changes in intracranial pressure, because the arachnoid granulations do not appear until roughly age three. (Until then, CSF is reabsorbed into small vessels in the subarachnoid space and beneath the ependyma lining the ventricles.) As in adults, if intracranial pressure becomes abnormally high, the ventricles will expand. But in infants, the cranial sutures have yet to fuse, so the skull can enlarge to accommodate the extra fluid volume. The result is a condition called **hydrocephalus**, or "water on the brain" (❱ Figure 45). Hydrocephalus may result from a blockage of the aqueduct of the midbrain or a constriction of the connection between the subarachnoid spaces of the cranial and spinal meninges. Untreated progressive hydrocephalus is usually fatal. Despite treatment, infants may have some degree of developmental problems. Successful treatment generally involves reduction of the intracranial pressure via the installation of a *shunt* that drains the excess CSF. (Shunts may also help adults with hydrocephalus.) For most cases a shunt is a lifetime requirement. The shunt may be removed if (1) further growth of the brain eliminates the blockage or (2) the intracranial pressure decreases after the arachnoid granulations develop when the child reaches three years of age.

Cerebellar Dysfunction

Cerebellar problems produce coordination disturbances that affect voluntary motor control. Cerebellar function can be altered permanently by trauma or a stroke or temporarily by drugs such as alcohol. In severe ataxia, balance problems are so

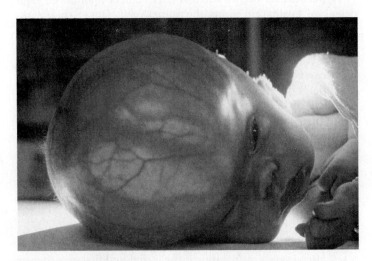

❱ **Figure 45 Hydrocephalus.** An infant with hydrocephalus, a condition caused by impaired circulation and removal of cerebrospinal fluid. The buildup of CSF leads to the distortion of the brain and enlargement of the infant cranium.

Nervous System

great that the individual cannot sit or stand upright. Less severe conditions cause an obvious unsteadiness and irregular patterns of movement. When standing or walking the individual assumes a wide-based stance, with feet placed wider than the hips, because this improves balance. The individual typically watches his or her feet to see where they are going and controls ongoing movements by intense concentration and voluntary effort. Without the cerebellar ability to adjust movements while they are occurring, the individual becomes unable to anticipate the course of a movement over time. Reaching for something becomes a major exertion, because the only information available must be gathered by sight or touch while the movement is taking place. Most commonly, a reaching movement ends with the hand overshooting the target. This inability to anticipate and stop a movement precisely is called **dysmetria** (dis-MET-rē-uh; *dys-*, bad + *metron*, measure). In attempting to correct the situation, the person usually overshoots again, this time in the opposite direction, and so on. The hand oscillates back and forth until either the object can be grasped or the attempt is abandoned. This oscillatory movement is known as an **intention tremor**.

Clinicians check for ataxia by watching an individual walk in a straight line; one test for dysmetria involves touching the tip of the index finger to the tip of the nose or the examiner's fingertip. Because many drugs impair cerebellar performance, the same tests are used by police officers to check drivers suspected of driving while under the influence of alcohol or other drugs.

Analysis of Gait and Balance

To check a person's gait and balance, the examiner may ask the person to walk a line, first as usual, then heel to toe. Heel-to-toe walking on a straight line will magnify any gait abnormalities or problems due to a loss of balance sensations. While the subject is walking, the examiner also watches how the feet are placed and how the arms swing back and forth. The pattern of limb movement during walking is normally regulated by the cerebral nuclei. Problems with these nuclei, as in Parkinson's disease, will upset the pace and rhythm of the associated movements.

Another test (the *Romberg*) involves standing with the feet together, first with the eyes open and then with the eyes closed. The Romberg test is used to check balance and equilibrium sensations because it reveals how much the individual is relying on visual information to fine-tune motor functions. If the person stands still with the eyes open, but sways or starts to fall with the eyes closed, there are problems with the balance pathways and perhaps the cerebellum as well. This result, called a *positive Romberg* sign, may indicate a stroke (or other cause of brain damage), multiple sclerosis, peripheral neuropathies, or any of several vestibular (internal ear) disorders.

The Basal Nuclei and Parkinson's Disease

The basal nuclei of the cerebrum are involved with the subconscious control of skeletal muscle tone and coordination of learned muscle patterns such as riding a bicycle. There are two discrete populations of neurons. One group stimulates motor neurons by releasing acetylcholine (ACh), and the other inhibits motor neurons by releasing gamma aminobutyric acid (GABA). Under normal conditions, the excitatory neurons remain inactive and the inhibitory neurons (the descending tracts), are primarily responsible for motor neuron activity. If they are damaged, the loss of direct inhibitory control leads to a generalized state of skeletal muscle contraction known as decerebrate (no cerebral cortex input) rigidity.

The excitatory neurons are quiet because they are continuously exposed to the inhibitory effects of the neurotransmitter dopamine. This compound is manufactured by neurons in the substantia nigra of the brain stem and is carried by the ascending tract by axoplasmic transport to synapses in the basal nuclei. If the dopamine-producing neurons are damaged, the uninhibited excitatory neurons become increasingly active. This increased activity produces the motor symptoms of **Parkinson's disease**. Parkinson's disease, which is both chronic and progressive, is characterized by tremor, stiffness, slowness of movement (bradykinesia), and poor coordination and balance. A pronounced increase in muscle tone leads to stiffness and slows the start of voluntary movements. These become hesitant and jerky, because a movement cannot occur until one muscle group manages to overpower its antagonists. A tremor represents a tug of war between antagonistic muscle groups that produces a background shaking of the limbs (resting tremor). Even changing one's facial expression requires intense concentration, and the individual acquires a blank, static expression. Finally, the positioning and preparatory adjustments normally performed automatically no longer occur, causing unsteadiness. Every aspect of each movement must be voluntarily controlled, and the extra effort requires intense concentration that can prove tiring and extremely frustrating. Other symptoms include a rigid posture and a slow, shuffling, forward-leaning walk. In the late stages other CNS effects, such as depression, hallucinations, and dementia, may appear.

Providing the basal nuclei with dopamine or dopamine agonists (compounds that mimic the effect of dopamine on the postsynaptic receptors) can significantly reduce the symptoms. Giving dopamine itself either orally or by injection is not effective, because the molecule cannot cross the blood–brain barrier. The oral drug L-dopa (levodopa) does cross the blood–brain capillaries and is then converted to dopamine. Unfortunately, with prolonged treatment, L-dopa becomes less effective. The therapeutic usefulness of L-dopa can be prolonged and enhanced by giving it in combination with other drugs, such as *amantadine* or *bromocriptine*. Amantadine accelerates dopamine release at synaptic terminals, and bromocriptine, a dopamine agonist, stimulates dopamine receptors on postsynaptic membranes. All drugs in current use reduce symptoms but do not slow or stop the progression of Parkinson's disease.

Deep Brain Stimulation (DBS), the electrical stimulation of the subthalamic nucleus via implanted wires, has symptomatically improved motor function in many patients. It is added primarily after drug treatment has become less effective. Other surgeries to alleviate Parkinson's symptoms focus on the destruction of areas within the basal nuclei or thalamus to control the motor symptoms of tremor and rigidity. Attempts to transplant tissues that produce dopamine or related compounds (fetal stem cells or adrenal cells) into the basal nuclei have not proven very beneficial. Most individuals with Parkinson's disease are elderly. The champion boxer Muhammad Ali developed Parkinson's in his forties, possibly as a result of repeated cranial trauma during his career. In 1988 a number of young adults abruptly developed unusually severe Parkinson's disease. Most cases were linked to a street drug rumored to be "synthetic heroin." In addition to the compound that produced the "high" sought by users, the drug contained a complex molecule contaminant with the abbreviated name **MPTP**. This accidental by-product of the synthetic process destroys neurons of the substantia nigra, eliminating the manufacture and transport of dopamine to the basal nuclei.

Aphasia

Aphasia, characterized by problems comprehending speech, reading, or expressing oneself verbally or in writing, results from damage to the language areas of the brain. *Global aphasia*, the loss of comprehension and expression of both verbal and written language, results from extensive damage to the general interpretive area or to the associated sensory tracts in the frontal and temporal lobes, especially if the dominant hemisphere (usually the left) is involved. There are several other forms of aphasia. **Expressive aphasia,** also known as *nonfluent aphasia*, affects the frontal lobe and can develop after a brief period of global aphasia. This condition is extremely frustrating for the individual, who can comprehend most language and knows how to respond, but lacks the word-finding ability necessary to produce the right combinations of sounds (and, worse, is fully aware of the problem). If motor control is damaged, mechanical speech problems, or *dysarthria*, occurs. In *fluent*, or *receptive*, *aphasia*, the person does not understand what is heard and neither makes sense while speaking nor realizes that his or her speech is incomprehensible. The individual words and sounds come easily, but they convey no meaning. Usually only the temporal lobe is involved, and no motor problems occur.

Lesser degrees and mixed types of aphasia (*anomic* or *amnesiac aphasia*) commonly follow a minor stroke. There is no initial period of global aphasia, and the individual can understand spoken and written words. The problems encountered with speaking or writing gradually fade. Many individuals with minor aphasia recover completely. Treatment involves working with the patient to read, write, follow directions, and repeat what they hear.

Disconnection Syndrome

The functional differences between the hemispheres become apparent if the corpus callosum is cut, a procedure sometimes performed to treat epileptic seizures that cannot be controlled by other methods. This surgery produces symptoms of **disconnection syndrome**. In this condition, the two hemispheres function independently, each "unaware" of visual, tactile, or auditory stimuli that are restricted to one hemisphere, or motor commands that involve the other. Immediately after surgery, competitive movements between the left and right hands may occur, reflecting the different goals and perspectives of the two hemispheres. Special testing conditions that restrict sensory input must be used to demonstrate interesting but subtle changes in mental abilities. For example, objects touched by the left hand can be recognized but not verbally identified, because the sensory information arrives at the right hemisphere but the speech center is on the left. The object can be verbally identified if felt with the right hand, but then the person cannot say whether it is the same object previously touched with the left hand. Sensory information from the left side of the body arrives at the right hemisphere and cannot reach the general

interpretive area. Thus, conscious decisions are made without regard to sensations from the left side.

Over time, compensatory mechanisms develop that provide communication between the cerebral hemispheres, such as speaking aloud, or passing objects from one hand to the other. Most persons with this syndrome display "social ordinariness" except for some memory problems. Individuals born without a functional corpus callosum do not have major sensory, motor, or intellectual problems unless other brain abnormalities are also present. In some way the CNS adapts to these situations, probably by increasing the amount of information transferred across the anterior commissure.

Seizures and Epilepsies

A *seizure* is a temporary disorder of cerebral function, where groups of neurons in the brain send out inaccurate signals. Depending on the site(s) involved, abnormal, involuntary movements, unusual sensations, emotions, inappropriate behavior, and/or unconsciousness may occur. The terms **epilepsy** and *seizure disorder* refer to more than 40 conditions characterized by a recurring pattern of seizures over extended periods. In roughly 50 percent of patients, no obvious cause can be determined. Known causes include infection, brain trauma, brain damage from metabolic problems, stroke, genetic factors, and poisoning.

There are many types of seizures often categorized as either partial (starting from a specific location) or generalized (where both cerebral hemispheres are involved at the start). Seizures of all kinds are usually accompanied by a marked change in the pattern of electrical activity that can be detected in an electroencephalogram. The change may begin in one portion of the cerebral cortex, but may thereafter spread to adjacent regions. The neurons at the site of origin of the change are abnormally sensitive. When they become active, they may facilitate and subsequently stimulate adjacent neurons. As a result, the abnormal electrical activity can spread across the entire cerebral cortex.

The extent of the cortical involvement determines the nature of the observed symptoms. A partial **seizure** affects a relatively restricted cortical area, producing sensory or motor symptoms or both. Because partial seizures often affect focal areas previously damaged by stroke, injury, or tumors, the individual generally remains conscious throughout the attack. If the seizure occurs within a portion of the primary motor cortex, the activation of pyramidal cells will produce uncontrollable movements. The muscles affected or the specific sensations experienced provide an indication of the precise region involved. In a **temporal lobe seizure**, the disturbance spreads to the sensory cortex and association areas, so the individual may experience unusual memories, sights, smells, or sounds as well. Involvement of the limbic system can also produce sudden emotional changes.

In a **generalized seizure**, the entire cortical surface is involved. Generalized seizures range from prolonged, major events to brief, almost unnoticed incidents. We will consider only two examples here: *grand mal* and *petit mal seizures*.

Some epilepsies have been linked to genetic abnormalities affecting the sodium and potassium channels of neurons, the "channelopathies." Seizures may involve powerful, uncoordinated muscular contractions that affect the face, eyes, and limbs as well as loss of consciousness. These are symptoms of a tonic-clonic, or convulsive seizure, often called a **grand mal seizure**.

During a grand mal attack, the cortical activation begins at a single focus and then spreads across the entire surface. There may be no warning, but some individuals experience a vague apprehension or awareness that a seizure is about to begin. A sudden loss of consciousness follows, and the individual drops to the floor as major muscle groups go into tonic contraction. The body remains rigid for several seconds before a rhythmic series of contractions occurs in the limb muscles. Incontinence may occur. After the attack subsides, the individual may appear disoriented or may sleep for several hours. Muscles or bones subjected to extreme stresses may be damaged, and the person will probably be sore for days after the incident.

Petit mal seizures are very brief (less than 10 seconds in duration) and involve few motor abnormalities. Typically, the individual loses consciousness suddenly, with no warning. It is as if an internal switch were thrown and the conscious mind turned off. Because the individual is "not there" for brief periods during petit mal attacks, the incidents are also known as *absence seizures*. During the seizure, small-motor activities, such as fluttering of the eyelids or trembling of the hands, may occur.

Petit mal seizures generally begin between ages 6 and 14, seem to involve neural networks between the thalamus and cerebral cortex, and may have a genetic component. They can occur hundreds of times a day, so the child lives each day in small segments separated by blank periods. The individual is aware of brief losses of consciousness that occur without warning, but may not seek help, either from embarrassment or on the assumption that such lapses are normal. He or she becomes extremely anxious about the timing of future attacks. However, the motor signs are so minor that they tend to go unnoticed by family members, and the psychological stress caused by the condition is in many cases overlooked. The initial diagnosis is typically made during counseling for learning problems. (You have probably taken an exam after you have missed 1 or 2 lectures out of 20. Imagine taking an exam after you have missed every third minute of every lecture.)

Both petit mal and grand mal forms of epilepsy can be treated with anticonvulsive drugs, such as barbiturates, phenytoin (*Dilantin*) or valproic acid (*Depakene*). More than 20 different anticonvulsants are available, with varying effectiveness, side effects, and other risks. Usually one drug suffices to control or eliminate seizures; however, the drug ethosuximide specifically suppresses petit mal seizures. Many people can discontinue medication after being seizure free for two or three years. This is less likely if multiple drugs are needed. If medication does not control severe seizures, various forms of brain surgery may be effective in doing so.

Trigeminal Neuralgia

New cases of **trigeminal neuralgia** (also named *Tic douloureux*) affect 1 individual out of every 25,000 in the United States each year. Sufferers have recurrent episodes of severe, almost totally debilitating, brief stabbing facial pain that arrives with a sudden, shocking intensity. The pain is along the sensory path of the maxillary and mandibular branches of the trigeminal nerve. Local irritation of the nerve, caused by pressure from a blood vessel or (less often) by a tumor or a stroke or other causes leads to focal demyelination of some axons within the nerve. With gaps in the electrical insulation of myelin, benign stimulation of some sensory nerves seems to overexcite nearby pain-sensitive nerves. On testing, trigeminal sensation is normal. If no other

causes of facial pain such as dental problems are present, a CT or MRI scan may be done. Medicines usually used for seizure control (another problem associated with inappropriate nerve activation) control the pain in about 80 percent of cases, but surgical procedures may be required. If pressure from a blood vessel seems to be the cause, delicate vascular surgery may help, but often the goal of the surgery is to destroy the afferent nerves that carry the pain sensations. This can be attempted by actually cutting the nerves, a procedure called a **rhizotomy** (*rhiza-*, root), or by injecting chemicals such as alcohol or phenol into the nerves at the foramina ovale and rotundum. The sensory fibers can also be destroyed by inserting an electrode and cauterizing the sensory nerve trunks as they leave the semilunar ganglion.

Cranial Nerve Tests

A number of different tests are used to test the condition of specific cranial nerves:

- The olfactory nerve (I) is assessed by asking the subject to distinguish among various odors.

- Cranial nerves II, III, IV, and VI are assessed when vision and movement of the eyes are checked. Vision (cranial nerve II) is tested separately. For eye motion, the person is asked to hold the head still and track the movement of the examiner's finger with the eyes. For the eyes to track the finger, the oculomotor muscles and associated cranial nerves must be functioning normally. For example, if the person cannot track with the right eye a finger that is moving from left to right, the right lateral rectus muscle or cranial nerve VI on the right side may be damaged.

- Cranial nerve V, which provides motor innervation to the muscles of mastication, can be checked by asking the person to clench the teeth. The jaw muscles are then palpated. If motor components of V on one side are damaged, the muscles on that side will be weak or flaccid. Sensory components of cranial nerve V can be tested by lightly touching areas of the forehead and side of the face.

- The facial nerve (VII) is checked by watching the muscles of facial expression or by asking the person to perform particular facial movements. Wrinkling the forehead, raising the eyebrows, pursing the lips, and smiling are controlled by the facial nerve. If a branch of VII has been damaged, the affected side will show muscle weakness or drooping. For example, the corner of the mouth may sag and fail to curve upward when the person smiles. Special sensory components of VII can be checked by placing solutions known to stimulate taste receptors on the anterior third of the tongue.

- The hearing function of the vestibulocochlear nerve (VIII) can be screened for problems by the "whispered voice test." Standing at arm's length behind the seated patient (out of sight), the clinician whispers three or four numbers or letters and asks the patient to repeat the sequence. If the patient responds correctly, hearing is considered normal; if the patient responds incorrectly, they are given a second try using different numbers and letters. Each ear is tested separately, blocking the external acoustic canal of the ear not being tested. (Other hearing tests are discussed on page 103.)

- The glossopharyngeal and vagus nerves (IX and X) can be evaluated by the person saying "ahh" or "ehh," and eliciting the gag reflex by touching the oropharynx with a tongue blade or swab. Examination of the soft-palate arches and uvula for normal movement is important.

- The accessory nerve (XI) can be checked by asking the person to shrug the shoulders. Atrophy of the sternocleidomastoid or trapezius muscles may also indicate problems with the accessory nerve.
- The hypoglossal nerve (XII) can be checked by having the person extend the tongue and move it from side to side.

▶ Neural Integration I: Disorders of Sensory and Somatic Pathways

Analyzing Sensory Disorders

A recurring theme of this *Manual* is that an understanding of how a system works enables you to predict how things might go wrong. You are already familiar with the organization and physiology of sensory systems, and we have previously discussed some of the most important clinical problems. Placing the entire array into categories provides an excellent example of a strategy that can be used to analyze any system in the body.

Every sensory system contains peripheral receptors, afferent fibers, ascending tracts, nuclei, and areas of the cerebral cortex. Any malfunction affecting the system must involve one of those components. Clinical diagnosis can be approached by an algorithm or flowchart process where a series of yes-or-no questions eliminates one possibility at a time until the nature of the problem becomes apparent. We will begin with a discussion of the events that convert an environmental stimulus into a sensation that can be interpreted by the CNS.

Transduction: A Closer Look

Transduction is the process that converts a stimulus into a series of action potentials in a sensory neuron. Those action potentials are interpreted by the CNS as a sensation. Transduction can be divided into three steps:

STEP 1: *A Stimulus Changes the Transmembrane Potential of the Receptor Membrane.* This change in the transmembrane potential is called a receptor potential.

The receptor potential may be a depolarization or a hyperpolarization. It is a graded potential change: The stronger the stimulus, the larger the receptor potential.

STEP 2: *The Receptor Potential Affects a Sensory Neuron.* A graded membrane depolarization that may, if sufficiently strong, lead to an action potential in a sensory neuron is called a generator potential. The usual receptors for the general senses (touch, pain, temperature, pressure, vibration, and position awareness) are the dendrites of sensory neurons (▶ **Figure 46a**). Receptor potentials in the dendrites produce generator potentials that can be large enough to trigger an action potential in the axon. In this case, when a sensory neuron itself acts as a receptor, the terms *receptor potential* and *generator potential* are used interchangeably.

Receptor potentials for the sensations of taste, hearing, equilibrium, and vision are provided by specialized receptor cells

that then communicate with sensory neurons across chemical synapses. Processing at the receptor cell varies the rate of neurotransmitter release at this synapse. In the case of taste, stimulation of receptor cells, triggers the release of neurotransmitters at the synapse that produce the generator potential in the sensory neuron (▶ **Figure 46b**). The larger the receptor cell potential, the more neurotransmitter is released and the greater is the generator potential of the sensory neuron.

STEP 3: *Action Potentials Travel to the CNS along an Afferent Fiber.* When a generator potential occurs in the sensory neuron, action potentials develop in the afferent fiber. The greater the depolarization produced by the generator potential, the higher the frequency of action potentials in the afferent fiber. The arriving information is then processed and interpreted by the CNS at the conscious and subconscious levels.

Pain Mechanisms, Pathways, and Control: A Closer Look

The sensory neurons that bring pain sensations into the CNS release the neurotransmitters *glutamate* and *Substance P.* These two neurotransmitters have very different but complementary effects. Glutamate produces an immediate depolarization of the postsynaptic membrane by binding to receptors, which stimulate the interneuron, and pain impulses ascend to the thalamus within the spinothalamic tracts.[1] The excitatory effects are restricted to the postsynaptic cell, because glutamate is rapidly reabsorbed by the synaptic knobs.

The net result is that the stimulated neurons become strongly facilitated, and pain sensitivity increases.

Substance P released by sensory neurons has a more widespread stimulatory effect. This neurotransmitter diffuses through the gray matter of the dorsal gray horn of the spinal cord, where it affects large numbers of interneurons. Substance P binds to membrane receptors and is brought into the cytoplasm by means of receptor-mediated endocytosis. The result is the stimulation of some interneurons, producing sensations of pain, and the facilitation of many other interneurons involved with pain pathways.

The ascending pain sensations are widely distributed. Most ascending fibers travel within the lateral spinothalamic tracts for projection to the primary sensory cortex. The thalamus also relays pain sensations to the cingulate gyrus, an emotional center of the limbic system. Some of the ascending fibers do not reach the thalamus, but synapse in the reticular formation or hypothalamus instead. Pain can thus influence CNS activities at both the conscious and subconscious levels. Endorphins and enkephalins are neuromodulators whose release inhibits activity along pain pathways in the brain. These compounds, structurally similar to morphine, are found in the limbic system, hypothalamus, and reticular formation. The pain centers in these areas also use Substance P as a neurotransmitter. Endorphins bind to the presynaptic membrane and prevent the release of Substance P, thereby reducing the conscious perception of pain, although the painful stimulus remains.

---

[1] AMPA receptors and the other NMDA receptors discussed were originally named after administered compounds that would bind to them. AMPA receptors bind alpha-amino-3-hydroxy-5-methyl-4-isoxazole propionic acid; NMDA receptors bind n-methyl-d-aspartate, a synthetic molecule related to glutamate.

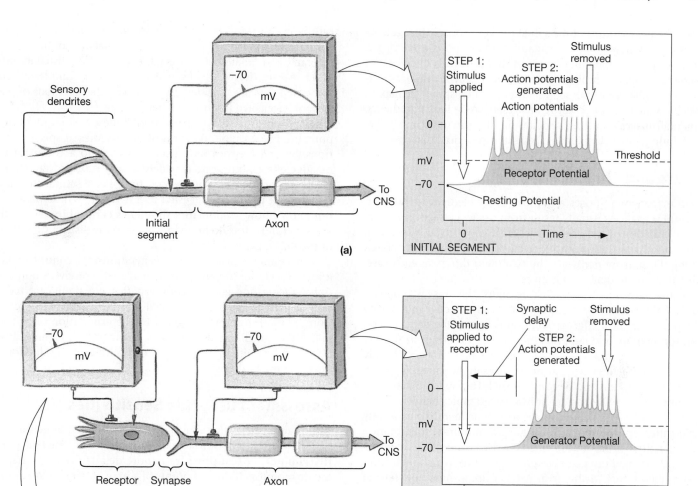

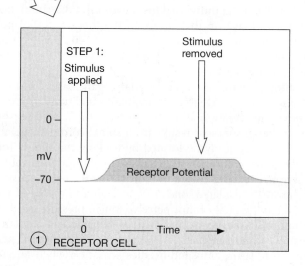

Figure 46 Transduction. (a) When a sensory neuron acts as a receptor, a stimulus that depolarizes the dendrites may bring the initial segment of the axon to threshold. The receptor and the neuron are the same cell, so the receptor potential is a generator potential. **(b)** For the senses of taste, equilibrium, hearing, and vision, the receptor cells are specialized cells that communicate with neurons across chemical synapses. (1) The receptor cell shows a receptor potential in response to stimulation. In this example, the receptor potential is a depolarization that accelerates the release of a neurotransmitter, and (2) the neurotransmitter produces a generator potential in the postsynaptic membrane.

Pain can be considered both a sensation and an emotion, and pain control involves more than simply dealing with receptor activation, or even activity along major pain pathways. Due to the facilitation that results from the release of either glutamate or Substance P, the level of pain experienced (especially chronic pain) can be out of proportion to the amount of painful stimuli or the apparent tissue damage. This effect may be one reason people

differ so widely in their perception of the pain associated with childbirth, headaches, or back pain. Such facilitation is also presumed to play a role in phantom limb pain: The sensory neurons may be inactive, but the hyperexcitable interneurons may continue to generate pain sensations. An interesting piece of evidence comes from clinical research: When general anesthesia shuts down the cerebral cortex and prevents conscious perception of pain, the interneurons of the spinal cord are not anesthetized. As a result, they may become facilitated, and this effect can exaggerate postoperative pain. In one recent study, roughly 20 percent of patients who had a limb amputated under general anesthesia experienced phantom limb pain. However, when spinal facilitation was prevented by "pre-emptive analgesia," using local anesthetics, epidural analgesia, and opiate or aspirin-like pain medicines before and during surgery, none of the patients experienced that pain.

Acute and Chronic Pain

Pain management poses a number of problems for clinicians. Painful sensations usually result from tissue damage or sensory nerve irritation; it can originate where it is perceived, be referred from another location, or represent a "false" signal generated along the sensory pathway. The treatment differs in each case, and an accurate diagnosis is an essential first step.

Acute pain is the result of tissue injury; the cause is apparent, and local treatment of the injury is typically effective in relieving the pain. The most effective solution is to stop the damage, end the stimulation, and suppress the painful sensations at the site of injury. Pain sensations are suppressed when topical or locally injected anesthetics inactivate nociceptors in the immediate area. Analgesic drugs can also be administered. They work in many different ways; we will consider only a few examples here.

Tissue injury results in damage to plasma membranes. A fatty acid called arachidonic acid escapes from injured membranes. In interstitial fluid, an enzyme called *cyclo-oxygenase* converts arachidonic acid molecules to prostaglandins, which then stimulate nociceptors in the area. Aspirin, ibuprofen, and related analgesics reduce inflammation and suppress pain by blocking the action of cyclo-oxygenase.

Chronic pain is persistent and more difficult to treat. It includes nociceptive pain from ongoing tissue damage, neuropathic pain from damage to the central or peripheral nervous system, and pain from mixed or undetermined causes. The American Chronic Pain Association (ACPA) defines chronic pain as pain that continues when it should not. Chronic pain in part reflects permanent facilitation of the pain pathways and the creation of a reverberating "pain memory." Complex psychological (pain as an emotion) and physiological (pain as a sensation) components are also involved. For example, many chronic-pain patients develop a tolerance for pain medications, and insomnia and depression are common complaints. Chronic pain can be helped by antidepressants, which affect neurotransmitter levels, and by antiseizure medicines. Counseling may help the person focus attention outward rather than inward; the outward focus can lessen the perceived level of pain and reduce the amount of pain medication required. Curiously, developing a second, acute source of pain, such as a herpes-zoster infection, can reduce the perception of preexisting chronic pain.

In some cases, chronic pain and severe acute pain can be suppressed by inhibition of the central pain pathway. Analgesics related to morphine reduce pain by mimicking the action of endorphins. The perception of pain may be altered, although the pain remains. For example, patients on morphine report being aware of painful sensations, but they are not distressed by them. Surgical steps can be taken to control severe pain; for instance, (1) the sensory innervation of an area can be destroyed by an electrical current, (2) the dorsal roots carrying the painful sensations can be cut (a *rhizotomy*), (3) the ascending tracts in the spinal cord can be severed (a *tractotomy*), or (4) thalamic or limbic centers can be stimulated or destroyed. These options, listed in order of increasing degree of effectiveness, surgical complexity, and associated risk, are considered only when other methods of pain control have failed to provide relief.

When used to control pain, the Chinese technique of acupuncture involves the insertion of fine needles at specific locations. Several theories have been proposed to account for the positive effects, but none is widely accepted or proven. Higher levels of spinal fluid endorphins have been found after acupuncture. It has been suggested that this may relieve pain. It is not known how acupuncture stimulates endorphin release; the acupuncture points do not correspond to the distribution of any of the major peripheral nerves.

Many other aspects of pain generation and control remain a mystery. Up to 30 percent of patients who receive a nonfunctional medication subsequently experience a significant reduction in pain. It has been suggested that this *placebo effect* results from endorphin release triggered by the expectation of pain relief. Although the medication has no direct effect, the indirect effect can be quite significant and complicates the evaluation of analgesic medications.

Assessment of Tactile Sensitivities

Regional sensitivity to light touch can be checked by gentle contact with a fingertip or a slender wisp of cotton. The **two-point discrimination test** provides a more detailed sensory map of tactile receptors. Two fine points of a bent paper clip or another object are applied to the skin surface simultaneously. The subject then describes the contact. When the points fall within a single receptive field, the individual will report only one point of contact. A normal individual loses two-point discrimination at 1 mm (0.04 in.) on the surface of the tongue, at 2–3 mm (0.08–0.12 in.) on the lips, at 3–5 mm (0.12–0.20 in.) on the backs of the hands and feet, and at 4–7 cm (1.6–2.75 in.) over the general body surface.

Vibration receptors are tested by applying the base of a tuning fork to the skin. Damage to an individual spinal nerve produces insensitivity to vibration along the paths of the related sensory nerves. If the sensory loss results from spinal-cord damage, the injury site can typically be located by walking the tuning fork down the spinal column, resting its base on the vertebral spines.

Descriptive terms are used to indicate the degree of sensitivity in the area considered. *Anesthesia* implies a total loss of sensation; the individual cannot perceive touch, pressure, pain, or temperature sensations in that area. *Hypesthesia* is a reduction in sensitivity, and *paresthesia* is the presence of abnormal sensations, such as the pins-and-needles sensation when an arm or leg "falls asleep" due to pressure on a peripheral nerve. (We discussed several types of *pressure neuropathies* that produce temporary paresthesia on page 85.)

Amyotrophic Lateral Sclerosis

Demyelinating disorders (see pp. 78–79) affect both sensory and motor neurons, producing losses in sensation and motor control. **Amyotrophic lateral sclerosis (ALS)** is a progressive disease that affects specifically motor neurons, leaving sensory neurons intact.

As a result, individuals with ALS experience a loss of motor control, but have no loss of sensation or intellectual function. Motor neurons throughout the CNS are destroyed. Neurons involved with the innervation of skeletal muscles are the primary targets.

Symptoms of ALS generally do not appear until the individual is over age 40. ALS occurs at a prevalence of 3–5 cases per 100,000 population worldwide. The disorder is more common among males than females. The pattern of symptoms varies with the specific motor neurons involved. When motor neurons in the cerebral hemispheres of the brain are the first to be affected, the individual experiences difficulty in performing voluntary movements and has exaggerated stretch reflexes. If motor neurons in other portions of the brain and the spinal cord are targeted, the individual experiences weakness, initially in one limb, but gradually spreading to other limbs and ultimately the trunk. When the motor neurons innervating skeletal muscles degenerate, a loss of muscle tone occurs. Over time, the skeletal muscles atrophy. The disease progresses rapidly, and the average survival after diagnosis is just 3–5 years. Because intellectual functions remain unimpaired, a person with ALS remains alert and aware throughout the course of the disease. This is one of the most disturbing aspects of the condition. Among well-known people who have developed ALS are baseball player Lou Gehrig and physicist Stephen Hawking.

The primary cause of ALS is uncertain; only 5–10 percent of ALS cases appear to have a genetic basis, with 5 percent of these genetic cases caused by a mutation in the superoxidase dismutase (*SOD1*) gene, which codes for a powerful antioxidant enzyme that normally reduces free radicals. At the cellular level, it appears that the underlying problem is at the postsynaptic membranes of motor neurons. It has been suggested that an abnormal receptor complex for the neurotransmitter glutamate in some way leads to the buildup of free radicals, such as NO (nitric oxide), that ultimately kill the neuron. Treatment with riluzole, a drug that suppresses the release of glutamate, has delayed the onset of respiratory paralysis and extended the life of ALS patients.

▶ Neural Integration II: The Autonomic Nervous System and Higher-Order Functions

Hypersensitivity and Sympathetic Function

Two interesting clinical conditions result from the disruption of normal sympathetic functions. In **Horner's syndrome**, the sympathetic postganglionic innervation to one side of the face becomes interrupted. The interruption may be the result of an injury, a tumor, or some progressive condition such as multiple sclerosis. The affected side of the face becomes flushed as vascular tone decreases. Sweating stops in the region, and the pupil on that side becomes markedly constricted. Other symptoms include a drooping eyelid and an apparent retreat of the eye into the orbit.

Primary Raynaud phenomenon, also called **Raynaud disease**, most commonly affects young women. In this condition, for unknown reasons, the sympathetic system temporarily orders excessive peripheral vasoconstriction of small arteries, usually in response to cold temperatures. The hands, feet, ears, and nose become deprived of their normal blood circulation and the skin in these areas changes color, becoming initially pale and then developing blue tones. A red color ends the cycle as normal blood flow returns. The symptoms may spread to adjacent areas as the disorder progresses. Most cases do not cause tissue damage, although in rare cases prolonged decreased blood flow may distort the skin and nails, even progressing to skin ulcers or the more extensive tissue death of dry gangrene.

Behavioral changes such as avoiding cold environments or wearing mittens and other protective clothing can usually reduce the frequency of occurrence. Stopping smoking and avoiding drugs that can cause vasoconstriction may also be beneficial. Drugs that prevent vasoconstriction (vasodilators) can be used if preventative steps prove ineffective. In some cases, putting one's hands in warm water upon entering a cold room may desensitize the ANS and suppress the responses of primary Raynaud syndrome.

Secondary Raynaud phenomenon is associated or possibly caused by an underlying problem. Trauma such as frostbite or the cumulative damage caused by the chronic use of vibrating machinery can cause symptoms of Raynaud syndrome. Symptoms may also appear in individuals with arterial diseases and in connective tissue disorders such as scleroderma, rheumatoid arthritis, and systemic lupus erythematosus (SLE).

A regional **sympathectomy** (sim-path-EK-to-mē) that affects the fibers that provide sympathetic innervation to the region may be permanent if sympathetic nerves are cut. Alternatively, injecting botulinus toxin can be temporary, lasting weeks to months. Under normal conditions, sympathetic tone provides the effectors with a background level of stimulation. After the elimination of sympathetic innervation, peripheral effectors may become extremely sensitive to norepinephrine and epinephrine. This hypersensitivity can produce extreme alterations in vascular tone and other functions after stimulation of the adrenal medullae. If the sympathectomy involves the postganglionic fibers, hypersensitivity to circulating norepinephrine and epinephrine may eliminate the beneficial effects. The prognosis improves if the preganglionic fibers are transected, because the ganglionic neurons will continue to release small quantities of neurotransmitter across the neuromuscular or neuroglandular junctions. This release keeps the peripheral effectors from becoming hypersensitive.

Biofeedback

Although conscious thought processes affect the ANS, you normally do not perceive the effect because visceral sensory information does not reach your cerebral cortex. Even when a conscious mental process triggers a physiological shift, such as a change in blood pressure, sweat gland activity, skin temperature, or muscle tone, the information never arrives at your primary sensory cortex. *Biofeedback* is an attempt to bridge this gap. In this technique, a person's autonomic processes are monitored, and a visual or auditory signal is used to alert the person when a particular change takes place. These signals let the individual know when ongoing conscious thought processes have triggered a desirable change in autonomic function. For example, when biofeedback is used to regulate blood pressure, a light or tone informs the person when blood pressure drops.

With practice, some people can learn to re-create the thoughts, sensations, or mood that will lower their blood pressure. This action is possible because the cerebral cortex (conscious thought),

Nervous System

thalamus (sensory feedback), and limbic system (emotions) all affect the hypothalamic centers that control autonomic activity.

Biofeedback techniques have been used to promote conscious control of blood pressure, heart rate, circulatory pattern, skin temperature, brain waves, and so forth. By reducing stress, lowering blood pressure, and improving circulation, these techniques can reduce the severity of clinical symptoms. In the process, they also lower the risks for serious complications, such as heart attacks or strokes, in individuals with high blood pressure. Unfortunately, not everyone can learn to influence their own autonomic functions, and the combination of variable results and expensive equipment has kept biofeedback unsuitable for widespread application.

Pharmacology and the Autonomic Nervous System

The treatment of many clinical conditions involves the manipulation of autonomic function. Drugs may be administered to counteract or reduce symptoms caused or aggravated by autonomic activities. These drugs are called **mimetic** if they mimic the activity of one of the normal autonomic neurotransmitters. Drugs that reduce the effects of autonomic stimulation by keeping the neurotransmitter from affecting postsynaptic membranes are known as **blocking agents**. Mimetic drugs commonly have advantages over natural neurotransmitters. For example, norepinephrine and epinephrine must be administered into the bloodstream by injection or infusion, and the therapeutic effects are short lived. In contrast, mimetic drugs can be administered in a variety of ways and their actions can be long-lasting. For example, sympathomimetic drugs can be swallowed, applied topically to reduce hemorrhaging, sprayed to reduce nasal congestion, inhaled to dilate the respiratory passageways, or put in drops to dilate the pupils. They can also be injected to elevate blood pressure and improve cardiac performance after severe blood loss or heart muscle failure. **Table 22** relates important mimetic drugs and blocking agents to specific autonomic activities.

Sympathomimetic drugs are used to treat a variety of disorders; we will consider only representative examples here. *Phenylpropanolamine* stimulates alpha receptors, causing a constriction of peripheral vessels and elevating blood pressure. Once widely used as a nasal decongestant and nonprescription diet pill, phenylpropanolamine has been linked to strokes in young women and removed from the market. Three other less powerful sympathomimetic drugs, phenylephrine, pseudoephedrine, and ephedrine stimulate beta-receptors. They have been used in many cold remedies (such as *Actifed* and *Sudafed*) that reduce nasal congestion and open respiratory passages. Undesirable side effects, such as jitteriness, anxiety, sleeplessness, and high blood pressure, result from the facilitation of CNS pathways and the stimulation of peripheral α and β_1 receptors. Pseudoephedrine is an ingredient used to make methamphetamine, an illegal, dangerous, and addictive stimulant. To avoid misuse, pseudoephedrine sale is monitored. Most cold remedies now contain phenylephrine instead. Ephedrine was used for asthma until more selective β_2 receptor agonists with fewer side effects were developed. Purveyors of "Natural" and "Health" supplements were able to market unregulated ephedrine-containing products for years as diet aids and stimulants, despite strokes and

deaths being attributed to their use. In 2004 the FDA was finally able to have them removed from the marketplace.

Sympathomimetic drugs that selectively target β_1 receptors, such as *dobutamine*, are especially valuable in increasing heart rate and blood pressure. They are used in some diagnostic heart tests and in severe heart failure (see pp. 136–137). Sympathomimetic drugs targeting β_2 receptors, such as *albuterol*, have been developed to treat the bronchial constriction that accompanies asthma attacks.

Sympathetic blocking agents that prevent a normal response to neurotransmitters or sympathomimetic drugs include alphablockers and beta-blockers. **Alpha-blockers** eliminate the peripheral vasoconstriction that accompanies sympathetic stimulation. The alpha-blockers include doxazosin, which selectively targets α_1 receptors and is used to reduce high blood pressure. The drug also reduces constriction of the prostatic urethra, so it is used to treat restricted urine flow in *benign prostatic hypertrophy*. **Beta-blockers** are effective and clinically useful for treating chronic high blood pressure and congestive heart failure, and for reducing the risk of death after a heart attack. In general, beta-blockers decrease the heart rate and force of contraction, reducing the strain on the heart and simultaneously lowering peripheral blood pressure. *Propranolol* and *metoprolol* are two of the most popular betablockers currently on the market. Propranolol affects both β_1 and β_2 receptors, so persons with asthma may experience difficulties in breathing as their respiratory passageways constrict. Metoprolol targets β_1 receptors almost exclusively, leaving the respiratory smooth muscles relatively unaffected.

Parasympathomimetic drugs may be used to increase activity along the digestive tract and encourage defecation and urination. *Physostigmine* and *neostigmine* are important parasympathomimetic, cholinesterase inhibiting drugs that work by blocking the action of acetylcholinesterase. Because this enzyme is rendered inoperative, levels of ACh within the synapses climb and parasympathetic activity is enhanced. The blocking agent *d-tubocurarine* blocks ACh neuromuscular transmission; the administration of physostigmine or neostigmine can counteract the resulting paralytic effects of this drug.

Parasympathetic blocking agents, such as *atropine*, inhibit the action of ACh on the muscarinic receptors at neuromuscular and neuroglandular junctions. These drugs have diverse effects, but they are typically used to control the diarrhea and cramps associated with various forms of food poisoning. The drug *Lomotil*, known as the "traveler's friend," can provide temporary relief from diarrhea for the duration of a plane flight home. Among its other effects, atropine causes an elevation of the heart rate due to a loss of parasympathetic tone. *Scopolamine* has similar effects on peripheral tissues, but it has greater influence on the CNS. Its most useful effects are promoting drowsiness, reducing nausea and salivation, and relieving anxiety. As a result, scopolamine may be given to a patient who is being prepared for surgery, prior to the administration of the anesthetic agent. (Scopolamine is also administered transdermally to control nausea and motion sickness.) Botulinus toxin also blocks release of ACh but requires local injection.

Amnesia

Amnesia, or memory loss, occurs suddenly or progressively, and recovery is complete, partial, or nonexistent, depending on the nature of the problem. In **retrograde amnesia** (*retro-*, behind),

Table 22	Drugs and the ANS		
Drug	**Mechanism**	**Action**	**Clinical Uses**
Sympathomimetic			
Phenylephrine (Neosynephrine)	Stimulates α_1 receptors	Elevates blood pressure; stimulates smooth muscle	As a topical nasal decongestant and to elevate low blood pressure
Clonidine	Stimulates α_2 receptors	Lowers blood pressure	Treatment of high blood pressure
Dobutamine	Stimulates β_1 receptors	Stimulates cardiac output	Treatment of heart failure; heart attack
Albuterol	Stimulates β_2 receptors	Dilate respiratory passages	Treatment of asthma
Ephedrine/pseudoephedrine	Stimulates α and β_1 receptors, and norepinephrine release at neuromuscular and neuroglandular junctions	Similar to effects of epinephrine	As a nasal decongestant and to elevate blood pressure or dilate respiratory passageways
Sympathetic blocking agents			
Doxazosin (Cardura)	Blocks α_1 receptors	Lowers blood pressure	Treatment of high blood pressure and benign prostatic hypertrophy
Propranolol (Inderal)	Blocks β_1 and β_2 receptors	Reduces metabolic activity in cardiac muscle, but may constrict respiratory passageways; slows heart rate	Treatment of high blood pressure; used to reduce heart rate and force of contraction in heart disease
Metoprolol	Blocks β_1 receptors	Reduces metabolic activity in cardiac muscle	Similar to those of Inderal, but has less of an effect on respiratory muscles
Parasympathomimetic			
Physostigmine, neostigmine	Block action of acetylcholinesterase	Increase ACh concentrations at parasympathetic neuromuscular and neuroglandular junctions	Stimulate digestive tract and smooth muscles of urinary bladder; treatment of myasthenia gravis
Pilocarpine	Stimulates muscarinic receptors	Similar to effects of ACh	Applied topically to cornea of eye to cause pupillary contraction; treats glaucoma
Parasympathetic blocking agents			
Atropine, related drugs	Block ACH action on muscarinic receptors	Inhibit ACh parasympathetic activity	Treatment of diarrhea; used to dilate pupils; raise heart rate, reduce secretions of digestive and respiratory tracts prior to surgery, treat accidental exposure to anticholinesterase drugs, such as pesticides or military nerve gases
Botulinus toxin (Table 20, p. 79)			

the individual loses memories of past events. Some degree of retrograde amnesia commonly follows a head injury; after a car wreck or a football tackle, many victims are unable to remember the moments preceding the accident. In **anterograde amnesia** (*antero-*, ahead), an individual may be unable to store additional memories, but earlier memories are intact and accessible. The problem appears to involve an inability to generate long-term memories. At least two drugs—*diazepam* (*Valium*) and *Halcion*—have been known to cause brief periods of anterograde amnesia. Brain injuries can cause more prolonged memory problems. A person with permanent anterograde amnesia lives

in surroundings that are always new. Magazines can be read, chuckled over, and reread a few minutes later with equal pleasure, as if they had never been seen before. Clinicians must introduce themselves at every meeting, even if they have been treating the patient for years.

Some degree of **post-traumatic amnesia (PTA)** commonly occurs after a head injury. The duration of the amnesia varies with the severity of the injury. This condition combines the characteristics of retrograde and anterograde amnesias; the individual can neither remember the immediate past nor consolidate recent memories of the present.

Nervous System

Altered States

A fine line separates normal from abnormal states of awareness, and many variations in these states are clinically significant. The state of **delirium** involves wild oscillations in the level of wakefulness. The individual has little or no grasp of reality and often experiences hallucinations. When capable of communication, the person seems agitated, confused, and unable to deal with situations and events. Delirium may develop (1) in individuals with very high fevers or some metabolic abnormalities; (2) in patients experiencing withdrawal from some drugs, such as alcohol; (3) after hallucinogenic drugs, such as LSD, are taken; (4) as a result of brain tumors affecting the temporal lobes; and (5) in the late stages of some infectious diseases, including syphilis and AIDS. The term **dementia** implies a more chronic state characterized by deficits in memory, language, reasoning, judgment, or personality. We shall discuss dementia in a later section.

Table 23 indicates the entire range of conscious and unconscious states, ranging from *delirium* through *coma*. Assessing the level of consciousness in an awake person involves noting the person's general alertness, patterns of speech, content of that speech (as an indication of ongoing thought processes), and general motor abilities. Orientation to three concepts—Who are you? What day is it? Where are you?—is a basic part of the *mental status* assessment. In an unconscious person, the *pupillary reflex* provides important information about the status of the brain stem. When light is shined into one eye, the pupils of both eyes should constrict. This reflex is coordinated by the superior colliculus in the mesencephalon. If the pupils are unreactive, or *fixed*, and this is not caused by a drug effect, serious brain damage has occurred.

A person's level of consciousness is often reported in terms of the *Glasgow Coma scale*, a classification system given in **Table 24**. When the rating is completed, a tally is taken. Someone with a total score of 7 or less is probably comatose. Low scores in more than one section indicate that the condition is relatively severe. A comatose individual with a score of 3–5 has probably suffered irreversible brain damage. One major exception is general anesthesia, which is essentially a medically induced temporary coma that facilitates surgery.

Sleep Disorders

Sleep disorders include abnormal patterns of REM or deep sleep, variations in the time of onset of sleep or the time devoted to sleeping, and unusual behaviors occurring while sleeping. Sleep disorders of one kind or another are very common, affecting an estimated 25 percent of the U.S. population at any time.

Many clinical conditions influence sleep patterns or are exaggerated by the autonomic changes that accompany the various stages of sleep. Major clinical categories of sleep disorders include *insomnias, hypersomnias,* and *sleep apnea*.

Insomnia has been defined as difficulty falling asleep, staying asleep, and poor/unrefreshing sleep. Sleep duration varies among people, and short refreshing sleep is not insomnia. Short-term insomnia lasting days to weeks is often a reaction to stress, trauma, or a short-term medical condition. Brief use of a sedative such as a benzodiazepine (Table 25) may help. Chronic insomnia lasting months may be either primary or secondary to other medical problems. Secondary insomnia usually responds to treatment of the underlying problem. Primary chronic insomnia is more difficult to treat due to possible genetic elements, prolonged serious stress, and learned maladaptive behavioral causes.

The following behaviors promote good sleep:

- Sleep only as much as necessary to feel rested, and then get out of bed.
- Maintain a regular sleep schedule (the same bedtime and wake time every day).
- Avoid caffeinated beverages after lunch.
- Adjust the bedroom environment (light, noise, temperature) so that you are comfortable before you lie down.
- Deal with concerns or worries before bedtime. Make a list of things to work on for the next day so anxiety is reduced at night.
- Exercise regularly, preferably four or more hours before bedtime.

Hypersomnia involves extremely long periods of otherwise normal sleep. The individual may sleep until noon and nap before

Table 23	States of Consciousness
Level or State	**Description**
Conscious States	
Normal consciousness	Aware of self and external environment; well oriented; responsive
Delirium	Disorientation, restlessness, confusion, hallucinations, and agitation, alternating with other conscious states
Confusion	Reduced awareness; easily distracted; easily startled by sensory stimuli; alternation between drowsiness and excitability; resembles minor form of delirium state
Dementia	Difficulties with spatial orientation, memory, language; changes in personality
Somnolence	Extreme drowsiness, but response to stimuli is normal
Chronic vegetative state	Conscious, but unresponsive; no evidence of cortical function
Unconscious States	
Asleep	Can be aroused by normal stimuli (light touch, sound, etc.)
Stupor	Can be aroused by extreme or repeated stimuli
Coma	Cannot be aroused and does not respond to stimuli (coma states can be further subdivided according to the effect on reflex responses to stimuli)

Table 24	The Glasgow Scale	
Area or Aspect Assessed	**Response**	**Score**
Motor abilities	None	1
	Decerebrate rigidity (individual is supine with extension of hips, knees, ankles in plantar flexion; arms adducted and elbows extended; forearms pronated; finger joints flexed)	2
	Decorticate rigidity (individual is supine with elbows, wrists, and fingers flexed and pressed against the chest; lower limbs rigidly extended with ankles in plantar flexion)	3
	Withdrawal reflex to stimulus	4
	Ability to pinpoint painful stimulus	5
	Ability to move body parts according to verbal request	6
Eyes	No reaction	1
	Opens eyes with painful stimulus	2
	Opens eyes upon verbal request	3
	Opens eyes spontaneously	4
Verbal ability	None	1
	Sounds emitted are not understandable	2
	Words used are not appropriate	3
	Able to speak, but is not oriented	4
	Converses and is oriented	5

dinner, despite an early retirement in the evening. The condition has physiological and psychological origins, and successful treatment may involve drug therapy, counseling, or both. Two important examples of hypersomnia are *narcolepsy* and *sleep apnea*.

Roughly 0.2–0.3 percent of the population has **narcolepsy**, characterized by dropping off to sleep at inappropriate times. The sleep lasts only a few minutes and is preceded by a period of muscle weakness. Any exciting stimulus, such as laughter or other strong emotion, may trigger an attack. (Interestingly, this condition may be inherited and occurs in other mammals; narcoleptic dogs will keel over into a sound sleep when presented with their favorite treats.) In some instances, the sleep period is followed by a brief period of amnesia. Drugs that block REM sleep, such as methylphenidate (*Ritalin*), may reduce or eliminate narcoleptic attacks.

In **sleep apnea**, a sleeping individual stops breathing for short periods, probably due to a disturbance or abnormality in the respiratory control mechanism or blockage of the upper airway. The problem can be exaggerated by various drugs (including alcohol), obesity, tonsillitis, and other medical conditions that either affect the CNS or cause intermittent obstruction of the respiratory passageways. Roughly half those who experience sleep apnea show simultaneous cardiac arrhythmias, but the link between the two is not understood. Often people are not aware of disturbed sleep, but will complain of daytime sleepiness. Treatment focuses primarily on alleviating potential causes; for example, overweight individuals are encouraged to lose weight, and enlarged tonsils are removed. In severe cases, air under pressure is administered through the nose to keep the upper airway open during sleep. This treatment is called *continuous positive airway pressure (CPAP)*.

Huntington's Disease

Huntington's disease is an inherited fatal disease marked by a progressive deterioration of muscle control and mental abilities

(*dementia*). About 15,000 Americans have this condition and another 75,000 to 100,000 have the abnormal gene and will eventually develop the disease. In Huntington's disease, the basal nuclei and frontal lobes of the cerebral cortex show degenerative changes. The basic problem is the destruction of ACh-secreting and GABA-secreting neurons in the basal nuclei and cerebral cortex, probably caused by excess amounts of the neurotransmitter glutamate. The first signs of the disease generally appear in early adulthood. As you would expect in view of the areas affected, the symptoms involve difficulties in performing voluntary movements, and experiencing involuntary patterns of movement called *chorea*. Over time a gradual decline in emotional control and intellectual ability progresses to dementia and eventually death.

An abnormality in an autosomal dominant gene located on chromosome 4 causes Huntington's disease. In people with the disease, this gene, which codes for the protein *huntingtin*, contains an increased number of repetitions of the nucleotide sequence CAG. This DNA segment appears to be unstable, and the number of repetitions can change from generation to generation. The duplication is thought to occur during gamete formation. The more repetitions, the earlier in life the symptoms and eventual death from Huntington's disease occur. The multiple copies of the CAG nucleotide sequence result in an abnormal form of huntingtin, which clumps in brain tissue and seems to contribute to cell death.

In 2008 the FDA approved Tetrabenazine (Xenazine), a dopamine-depleting drug previously used for schizophrenia. It significantly reduces the abnormal movements of chorea, but disease progression continues and there is no effective treatment for Huntington's disease. The children of a person with Huntington's disease have a 50 percent chance of receiving the gene and developing the condition. Genetic testing can be done either prenatally or postnatally to confirm the disease. For people born to affected families, this test can indicate whether or not they will develop the disease. While the test relieves uncertainty, it reveals a virtual early death sentence to those who test positive, and not all at-risk individuals choose to take the test.

Pharmacology and Drug Abuse

Although accurate and comparable statistics are hard to find, the prescription drug industry in the United States certainly qualifies as "big business": Yearly sales grew to $263 billion in 2011, and 354 billion prescriptions were filled in 2008. In 2005, when drug sales totaled $163 billion, $29.8 billion were spent on advertising and samples.

With that level of funding available, it's not surprising that television viewers are exposed to a dazzling array of advertisements for cold medicines, diet pills, headache remedies, anxiety relievers, and sleep promoters. Although roughly one-third of the drugs sold affect the CNS, few consumers have any understanding of how these drugs exert their effects. Despite an overwhelming interest in medicinal and "recreational" (typically illicit) drugs, we seldom encounter accurate information about their mechanisms of action or the associated hazards, except when speed-read by voice-over actors at the end of commercials. Unfounded and inaccurate rumors are quite common; the concept that "natural" drugs are safer or more effective than "artificial" (synthetic) drugs is an example of a potentially dangerous, but popular, misconception.

The major categories of drugs that affect CNS function are presented in **Table 25**. Considerable overlap exists among these categories. For example, any drug that causes heavy sedation will also depress the perception of pain and alter mood at the same time. Well-known prescription, nonprescription, and illicit drugs are included in the table where appropriate. Some of the nonprescription drugs, most of the prescription drugs, and all of the illegal CNS-active drugs are prone to abuse because tolerance and addiction occur.

Sedatives lower the general level of CNS activity, reduce anxiety, and have a calming effect. **Hypnotic** drugs further depress activity, producing drowsiness and promoting sleep. All levels of CNS activity are reduced, so the effects can range from a mild relaxation to a general anesthesia, depending on the drug and the dosage administered.

Analgesics, provide relief from pain. Some, like aspirin, act on the periphery by reducing the source of the painful stimulation. (Aspirin slows the release of prostaglandins, which promote inflammation and stimulate pain receptors.) Others, especially the drugs structurally related to opium, target the CNS processing of pain sensations. The compounds bind to receptors on the surfaces of CNS neurons, notably in the basal nuclei, limbic system, thalamus, hypothalamus, midbrain, medulla oblongata, and spinal cord. Some opium derived drugs, such as morphine and its relatives, mimic the activity and structure of endorphins already present in the CNS.

Psychotropics are used to produce changes in mood and emotional state. Certain forms of **depression** have been shown to result from the inadequate production of the neurotransmitters norepinephrine, dopamine, or serotonin in key nuclei of the brain. Overexcitement, or **mania**, has been correlated with an overproduction of the same compounds. Enhancing or blocking the production of these neurotransmitters can often provide the mental stabilization needed to alleviate mood disorders.

Antipsychotics reduce the hallucinations and behavioral or emotional extremes that characterize the severe mental disorders known as psychoses, but they leave other mental functions relatively intact. Phenothiazines, notably chlorpromazine (*Thorazine*), were initially used to control nausea and vomiting and to promote sedation and relaxation. However, they ultimately proved more useful in controlling psychotic behavior.

The term *tranquilizer* was first applied to the early antipsychotic drugs, such as *Reserpine* (which was also used to treat hypertension). Later, the category was expanded to include "minor tranquilizers," such as *Valium*, whose effects more closely resemble those of sedatives. Eventually, it became apparent that

Table 25	A Simple Classification of Drugs That Affect the CNS		
		Examples	
Category (Common name)	**Actions**	**Prescription**	**Nonprescription**
Sedatives and hypnotics ("downers")	Depress CNS activity; promote sleep, reduce anxiety, create calm; those used to depress seizures are considered "anticonvulsants"; can be addictive	Barbiturates (phenobarbital, Nembutal, Amytal); benzodiazepines (Librium, Valium)	Sominex, Nytol, Benadryl, alcohol
Analgesics ("painkillers")	Relieve pain at source or along CNS pathways; often addictive	Opiates (morphine, Demerol, codeine)	Aspirin, acetaminophen (Tylenol), Ibuprofen
Psychotropics ("mood changers")	Alter CNS function and change mental state or mood; many are addictive	Antipsychotics (chlorpromazine, Risperdal); cocaine	Caffeine, alcohol, Kava
	Antidepressants increase levels of some neurotransmitters, including serotonin and norepinephrine	Prozac, Effexor, nortriptyline	St. John's Wort
	Facilitation of GABA neurotransmitter release	Antianxiety drugs (Librium, Valium)	
	Alters sodium transport in nerve and muscle cells; specific mechanism of action remains unknown	Mood stabilizers (lithium)	
Anticonvulsants	Inhibit spread of cortical stimulation	Dilantin; also sedative–hypnotics	
Stimulants	Facilitate CNS activity; some are addictive; typically affect cardiovascular system	Ritalin, amphetamines	Diet pills, Sudafed, Actifed, Xanthines (caffeine)

"tranquilization" was more useful as a descriptive term than as a classification for specific drugs. For example, a tranquilizing effect can be produced by sedatives (alcohol, Valium), hypnotics (barbiturates), and antipsychotics.

The first **anticonvulsants**, used to control seizures, were sedatives such as phenobarbital. Over time, however, patients required larger and larger doses to achieve the same level of effect. This phenomenon, called **tolerance**, commonly appears when any CNS-active drug is administered. *Dilantin* and related drugs have powerful anticonvulsant effects without producing tolerance.

Stimulants facilitate activity in the CNS. Few stimulants are used clinically, as they may trigger convulsions or hallucinations. Several of the ingredients in coffee and tea are CNS stimulants, caffeine being the most familiar example. **Amphetamines** prompt the release of norepinephrine, stimulating the respiratory and cardiovascular control centers and elevating muscle tone to the point at which tremors may begin. Amphetamines also stimulate dopamine and serotonin release in the CNS and facilitate cranial and spinal reflexes. Attention deficit/hyperactivity disorders (ADHDs) and narcolepsy may respond to stimulant therapy.

Drug abuse has a very hazy definition, implying that the individual voluntarily uses a drug in some inappropriate manner. Any drug use that violates medical advice, prevalent social mores, or common law can be included within this definition. **Drug addiction** refers to an overwhelming dependence and compulsion to use a specific drug, despite the medical or legal risks involved. If use of the drug is stopped or prevented, the individual will typically suffer physical and psychological symptoms of **withdrawal**. Some drugs can be beneficial with responsible use, but because they are addictive, users experience withdrawal symptoms. Caffeine, which with moderate use has been shown to safely increase alertness, can cause withdrawal symptoms, mainly headaches, that last for 1 to 7 days after stopping use. The physiological foundations of drug dependence and addiction vary with the compound considered. In general, the mechanisms are poorly understood.

The dangers inherent in recreational drug use have been repeatedly overlooked or ignored. Until recently, the abuse of common addictive drugs such as alcohol and nicotine was considered relatively normal or at least excusable. The risks of alcohol abuse can hardly be overstated; it has been estimated that 40 percent of young men have problems with alcohol consumption, and 7 percent of the adult population show signs of alcohol abuse or addiction. Long-term abuse increases the risks of diabetes, liver and kidney disorders, cancer, cardiovascular disease, and digestive system malfunctions. In addition, CNS disturbances commonly lead to accidents, due to poor motor control, and violent behaviors, due to interference with normal emotional balance and analytical function. Medications used for other problems, such as those for hypertension or even antibiotics, can also have CNS effects, ranging from drowsiness to hallucinations or seizures. These side effects remind us of the importance of overall homeostasis and our inability to separate the mind from the body.

Alzheimer's Disease

Alzheimer's disease is a chronic, progressive illness characterized by memory loss and impairment of higher-order cerebral functions including abstract thinking, judgment, and personality. It is the most common cause of **dementia**. Symptoms may appear at 50–60 years of age or later, although the disease occasionally affects younger individuals. Alzheimer's disease has widespread impact. An estimated 5.1 million people in the United States have Alzheimer's—including roughly 5 percent of those from age 65 to 70, with the number doubling for every 5 years of aging until nearly 50 percent of those over age 85 have some form of the condition. Over 230,000 victims require nursing home care, and Alzheimer's disease is the sixth leading cause of death.

Most cases of Alzheimer's disease are associated with large concentrations of neurofibrillary tangles and plaques in the nucleus basalis, hippocampus, and parahippocampal gyrus. These brain regions are directly associated with memory processing. It remains to be determined whether these deposits cause Alzheimer's disease or are secondary signs of ongoing metabolic alterations with an environmental, hereditary, or infectious basis.

The tangles are intracellular masses of abnormal microtubular proteins. The plaques are extracellular deposits of protein fragments, primarily **amyloid beta** (also known as beta-amyloid) **protein**. In Down's syndrome (which results from an extra copy of chromosome 21 and is discussed further on page 230) and in some inherited forms of Alzheimer's disease, mutations affecting genes on either chromosome 21 or a small region of chromosome 14 lead to excessive production and deposition of amyloid beta protein and increased risk of the early onset of the disease. Other genetic factors certainly play a major role. The late-onset form of Alzheimer's disease has been traced to the apolipoprotein E gene on chromosome 19 that codes for proteins involved in cholesterol transport.

Diagnosis involves excluding metabolic and anatomical conditions that can mimic dementia, a detailed history and physical exam, and an evaluation of mental functioning. Initial symptoms are subtle: moodiness, irritability, depression, and a general lack of energy. These symptoms are often ignored, overlooked, or dismissed. Elderly relatives are viewed as eccentric or irascible and are humored whenever possible.

As the condition progresses, however, it becomes more difficult to ignore or accommodate. An individual with Alzheimer's disease has difficulty making decisions, even minor ones. Mistakes—sometimes dangerous ones—are made, through either bad judgment or forgetfulness. For example, the person might light the gas burner, place a pot on the stove, and go into the living room. Two hours later, the pot, still on the stove, melts and starts a fire.

As memory losses continue, the problems become more severe. The individual may forget relatives, his or her home address, or how to use the telephone. The memory loss commonly starts with an inability to store new long-term memories, followed by the loss of recently stored memories. Eventually, basic long-term memories, such as the sound of the individual's own name, are forgotten. The loss of memory affects both intellectual and motor abilities, and a person with severe Alzheimer's disease has difficulty performing even the simplest motor tasks. Although by that time victims are relatively unconcerned about their mental state or motor abilities, the condition can continue to have devastating emotional effects on the immediate family.

Individuals with Alzheimer's disease show brain shrinkage with a pronounced decrease in the number of cortical neurons, especially in the frontal and temporal lobes. This loss is correlated with inadequate ACh production in the *nucleus basalis* of the cerebrum. Axons leaving that region project throughout the cerebral cortex; when ACh production declines, cortical function deteriorates.

There is no cure for Alzheimer's disease, but a few medications and supplements may slow its progress in some patients and delay the need for nursing home care. The antioxidants vitamin E, and perhaps ginkgo biloba, and the B vitamins of folate, B_6, and B_{12} may help some patients and may delay or prevent the disease. Both anticholinesterase inhibitors that improve ACh levels, and a medicine that partially blocks the NMDA neurotransmitter receptor and thus increases glutamate levels also give some additional benefit. Other medicines may help control some of the difficult behaviors such as agitation, anxiety, aggression, and insomnia that often accompany the disease. Various toxicities and side effects determine what combination of drugs is used. Aging is the single largest risk factor for Alzheimer's, with ill health due to diabetes or cardiovascular disease contributing as well. There are many studies under way that aim to find effective treatments or preventative measures.

▶ Disorders of the Special Senses

Otitis Media and Mastoiditis

Acute otitis media is an infection of the middle ear, frequently of bacterial origin. It commonly occurs in infants and children and is occasionally seen in adults. The middle ear, usually a sterile, air-filled cavity, becomes fluid-filled and infected by pathogens that arrive via the auditory tube, often during an upper respiratory infection. If caused by a virus, otitis media may resolve in a few days without use of antibiotics. This "watchful waiting" is most appropriate where medical care is readily available; the pain is reduced by analgesics, and the use of decongestants helps drain the stagnant clear mucus produced in response to mucosal swelling.

If bacteria become involved, symptoms worsen and the mucus becomes cloudy with the bacteria and active or dead neutrophils. Severe otitis media must be promptly treated with antibiotics. As pus accumulates in the middle ear cavity, the tympanic membrane becomes painfully distorted, and in untreated cases it will often rupture, producing a characteristic drainage from the external acoustic canal. The infection may also spread to the mastoid air cells. Chronic mastoiditis, accompanied by drainage through a perforated eardrum and scarring around the auditory ossicles, is a common cause of hearing loss in areas of the world without access to medical treatment. In developed countries, it is rare for otitis media to progress to the stage at which rupture of the tympanic membrane or infection of the adjacent mastoid bone occurs.

Serous otitis media (SOM) involves the accumulation of clear, thick, gluelike fluid in the middle ear. The condition, which can follow acute otitis media or can result from chronic nasal infection and allergies, causes conductive hearing loss. Affected toddlers may have delayed speech development as a result. Treatment may involve decongestants, antihistamines, and, in some cases,

prolonged antibiotic treatment. Nonresponsive cases and recurrent otitis media may be treated by myringotomy (drainage of the middle ear through a surgical opening in the tympanic membrane) and the placement of a temporary tube in the membrane. As toddlers grow, the auditory tube enlarges, allowing better drainage during upper respiratory infections, so both forms of otitis media become less common.

Vertigo, Motion Sickness, and Ménière's Disease

The term **vertigo** describes an inappropriate sense of motion, usually a spinning sensation. This meaning distinguishes it from "dizziness," a sensation of light-headedness and disorientation. Vertigo can result from abnormal conditions in, or stimulation of, the internal ear or from problems elsewhere along the sensory pathway that carries equilibrium sensations. It can accompany CNS or other infections, and many people experience vertigo when they have a high fever.

Any event that sets endolymph in motion can stimulate the equilibrium receptors and produce vertigo. Placing an ice pack in contact with the area over the mastoid process of the temporal bone or flushing the external acoustic canal with cold water may chill the endolymph in the outermost portions of the labyrinth and establish a temperature-related circulation of fluid. A mild and temporary vertigo is the result. The consumption of excessive quantities of alcohol or exposure to certain drugs can also produce vertigo by changing the composition of the endolymph or disturbing the hair cells of the internal ear. Other causes of vertigo include viral infection of the vestibular nerve and damage to the vestibular nucleus or its tracts. Acute vertigo can also result from damage caused by abnormal endolymph production, as in Ménière's disease. Probably the most common cause of vertigo is *motion sickness*.

The exceedingly unpleasant signs and symptoms of **motion sickness** include headache, sweating, flushing of the face, nausea, vomiting, and various changes in mental perspective. (Sufferers may go from a state of giddy excitement to almost suicidal despair in a matter of moments.) It has been suggested that the condition results when central processing stations, such as the tectum of the mesencephalon, receive conflicting sensory information. Why and how these conflicting reports result in nausea, vomiting, and other symptoms are not known. Sitting below decks on a moving boat or reading in a car or airplane tends to provide the necessary conditions. Your eyes (which are tracking lines on a page) report that your position in space is not changing, but your semicircular ducts report that your body is lurching and turning. To counter this effect, seasick sailors watch the horizon rather than their immediate surroundings, so that their eyes will provide visual confirmation of the movements detected by their internal ears. It is not known why some individuals are almost immune to motion sickness, whereas others find travel by boat or plane almost impossible. Visual and equilibrium receptors are both involved; roughly half of the astronauts on the space shuttle suffer from "space sickness" upon glancing out a window and seeing the Earth appearing at an unexpected angle. Fortunately for most of them, the nervous system adapts quickly, and space sickness is seldom a problem after a day or two in orbit.

Drugs commonly administered to prevent motion sickness include dimenhydrinate (*Dramamine*), scopolamine, and promethazine. These compounds appear to depress activity at the vestibular nuclei. Sedatives, such as prochlorperazine (*Compazine*), may also be effective. Scopolamine can be administered across the skin surface by using an adhesive patch (*Transderm Scop*).

In **Ménière's disease**, distortion of the membranous labyrinth of the internal ear by high fluid pressures may rupture the membranous wall and mix endolymph and perilymph. The receptors in the vestibule and semicircular canals then become highly stimulated. The individual may be unable to start a voluntary movement because he or she is experiencing intense spinning or rolling sensations. In addition to the vertigo, the person may "hear" unusual sounds as the cochlear receptors are activated and some episodic, fluctuating hearing loss may occur.

Hearing Deficits

Conductive deafness results from conditions in the outer or middle ear that block the normal transfer of vibrations from the tympanic membrane to the oval window. An external acoustic canal plugged with accumulated wax or trapped water can cause a temporary hearing loss. Scarring or perforation of the tympanic membrane and immobilization of one or more of the auditory ossicles are more serious causes of conductive hearing loss.

In **nerve deafness**, the problem lies within the cochlea or somewhere along the auditory pathway. The vibrations reach the oval window and enter the perilymph, but either the receptors cannot respond or their response cannot reach its central destinations. Causes of nerve deafness include the following:

- Very loud (high-intensity) sounds can produce nerve deafness by breaking stereocilia off the surfaces of the hair cells. (The reflex contraction of the tensor tympani and stapedius muscles in response to a dangerously loud noise occurs in less than 0.1 second, but this may not be fast enough to prevent damage.) As trauma occurs repeatedly over a lifetime, with every loud noise, high-frequency hearing loss gradually develops, and hearing losses are common among older individuals. Wearing ear protection around loud machinery, gunfire, and music is advised.

- Drugs such as the aminoglycoside antibiotics (*neomycin* and *gentamicin*) can diffuse into the endolymph and kill hair cells. Because hair cells and sensory nerves can also be damaged by bacterial infection, the potential side effects must be balanced against the severity of infection.

Testing and Treating Hearing Deficits

In the most common hearing test, an individual listens to sounds of varying frequency and intensity generated at irregular intervals. A record is kept of the responses, and the graphed record, or *audiogram*, is compared with that of an individual with normal hearing (❯ **Figure 47a**). The audiogram is a fairly precise test but can be time consuming to perform. **Bone conduction tests** are used to discriminate between conductive deafness and nerve deafness. If you put your fingers in your ears and talk quietly, you can still hear yourself because the bones of the skull conduct the sound waves to the cochlea, bypassing the middle ear. In a *Rinne test*, the examiner first places the handle of a vibrating tuning fork (256 or 512 Hz) against one mastoid process to test bone conduction. When the patient no longer hears the sound, the still-vibrating tuning fork is held in the air at the entrance to the external acoustic canal on that side. Normally the subject still hears a sound with air conduction (AC) after bone conduction (BC) ceases. In a *Weber test* the vibrating tuning fork (256 to 512 Hz) is held against the top of the skull in the center of the head. With normal hearing both ears hear the sound equally (normal Weber), and in both ears AC is better than BC (normal Rinne). If the Weber test sound is heard louder in one ear, known as lateralized to that side, and if the Rinne test on that ear shows AC > BC, there is hearing loss due to nerve damage on the opposite (non-lateralized) side. If the Weber lateralizes and the Rinne shows BC > AC on that side, there is conductive hearing loss on the lateralized side. Because patients may have difficulty describing sound strength, further, more precise testing by audiogram is indicated if the hearing loss is significant.

Several effective treatments exist for conductive deafness. A hearing aid overcomes the loss in sensitivity by increasing the volume of received sounds. Surgery may patch the tympanic membrane or free damaged or immobilized auditory ossicles. Artificial ossicles may also be implanted if the original ones are damaged beyond repair.

Few treatments are available for nerve deafness. Mild conditions may be overcome by the use of a hearing aid if some functional hair cells remain. In a **cochlear implant**, a small receiving device is inserted beneath the skin behind the mastoid process (❯ **Figure 47b**). Small wires are run through the round window to reach inside the cochlear nerve, and impulses stimulate different areas of the nerve, which "hears" a sound and stimulates the nerve directly. The implant receives information through the skin from an external battery-powered, magnetically attached device that includes a microphone, to detect sound, and a speech processor, which selects and arranges sounds into information that is transmitted through the skin to the implanted receiver. Increasing the number of wires (channels) and varying their implantation sites within the cochlea make it possible to create a number of different frequency sensations. Those sensations do not approximate normal hearing: There is as yet no way to target the specific afferent fibers responsible for the perception of a particular sound. Instead, a random assortment of afferent fibers are stimulated, and the individual learns to recognize the meaning and probable origin of the perceived sound. Congenitally deaf children given implants have learned normal speech. Older persons who have lost hearing and cannot hear with external hearing aids are also benefitting from cochlear implants. Multiple factors affect the performance of cochlear implants including age at implantation, existing oral or sign language skills, consistency of use, education, and device programming.

A new approach involves inducing the regeneration of hair cells of the organ of Corti. Researchers working with mammals other than humans have been able to induce hair cell regeneration both in cultured hair cells and in live animals. This is a very exciting area of research, and there is hope that it may ultimately lead to an effective treatment for human nerve deafness.

Nervous System

Nervous System

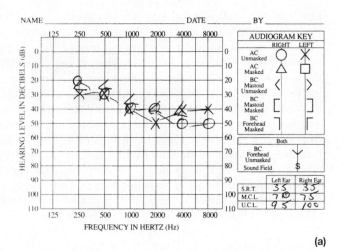

(a)

External Components
Transmitter system
Sound processor
Microphone

Internal Components
Implanted receiver
Electrode system

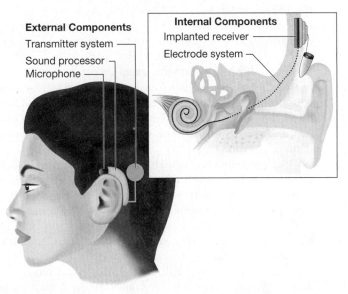

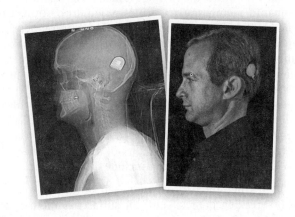

(b)

▶ **Figure 47** **Hearing Disorders. (a)** An audiogram. **(b)** A cochlear implant.

► The Endocrine System and Its Disorders

The endocrine system provides long-term regulation and adjustment of homeostatic mechanisms and a variety of body functions. For example, the endocrine system is responsible for the regulation of fluid and electrolyte balance, cell and tissue metabolism, growth and development, and reproductive functions. The endocrine system also assists the nervous system in responding to stressful stimuli.

The classic endocrine system is composed of nine major endocrine glands and several other organs, such as the heart and kidneys, that have nonendocrine functions as well. The hormones secreted by these endocrine organs are distributed by the circulatory system to target tissues throughout the body. Each hormone (from the Greek phrase *"to set in motion"*) affects a specific set of target tissues that may differ from those affected by other hormones. The selectivity is based on the presence or absence of hormone-specific receptors in the target cell's plasma membrane, cytoplasm, or nucleus. As researchers learn more about how cells interact, they are discovering that tissues contain a variety of molecules secreted by cells to affect their neighbors and coordinate tissue activities. The term *local hormones* has been used to describe these molecules. Although it was initially thought that their effects were limited to their tissues of origin, it is now clear that many have more widespread impact. The central nervous system and the gastrointestinal tract both produce many peptide hormones. This recognition of many more compounds with hormonal behavior and the multiple interactions of hormones with various organ systems is under active study. For the purposes of this text we will confine ourselves to the "traditional" hormones of the endocrine system.

A Classification of Endocrine Disorders

Homeostatic regulation of circulating hormone levels primarily involves negative feedback control mechanisms. The feedback loop features an interplay between the endocrine organ and its target tissues. An endocrine gland may release a particular hormone in response to one of three types of stimuli:

1. **Some hormones are released in response to variations in the concentrations of specific substances in body fluids.** Parathyroid hormone, for example, is released when calcium levels decline.

2. **Some hormones are released only when the gland cells receive hormonal instructions from other endocrine organs.** For example, the rate of production and release of triiodothyronine (T_3) and tetraiodothyronine (T_4, thyroxine) by the thyroid gland is controlled by thyroid-stimulating hormone (TSH) from the anterior pituitary gland (adenohypophysis). The secretion of TSH is in turn regulated by the release of thyrotropin-releasing hormone (TRH) from the hypothalamus.

3. **Some hormones are released in response to neural stimulation.** The release of epinephrine and norepinephrine from the adrenal medullae during sympathetic activation is an example.

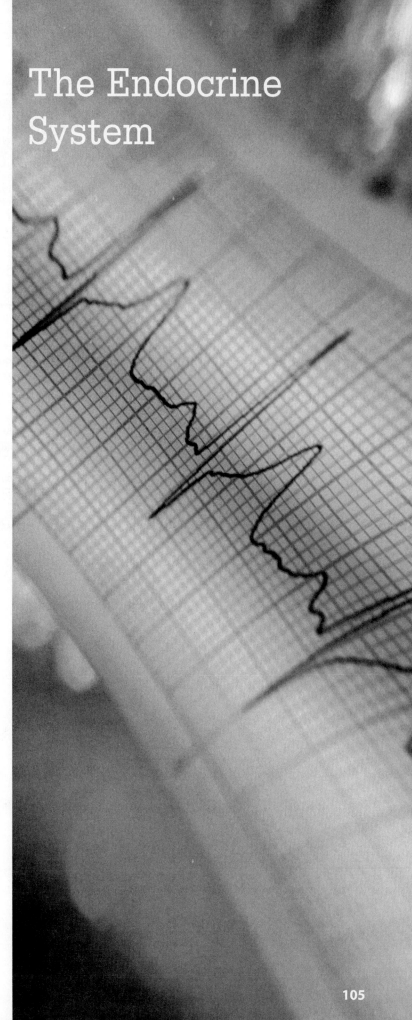

The Endocrine System

▶ Figure 48 Representative Disorders of the Endocrine System.

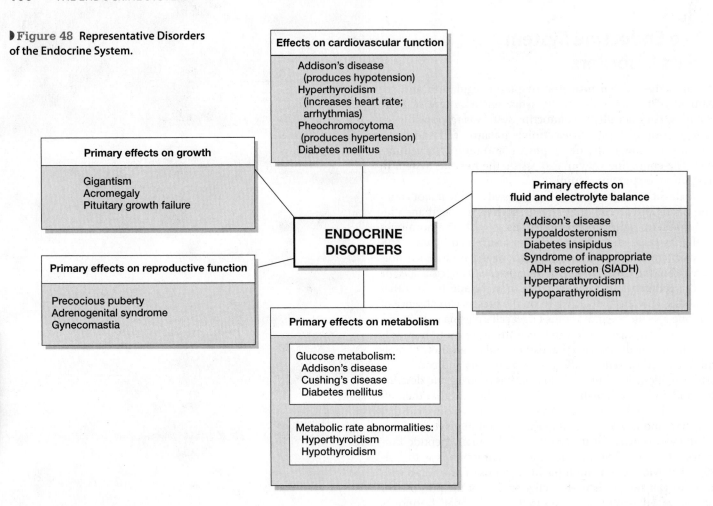

Effects on cardiovascular function

Addison's disease
(produces hypotension)
Hyperthyroidism
(increases heart rate;
arrhythmias)
Pheochromocytoma
(produces hypertension)
Diabetes mellitus

Primary effects on growth

Gigantism
Acromegaly
Pituitary growth failure

Primary effects on reproductive function

Precocious puberty
Adrenogenital syndrome
Gynecomastia

ENDOCRINE DISORDERS

Primary effects on fluid and electrolyte balance

Addison's disease
Hypoaldosteronism
Diabetes insipidus
Syndrome of inappropriate
ADH secretion (SIADH)
Hyperparathyroidism
Hypoparathyroidism

Primary effects on metabolism

Glucose metabolism:
Addison's disease
Cushing's disease
Diabetes mellitus

Metabolic rate abnormalities:
Hyperthyroidism
Hypothyroidism

Endocrine disorders can therefore develop due to abnormalities in the endocrine gland, the endocrine or neural regulatory mechanisms, or the target tissues. ▶ Figure 48 provides an overview of the major classes of endocrine disorders. In the discussions that follow, we will first consider *primary disorders* that originate in an endocrine gland itself and that may result in hormone overproduction (*hypersecretion*) or underproduction (*hyposecretion*). We will use the thyroid gland as an example.

Disorders Due to Endocrine Gland Abnormalities

Most endocrine disorders are the result of problems within the endocrine gland itself. Causes of hyposecretion include the following:

- **Metabolic factors.** Hyposecretion may result from a deficiency in some key substrate needed to synthesize the hormone in question. For example, primary hypothyroidism can be caused by inadequate dietary iodine levels or by exposure to drugs that inhibit iodine transport or utilization at the thyroid gland.

- **Physical damage.** Any condition that interrupts the normal circulatory supply or that physically damages the endocrine cells may cause them to become inactive immediately or after an initial surge of hormone release. If the damage is severe, the gland can become permanently inactive. For instance, temporary or permanent primary hypothyroidism can result from infection or inflammation of the gland (*thyroiditis*), from the

interruption of normal blood flow, or from exposure to radiation as part of treatment for cancer of the thyroid gland or adjacent tissues. The thyroid gland can also be damaged in an *autoimmune disorder* that results in the production of antibodies that attack and destroy normal follicle cells.

- **Congenital disorders.** An individual may be unable to produce normal amounts of a particular hormone because (1) the gland itself is too small, (2) the required enzymes are abnormal, (3) the receptors that trigger secretion are relatively insensitive, or (4) the gland cells lack the receptors normally involved in stimulating secretory activity.

Disorders Due to Endocrine or Neural Regulatory Mechanism Abnormalities

Endocrine disorders can result from problems with other endocrine organs involved in the negative feedback control mechanism. For example,

- **Secondary hypothyroidism** can be caused by inadequate TSH production at the pituitary gland or by inadequate TRH secretion at the hypothalamus.

- **Secondary hyperthyroidism** can be caused by excessive TRH or TSH production, such as the secondary hyperthyroidism that occurs in individuals with TSH-secreting tumors of the pituitary gland.

Disorders Due to Target Tissue Abnormalities

Endocrine abnormalities can also be caused by the presence of abnormal hormonal receptors in target tissues. In such a case, the gland involved and the regulatory mechanisms are normal, but the peripheral cells are unable to respond to the circulating hormone. Perhaps the best example of this type of abnormality is *type 2 diabetes*, in which peripheral cells do not respond normally to insulin.

Symptoms and Diagnosis of Endocrine Disorders

Knowledge of the individual endocrine organs and their functions makes it possible to predict the symptoms of specific endocrine disorders. For example, thyroid hormones increase basal metabolic rate, body heat production, perspiration, and heart rate. An elevated metabolic rate, increased body temperature, weight loss, nervousness, excessive perspiration, and an increased or irregular heartbeat are symptoms of hyperthyroidism. Conversely, a low metabolic rate, decreased body temperature, weight gain, lethargy, dry skin, and a reduced heart rate accompany advanced hypothyroidism.

The next step in the diagnosis of an endocrine disorder, after obtaining a patient's medical history, is the physical examination. Several disorders produce characteristic physical signs that reflect abnormal hormone activities. Examples include:

- **Cushing's disease and syndrome**, which result from an oversecretion of glucocorticoids by the adrenal cortex, either in response to pituitary hypersecretion of ACTH (disease) or from adrenocortical hyperactivity (syndrome). As the condition progresses, the normal pattern of fat distribution in the body shifts. Adipose tissue accumulates in the abdominal area, the lower cervical area (causing a "buffalo hump"), and the face (producing a "moonface"), while the limbs become relatively thin.

- **Acromegaly**, which results from the oversecretion of growth hormone in adults, usually from a pituitary tumor. In this condition, the facial features become distorted due to excessive cartilage and bone growth, and the lower jaw protrudes, a sign known as *prognathism*. The hands and feet also become enlarged.

- **Adrenogenital syndrome**, which results from the oversecretion of androgens by the adrenal glands in females. Hair growth patterns change to resemble that of males, and *hirsutism* (p. 47) develops.

- **Hypothyroidism**, which can produce a distinctively enlarged thyroid gland, or *goiter* as the thyroid gland fails to produce adequate amounts of thyroid hormones.

- **Hyperthyroidism**, which can also produce a goiter while making excess amounts of thyroid hormones and, for unknown reasons, cause protrusion of the eyes (*exophthalmos*).

These signs are very useful, but many other signs and symptoms related to endocrine disorders are less definitive. For example, *polyuria*, or increased urine production, can result from hyposecretion of ADH (*diabetes insipidus*) or the hyperglysuria caused by *diabetes mellitus*; a symptom such as hypertension (high blood pressure) can be caused by a variety of cardiovascular or endocrine problems. Diagnostic decisions are based on measurement of hormone concentrations, if available, and other tests, which can confirm the presence of an endocrine disorder by detecting abnormal levels of circulating hormones or metabolic products resulting from hormone action. Often followed by tests that determine whether the primary cause of the problem lies with the endocrine gland, the regulatory mechanism(s), or the target tissues. Often, it is a pattern of several different test results that leads to the diagnosis. **Table 26** provides an overview of important tests used in the diagnosis of endocrine disorders.

Growth Hormone Abnormalities

Growth hormone stimulates muscular and skeletal development. Until the epiphyseal cartilages close, it causes an increase in height, weight, and muscle mass. In extreme cases of GH overproduction, *gigantism* can result. In **acromegaly** (*akron*, great), an excessive amount of GH is released after puberty, when most of the epiphyseal cartilages have already closed. Cartilages and small bones respond to the hormone, however, resulting in abnormal growth of the hands, feet, lower jaw, skull, and clavicle. ❱**Figure 49** shows an acromegalic individual with prominent brows and mandible.

Children who are unable to produce adequate concentrations of GH have *pituitary growth failure*. The steady growth and maturation that typically precede and accompany puberty do not occur in these individuals, who have short stature, slow epiphyseal growth, and larger-than-normal adipose tissue reserves.

Normal growth patterns can be restored by the injected administration of GH. Before the advent of gene splicing and recombinant DNA techniques, GH had to be carefully extracted and purified from the pituitary glands of cadavers, at considerable expense and risk of infectious disease. Now genetically manipulated bacteria are used to produce human GH in commercial quantities. The current availability of purified human growth hormone has led to its use under medically questionable circumstances. GH is considered a performance-enhancing drug and banned by WADA (the World Anti-Doping Agency) for competitive athletes.

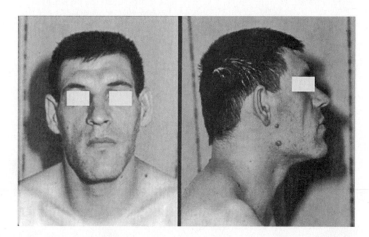

❱**Figure 49 Acromegaly.** Acromegaly results from the overproduction of growth hormone after the epiphyseal cartilages have closed. Bone shapes change, and cartilaginous areas of the skeleton enlarge. Notice the broad facial features with enlarged brows and lower jaw.

Endocrine System

Table 26	Representative Diagnostic Procedures for Disorders of the Endocrine System	
Diagnostic Procedure	**Method and Result**	**Representative Uses**
Pituitary Gland		
Wrist and hand x-ray film	Standard x-rays of epiphyseal cartilages for estimation of "bone age," based on the time of closure of epiphyseal cartilages	Compares a child's bone age and chronological age; a bone age greater than two years behind the chronological age suggests possible growth hormone deficiency with hypopituitarism or pituitary growth failure
X-ray study of sella turcica	Standard x-ray of the sella turcica, which houses the pituitary gland	Determine with increasing accuracy (and cost) the size of the pituitary gland; detect pituitary tumors
CT scan of pituitary gland	Standard cross-sectional CT; contrast media may be used	
MRI of pituitary gland	Standard MRI	
Thyroid Gland		
Thyroid scanning	A dose of radionucleotide accumulates in the thyroid, giving off detectable radiation captured to create an image of the thyroid	Determines size, shape, and abnormalities of the thyroid gland; detects presence of nodules and/or tumors: hyperactive or hypoactive areas; may determine cause of a mass in the neck
Ultrasound examination of thyroid	Sound waves reflected off internal structures are used to generate a computer image	Detects thyroid cysts or tumors, enlarged lymph nodes, or abnormalities in the shape or size of the thyroid gland
Radioactive iodine uptake (RAIU) test	Radioactive iodine is ingested and trapped by the thyroid; detector determines the amount of radioiodine taken up over a period of time	Determines hyperactivity or hypoactivity of the thyroid gland. Frequently done at same time as thyroid scan
Parathyroid Glands		
Ultrasound examination of parathyroid glands	Standard ultrasound	Determines structural abnormalities of the parathyroid gland, such as enlargement
Adrenal Glands		
Ultrasound of adrenal gland	Standard ultrasound	Determines abnormalities in adrenal gland size or shape; may detect tumors
CT scan of adrenal gland	Standard cross-sectional CT	Determines abnormalities in adrenal gland size or shape; may detect tumor (pheochromocytoma)
Adrenal angiography	Injection of radiopaque dye for examination of the vascular supply to the adrenal gland	Detects tumors and hyperplasia
Laboratory Test	**Normal Values in Blood Plasma or Serum**	**Significance of Abnormal Values**
Pituitary Gland		
Growth hormone (GH)	0.5–17 ng/mL	<10 ng/mL on stimulation for GH suggests deficiency (found in hypopituitarism and pituitary growth failure)
Plasma ACTH	Adults, morning: 20–80 pg/mL Adults, late afternoon: 10–40 pg/mL	Elevated levels of ACTH could indicate hypersecretion from either a pituitary or other tumor or adrenocortical deficiency (Addison's disease) with a compensatory elevation of ACTH; decreased levels suggest pituitary hyposecretion, or adrenocortical hyperfunction (Cushing's syndrome), or carcinoma of the adrenal gland
Serum TSH	Adults: <0.4–5.0 mU/mL	Elevated levels indicate pituitary hypersecretion of TSH or primary hypothyroidism; decreased levels suggest pituitary hyposecretion of TSH or hyperthyroidism
Serum LH	Premenopausal females: 3–30 mLU/mL Females, mid-menstrual cycle: 30–100 mLU/mL	Mid-menstrual cycle peak triggers ovulation; home urine tests can predict ovulation; decreased levels occur in pituitary hyposecretion or adrenal tumors
Laboratory Test	**Normal Values in Blood Plasma or Serum**	**Significance of Abnormal Values**
Thyroid Gland		
Free serum T_4	Adults: 1.0–2.3 ng/dL	Elevated levels occur in hyperthyroidism; decreased levels occur in hypothyroidism

(continued)

Table 26	Representative Diagnostic Procedures for Disorders of the Endocrine System (*Continued*)	
Diagnostic Procedure	**Method and Result**	**Representative Uses**
Calcitonin (serum)	Adult males: 3–26 pg/mL Adult females: 2–27 pg/mL	Elevated levels occur in carcinoma of the thyroid gland
Antithyroglobulin antibodies	Negative to 1:20	High titers occur in autoimmune disorders such as Graves disease
TSH	0.4–6 mU/mL	If the TSH is high and T_4 is low, primary thyroid gland hypothyroidism is diagnosed; if TSH and T_4 are low, hypothyroidism secondary to pituitary hyposecretion is diagnosed If TSH is low and T_4 is high, primary thyroid gland hyperthyroidism is diagnosed; if TSH and T_4 are high, hyperthyroidism secondary to pituitary hypersecretion is diagnosed
Parathyroid Glands		
Serum parathyroid hormone (PTH)	Adults: 10–60 pg/mL	Increased levels occur in hyperparathyroidism and cause hypercalcemia; decreased levels occur in hypoparathyroidism
Serum phosphorus	Adults: 3–4.5 mg/dL	Increased levels occur in hypoparathyroidism; decreased levels occur in hyperparathyroidism
Serum calcium	Adults: 8.5–10.5 mg/dL	Decreased levels occur in hypoparathyroidism; increased levels occur in hyperparathyroidism
Adrenal Glands		
Plasma cortisol	Adults, morning: 5–23 mg/dL Adults, afternoon: 3–13 mg/dL	Increased levels occur in adrenal hypersecretion (Cushing's syndrome) from a pituitary or other tumor, stress, and steroid use; decreased levels in Addison's disease (adrenocortical deficiency) and pituitary hypofunction
Serum aldosterone	Adults: 4–30 ng/dL supine: decreased <1 ng/dL, elevated >9 ng/dL	Increased levels occur in dehydration, hyperactivity of adrenal glands, and hyponatremia; decreased levels occur in hypernatremia and adrenal hypoactivity
Serum sodium	Adults: 135–145 mEq/L	Increased levels occur in dehydration; decreased levels occur with adrenocortical insufficiency
Serum potassium	Adults: 3.5–5.0 mEq/L	Increased levels occur with hypoactivity of adrenal glands and hypoaldosteronism; decreased levels occur with hyperactivity of adrenal glands and aldosteronism
Urine hydroxycorticosteroid (detects breakdown products of cortisol)	Adults: 2–12 mg/24 hour	Increased levels occur in Cushing's syndrome; decreased levels occur in Addison's disease
Urine ketosteroids (detects metabolites of androgens)	Adults: 5–25 mg/24 hour	Increased levels occur in adrenal hyperactivity and pituitary hypersecretion; decreased levels occur in adrenocortical deficiency and pituitary hyposecretion
Pancreas		
Fasting glucose (plasma)	<99 mg/dL 100–125 mg/dL >126 mg/dL	Normal Impaired fasting glucose (pre-diabetes) Diabetes
Oral glucose tolerance test	<139 mg/dL 140–199 mg/dL >200 mg/dL	Normal (2-hour duration) Impaired fasting glucose (pre-diabetes) Diabetes
Serum insulin	Adults: 5–25 mU/mL	Increased levels occur in early type 2 diabetes and obesity; decreased levels occur in type 1 diabetes
Glycosylated hemoglobin (Alc)	Adults: 3.8%–6.4%	Increased levels occur in diabetes mellitus and are associated with serious chronic health complications

Endocrine System

Endocrine System

GH is now being praised as an "antiaging" miracle cure. Although GH injections do slow or even reverse the losses in bone and muscle mass that accompany aging, continuous use is required to maintain these benefits, and exercise training was more effective in increasing muscle function. Little is known about adverse side effects that may accompany long-term use of the hormone in adults, but concerns about increases in diabetes and cancer have been raised. GH is also being sought by some parents of short but otherwise healthy children. These parents view short stature as a handicap that merits treatment, and what height is considered "too short" varies with the local population norms. Whether we are considering GH treatment of adults or children, it is important to remember that GH and the somatomedins affect many different tissues and have widespread metabolic effects. For example, children exposed to GH may grow faster, but their body fat content declines drastically, and sexual maturation is delayed. The decline is associated with metabolic changes in many organs. The range and significance of these metabolic side effects are now the subjects of long-term studies. Since growth hormone is a protein that is digested and rendered ineffective if taken orally, diet supplements advertised as supplying or "boosting" growth hormone are probably (and fortunately) ineffective.

Thyroid Gland Disorders

Normal production of thyroid hormones controls the background rate of cellular metabolism. These hormones exert their primary effects on metabolically active tissues and organs, including skeletal muscles, the liver, and the kidneys. The inadequate production of thyroid hormones is called **hypothyroidism**.

Hypothyroidism typically results from some problem that involves the thyroid gland, rather than a problem with the pituitary production of TSH. In primary hypothyroidism, TSH levels are elevated, because the pituitary gland attempts to stimulate thyroid activity, but levels of T_3 and T_4 are depressed. In a fetus or infant, hypothyroidism produces *infantile hypothyroidism* (▶ Figure 50a), a condition marked by inadequate skeletal and nervous system development and a metabolic rate as much as 40 percent below normal levels. Unless detected (all 50 U.S. states screen newborns) and treated soon after birth, significant mental retardation may occur. Prior to birth, normal maternal thyroid hormones protect the fetus. The condition affects approximately 1 birth out of every 5000. Hypothyroidism developing later in childhood retards growth and delays puberty, but has less affect on mental development.

The signs and symptoms of adult hypothyroidism, collectively known as *myxedema* (miks-e-DĒ-muh), include subcutaneous swelling, dry skin, hair loss, low body temperature, muscular weakness, and slowed reflexes. Adults with hypothyroidism are lethargic and unable to tolerate cold temperatures. Hypothyroidism may also be associated with enlargement of the thyroid gland, producing a distinctive swelling called a goiter (▶ Figure 50b). The enlargement generally indicates an increased thyroid follicle size, but thyroxine release may be increased or decreased, depending on the cause of the goiter. Most goiters develop when the thyroid gland is unable to synthesize and release adequate amounts of thyroid hormones, and grows under increased TSH stimulation. Thyroglobulin production accelerates and the thyroid follicles enlarge. One type of goiter occurs if the thyroid fails to obtain enough iodine to meet its synthetic requirements. Goiters from inadequate dietary iodide are very rare in the United States, in part due to the addition of iodine to table salt, but these conditions can be relatively common in some countries, especially landlocked ones (seafood is a good source of iodine). Administering iodine may not solve the problem entirely: The sudden availability of iodine can temporarily produce symptoms of hyperthyroidism as the stored thyroglobulin becomes available.

In the absence of iodine deficiency, the usual therapy for hypothyroidism and/or goiter involves the administration of synthetic thyroid hormone, T_4 (thyroxine), which has a negative feedback effect on the hypothalamus and pituitary gland, thus inhibiting the production of TSH. Over time, the resting thyroid may return to its normal size. Treatment of chronic hypothyroidism, such as the hypothyroidism that follows radiation exposure or autoimmune thyroiditis, generally involves the administration of thyroxine to maintain normal blood concentrations.

Hyperthyroidism, or *thyrotoxicosis*, occurs when thyroid hormones are produced in excessive quantities, and may be associated with a slight goiter (enlargement of the thyroid). The metabolic

▶ **Figure 50** **Thyroid Disorders. (a)** Infantile hypothyroidism results from thyroid hormone insufficiency in infancy. **(b)** An enlarged thyroid gland, or goiter, can be associated with thyroid hyposecretion due to iodine insufficiency.

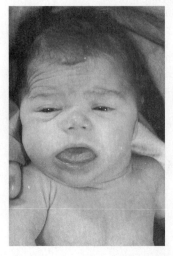

(a) (b)

rate climbs, and the skin may be flushed and moist with perspiration. Blood pressure and heart rate increase, and the heartbeat may become irregular. The effects on the central nervous system make the individual restless, excitable, and subject to insomnia and shifts in mood and emotional states. Despite the drive for increased activity, the person has limited energy and fatigues easily. *Graves disease* is a form of hyperthyroidism that develops when antibodies are produced that attack the thyroid gland. This autoimmune condition results in the release of excessive amounts of thyroid hormones, accompanied by goiter formation and other signs of hyperthyroidism. Protrusion of the eyes, or **exophthalmos** (eks-ahf-THAL-mōs), may appear with inflammation and increased fatty tissue behind the eye. Graves disease has a genetic autoimmune basis and affects many more women than it does men. Treatment may involve the use of antithyroid drugs, the surgical removal of portions of the thyroid, or the destruction of part of the gland by ingesting radioactive iodine.

Hyperthyroidism may also result from other causes of thyroid inflammation or, rarely, thyroid tumors. In extreme cases, the individual's metabolic processes accelerate out of control. During a *thyrotoxic crisis*, or "thyroid storm," the individual experiences a high fever, a rapid heart rate, and the dangerous malfunctioning of a variety of physiological systems.

Disorders of Parathyroid Function

When the parathyroid glands secrete inadequate or excessive amounts of parathyroid hormone, calcium concentrations move outside normal homeostatic limits. Inadequate parathyroid hormone production, a condition called **hypoparathyroidism**, leads to low Ca^{2+} concentrations in body fluids. The most obvious symptoms involve neural and muscle tissues: The nervous system becomes more excitable, and the affected individual may experience *hypocalcemic tetany*, a dangerous condition characterized by prolonged muscle spasms that initially involve the limbs and face.

Hypoparathyroidism with hypocalcemia can develop after neck surgery, especially a thyroidectomy, if the parathyroid glands are damaged. In many other cases, the primary cause of the condition is uncertain. Parathyroid hormone (PTH) is extremely costly and is not used to treat hypoparathyroidism. As an alternative, a dietary combination of vitamin D and calcium is used to elevate body fluid calcium concentrations, because vitamin D stimulates the absorption of calcium ions across the lining of the digestive tract.

In **hyperparathyroidism**, Ca^{2+} concentrations become abnormally high. Calcium salts in the skeleton are mobilized, and over time bones are weakened. On x-rays, the bones have a light, airy appearance, because the dense calcium salts no longer dominate the tissue. Central nervous system (CNS) function is depressed, thinking slows, memory is impaired, and the individual often experiences emotional swings and depression. Nausea and vomiting occur, and in severe cases the person becomes comatose. Muscle function deteriorates, and skeletal muscles become weak. Other tissues are typically affected as calcium salts crystallize in joints, tendons, and the dermis; calcium deposits may produce masses called *kidney stones*, which block filtration and conduction passages in the kidneys and urinary tract. Hyperparathyroidism most commonly results from a tumor of the parathyroid gland. Treatment involves the surgical removal of the overactive tissue. Fortunately, humans have four parathyroid glands, and the

secretion of even a portion of one gland can maintain normal calcium concentrations.

Disorders of the Adrenal Cortex

Clinical problems related to the adrenal gland vary with the zone involved. The problems may result from changes in the functional capabilities of the adrenal cells (primary conditions) or disorders that affect the regulatory mechanisms (secondary conditions).

In **hypoaldosteronism**, the zona glomerulosa fails to produce enough aldosterone, generally either as an early sign of adrenal insufficiency or because the kidneys are not releasing adequate amounts of renin. Low aldosterone levels lead to excessive losses of water and sodium ions at the kidneys, and the water loss in turn leads to low blood volume and a fall in blood pressure. The resulting changes in electrolyte concentrations, including *hyperkalemia* (high extracellular K^+ levels), affect transmembrane potentials, eventually causing dysfunctions in neural and muscular tissues.

Hypersecretion of aldosterone results in **aldosteronism**, or *hyperaldosteronism*. Under continued aldosterone stimulation, the kidneys retain sodium ions in exchange for potassium ions (K^+) that are lost in urine. Hypertension and hypokalemia occur as extracellular potassium levels decline, increasing the concentration gradient for potassium ions across plasma membranes. This increase leads to an acceleration in the rate of potassium diffusion out of the cells and into interstitial fluids. The reduction in intracellular and extracellular potassium levels eventually interferes with the function of excitable membranes, especially cardiac muscle cells, neurons, and kidney cells.

Addison's disease or adrenocortical deficiency, (▶Figure 51a) results from inadequate stimulation of the zona fasciculata by the pituitary hormone ACTH (adrenocorticotropic hormone) or, more commonly, from the inability of the adrenal cells to synthesize the necessary hormones, generally from adrenal cell loss caused by autoimmune problems or infection. Affected individuals produce insufficient levels of glucocorticoids and mineralocorticoids. They become weak and lose weight, owing to a combination of appetite loss, hypotension, and hypovolemia. They cannot adequately mobilize energy reserves, and their blood glucose concentrations may fall sharply within hours after a meal. Stresses cannot be tolerated, and a minor infection or injury can lead to a sharp and even fatal decline in blood pressure. A particularly interesting sign of adrenal insufficiency is the increased production of the pigment melanin in the skin. The ACTH molecule and the melanocyte-stimulating hormone (MSH) molecule are similar in structure, and at high concentrations ACTH stimulates the MSH receptors on melanocytes.

Addison's disease has vague symptoms and diagnosis may require measuring the amount of cortisol produced in response to injecting synthetic ACTH. Treatment consists of replacement corticosteroid and mineralocorticoid drugs (cortisone, prednisone, fludrocortisone, and others). However, chronic use or higher doses of corticosteroids to treat inflammatory conditions, such as rheumatoid arthritis or asthma, carries the risk of suppressing ACTH secretion and *causing* a secondary form of Addison's disease if the corticosteroid treatment is stopped. President John F. Kennedy had Addison's disease, possibly from corticosteroid treatment of another disease. For this reason corticosteroids are used for these conditions only when they are unresponsive to other treatments, and tapered, rather than stopped abruptly, after long-term use.

Endocrine System

▶Figure 51 **Adrenal Abnormalities. (a)** Addison's disease is caused by hyposecretion of adrenal corticosteroids, especially glucocorticoids. Pigment changes result from stimulation of melanocytes by ACTH, which is structurally similar to MSH. **(b)** Cushing's syndrome is caused by hypersecretion of glucocorticoids. Lipid reserves are mobilized and adipose tissue accumulates in the cheeks and at the base of the neck.

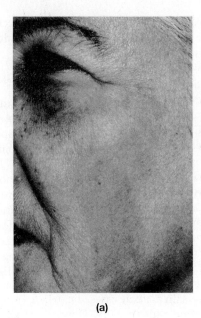

(a)

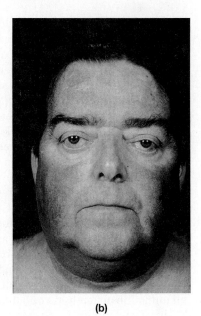

(b)

Cushing's syndrome (▶Figure 51b) results from adrenal overproduction of glucocorticoids. Hypersecretion of ACTH either by the pituitary or other tumor is the usual cause. The symptoms resemble those of a protracted and exaggerated response to stress. Glucose metabolism is suppressed, lipid reserves are mobilized, and peripheral proteins are broken down. Lipids and amino acids are mobilized in excess of the existing demand. The energy reserves are shuffled around, and the distribution of body fat changes. Adipose tissues in the cheeks and around the base of the back of the neck ("buffalo hump") become enlarged at the expense of other areas, producing a "moon-faced" appearance. The demand for amino acids falls most heavily on the skeletal muscles, which respond by breaking down their contractile proteins. This response reduces muscular power and endurance. The skin becomes thin and may develop *striae*, or stretch marks.

If the primary cause of Cushing's disease is ACTH oversecretion at the anterior lobe of the pituitary gland, the most common source is a *pituitary adenoma* (a benign tumor of glandular origin). Microsurgery can be performed through the nose and the sphenoid bone to remove the adenomatous tissue. Some oncology centers use pituitary radiation rather than surgery. Several pharmacological therapies act at the hypothalamus, rather than at the pituitary gland, to prevent the release of corticotropin-releasing hormone (CRH), which stimulates ACTH secretion. The drugs used are serotonin antagonists, gamma-aminobutyric acid (GABA) transaminase inhibitors, or dopamine agonists. Alternatively, a bilateral adrenalectomy (the removal of the adrenal glands) can be performed, but further complications may arise as the adenoma in the pituitary gland enlarges. Cushing's syndrome also results from the production of ACTH outside the pituitary gland; for example, the condition may develop with one form of lung cancer (oat cell carcinoma). The removal of the causative tumor in some cases relieves the symptoms.

As mentioned in the discussion of Addison's disease, the chronic administration of large doses of steroids is sometimes required to treat severe asthma, arthritis, and certain cancers or to prevent transplanted organs from being rejected. Prolonged use of such large doses can produce symptoms similar to those of Cushing's syndrome, which provides another reason to avoid such treatment if at all possible or to use the lowest dose for the shortest time possible.

The zona reticularis ordinarily produces a negligible amount of androgens. If a tumor forms there, androgen secretion may increase dramatically, producing symptoms of **adrenogenital syndrome**. In women, this condition leads to the gradual development of male secondary sex characteristics, including body and facial hair patterns, adipose tissue distribution, and muscle development. Tumors affecting the zona reticularis of males can result in the production of large quantities of estrogens. Affected males develop enlarged breasts (called **gynecomastia**; *gynaikos*, woman + *mastos*, breast), and in some cases, other female secondary sex characteristics.

Disorders of the Adrenal Medulla

The overproduction of epinephrine by the adrenal medullae may reflect chronic sympathetic activation. A **pheochromocytoma** (fē-ō-krō-mō-sī-TŌ-muh) is a tumor that produces catecholamines in massive quantities. The tumor generally develops within an adrenal medulla, but may also involve other sympathetic ganglia. The most dangerous symptoms are episodes of rapid and irregular heartbeat and high blood pressure; other symptoms include uneasiness, sweating, blurred vision, and headaches. The condition is rare, and surgical removal of the tumor is the most effective treatment.

Light and Behavior

Exposure to sunlight can do more than induce a tan or promote the formation of vitamin D_3. Evidence indicates that daily light–dark cycles have widespread effects on the central nervous system, with melatonin playing a key role. Several studies have indicated that residents of temperate and higher latitudes in the Northern Hemisphere undergo seasonal changes in mood and activity patterns. These people feel most energetic from June through September, and they experience relatively low spirits from December through March. (The opposite situation occurs in the Southern Hemisphere, where winter and summer are reversed relative to the Northern Hemisphere.) The degree of seasonal variation differs from individual to individual: Some people display no symptoms; other people are affected so severely that they seek medical attention. The observed symptoms are called **seasonal affective disorder (SAD)**. Individuals with SAD experience depression and lethargy and have difficulty concentrating. They tend to sleep for long periods, perhaps 10 hours or more a day. They may also go on eating binges and crave carbohydrates.

Melatonin secretion appears to be regulated by exposure to sunlight, not simply by exposure to light. Normal interior lights are apparently not strong enough or do not release the right mixture of light wavelengths to depress melatonin production. Because many people spend very little time outdoors in the winter, melatonin production increases then; the depression, lethargy, and concentration problems appear to be linked to elevated melatonin levels in blood. In experiments, comparable symptoms can be produced in a healthy individual by an injection of melatonin.

Many individuals with SAD are successfully treated by exposure to bright lights that produce full-spectrum light. Experiments are under way to define exactly how intense the light must be and to determine the minimal effective time of exposure. Some people have been using melatonin obtained (in varying doses and purity) from health-food stores to treat insomnia and jet lag. Because the health-food market is unregulated and few, if any, controlled studies have been done, it remains unclear whether melatonin is truly an effective therapy for sleep disorders or whether it aggravates SAD and depression.

Diabetes Mellitus

Diabetes mellitus is a condition caused by inadequate production of, or sensitivity to, insulin. In the absence of insulin, blood glucose levels skyrocket, yet peripheral tissues become glucose-starved because they are unable to transport glucose into the cytoplasm. The two forms of diabetes mellitus are type 1 (5 percent of cases) and type 2 (90–95 percent of cases). Type 2 diabetes affects almost 8 percent of the U.S. population (24 million people), and the incidence is increasing as obesity becomes more common. Precise glucose control has been shown to reduce and delay the onset of the serious chronic complications of diabetes. These complications include accelerated coronary artery disease, kidney failure, and microvascular complications that are the leading cause of blindness and foot amputations in the United States. The most common examples include the following:

1. Vascular changes at the retina, including proliferation of capillaries and hemorrhaging, often causing partial or total blindness. This condition is called **diabetic retinopathy**.

2. Changes occur in the clarity of the lens, producing cataracts.

3. Small hemorrhages and inflammation at the microvasculature of the kidneys cause degenerative changes that can lead to kidney failure. This condition is called **diabetic nephropathy**.

4. A variety of neural problems appear, including nerve palsies, paresthesias, and autonomic dysfunction, including male erectile dysfunction. These disorders, collectively termed **diabetic neuropathy**, are probably related to disturbances in the blood supply to neural tissues.

5. Degenerative changes in cardiac circulation can lead to accelerated coronary artery disease and early heart attacks. For a given age group, heart attacks are three to five times more likely in diabetic individuals.

6. Other peripheral changes in the vascular system can disrupt normal circulation to the limbs. For example, a reduction in blood flow to the feet can lead to tissue death, ulceration, infection, and loss of toes or a major portion of one or both feet.

Insulin-Dependent Diabetes Mellitus

Type 1 diabetes most commonly occurs in individuals under 40 years of age. Because it often appears in childhood, and usually requires insulin treatment, it has been called **juvenile-onset diabetes** and **insulin-dependent diabetes mellitus (IDDM)**. The primary cause of **type 1 diabetes** is the autoimmune destruction of the beta cells of the pancreatic islets. It is thought to be caused by an infectious or environmental trigger in a genetically susceptible person. In most cells, glucose transport cannot occur in the absence of insulin. When insulin concentrations decline, cells can no longer absorb glucose; tissues remain glucose starved, despite the presence of adequate or even excessive amounts of glucose in the bloodstream. After a meal rich in glucose, blood glucose concentrations may become so elevated that the kidney cells cannot reclaim all the glucose molecules that enter the urine. The high urinary concentration of glucose limits the ability of the kidneys to conserve water, so the individual urinates frequently and may become dehydrated. The chronic hyperglycemia and dehydration leads to disturbances of neural function (blurred vision, tingling sensations, disorientation, and fatigue) and muscle weakness.

Despite high blood concentrations, glucose cannot enter endocrine tissues, and the endocrine system responds as if glucose were in short supply. Pancreatic alpha cells release glucagon, and glucocorticoid production accelerates. Peripheral tissues then break down lipids and proteins to obtain the energy needed to continue functioning. The breakdown of large numbers of fatty acids promotes the generation of molecules called **ketone bodies**. These small molecules are metabolic acids whose accumulation in large numbers can cause a dangerous reduction in blood pH. This condition, called **ketoacidosis**, may trigger vomiting. In severe cases, it can progress to coma and death.

If the individual survives (an impossibility without insulin therapy), long-term treatment involves a combination of dietary control, monitoring of blood glucose levels several times a day, and the administration of insulin. The treatment is complicated by the fact that tissue glucose demands vary with food eaten, physical activity, emotional state, stress, and other factors that are hard to assess or predict. Dietary control, including the regulation of the type of food, time of meals, and amount consumed, can help reduce oscillations in blood glucose levels.

Modern insulin comes in many forms, with varying durations of activity. Single or, more commonly, multiple injections throughout the day are guided by blood glucose measurements in an attempt to approach normal homeostatic glucose regulation. At present, this testing involves needle sticks and finger pricks, but visual glucose monitors are now available and multiple insulin injections can be avoided by using an **insulin pump** connected by tubing and a small catheter and needle to the subcutaneous tissue which can give a continuous infusion. A powdered insulin inhaled through the mouth for absorption in the lungs is now FDA approved, and nasal sprays, pills, and insulin patches are being researched. However, it remains difficult to maintain stable and normal blood glucose levels over long periods, even with an insulin pump.

Since 1990, pancreas transplants have been used to treat diabetes in the United States. The procedure is generally limited to gravely ill patients already undergoing kidney transplantation. The graft success rate over five years is roughly 50 percent, and the procedure is controversial. Pancreatic islet transplantation has recently shown promise, but it requires a large number of islets (two cadaver organs are needed to serve as source), and, as with all transplants, suppression of the immune system to prevent rejection is necessary.

Because of its autoimmune nature, attempts have been made to prevent type 1 diabetes with *azathioprine* (*Imuran*), a drug that suppresses the immune system. This treatment is potentially dangerous, however, because compromising immune function indefinitely increases the risk of acquiring serious infections or of developing cancer.

Non-Insulin-Dependent Diabetes Mellitus

Non-insulin-dependent diabetes mellitus (NIDDM), or **type 2 diabetes**, until recently usually affected obese individuals over 40 years of age. Because of the age factor, this condition is also called **maturity-onset diabetes**, although with more childhood obesity, significant numbers of adolescents are developing the disease. Approximately 2 million new cases are diagnosed each year in the United States alone.

In type 2 diabetes, insulin levels are normal or elevated, but peripheral tissues no longer respond normally. Because most patients are overweight, treatment consists of weight loss and dietary restrictions that may elevate insulin production and tissue response. The drug *metformin* (*Glucophage*) lowers plasma glucose concentrations, primarily by reducing glucose synthesis and release at the liver. The use of metformin in combination with other drugs that affect glucose metabolism promises to improve the quality of life for many type 2 diabetes patients.

A diagnosis of diabetes mellitus is based on two observations: a high fasting blood glucose and the persistence of an elevated blood glucose level two hours after drinking a fixed amount of glucose. These criteria have largely replaced the six-hour glucose tolerance test, which involved having the patient drink 75 to 100 grams of glucose and then testing the blood glucose level multiple times over four to six hours. Careful control to avoid high glucose levels and to keep the long-term marker (hemoglobin A1C) at a low level has been shown to reduce the risk of chronic kidney, eye, and cardiovascular complications.

END-OF-UNIT CLINICAL PROBLEMS

1. Jane, 33, awakens one morning with blurred vision and pain in her right eye. She sees her family physician, who determines that her sclera, conjunctiva, cornea, and ocular tension are normal but that her visual acuity is markedly reduced in that eye. An ophthalmologist is consulted and, after an exam, diagnoses optic neuritis. Jane is then referred to a neurologist, who learns that she had an earlier, previously forgotten episode of weakness and clumsiness of her left arm. That problem persisted for only two days, and Jane attributed it to lifting heavy suitcases on a trip. An MRI of her brain reveals several plaque-type lesions in the white matter of the brain but no other abnormalities. A lumbar puncture is performed, and the results are as follows:

   Pressure: 150 mm H_2O

   Color: clear, colorless

   Glucose: 50 mg/dL

   Protein: 50 mg/dL

   Cells: no RBCs, WBC count of 6/mm³

   IgG: high ratio of IgG to other proteins

   Culture: no bacteria

   What is the preliminary diagnosis?

2. Sandra is 50 years old. She has gained weight over the last year but has felt too tired and weak to exercise. She takes no medicine, but has tried taking vitamins with no response. During a physical examination, her physician notes the following signs:

   rounded "moon-shaped" face

   obesity of the trunk

   slender limbs

   muscular weakness

   blood pressure: 140/95 (normal = 120/80)

   On the basis of these signs, which of the following disorders would you suspect?

   a. Addison's disease

   b. Cushing's syndrome

   c. pheochromocytoma

   d. hypoaldosteronism

   Diagnostic and laboratory tests are ordered, and an appointment with an endocrinologist is arranged. A few of the pertinent test results are listed below:

   Plasma cortisol levels: 40 g/dL (taken at 7:00 a.m.)

   Plasma ACTH: 100 pg/mL (taken at 7:00 a.m.)

   X-ray of skull: erosion of the sella turcica

   MRI of pituitary gland: abnormal mass detected

   What is the probable cause of Sandra's disorder?

3. Mr. Johnson noticed pain on the left side of his chest under his arm two days after he cleaned out his garage. He thought it was from the lifting he had done and expected it to fade in a day or two. The next day the pain was stronger, and he couldn't sleep well that night. In the morning his wife noticed a red, bumpy rash in two patches in the area of pain. When he saw his doctor that afternoon, there were clear blisters in the rash area and more red patches were appearing. All of the patches

were in a continuous band, three inches wide, that extended from the posterior midline across the left side of his back and onto the anterior chest, roughly following the curve of the ribs. The doctor asked about childhood illnesses and confirmed that Mr. Johnson once had chicken pox. The physician then prescribed a medication.

What does the location and pattern of the rash suggest? What anatomical structures are involved? What is the probable diagnosis?

NOTES:

Answers to these problems can be found on page 241.

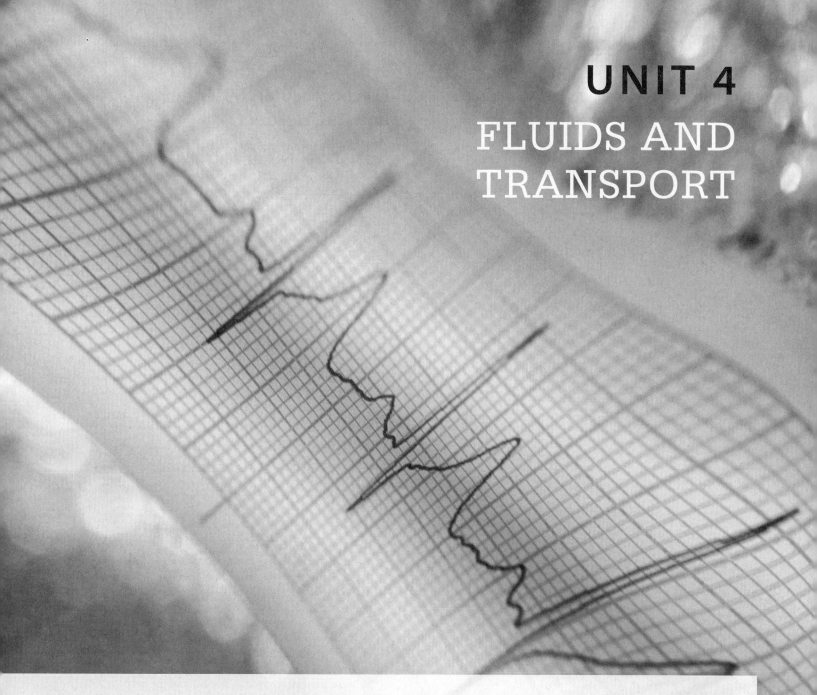

UNIT 4
FLUIDS AND TRANSPORT

The body is in constant chemical communication with its external environment. Nutrients are absorbed across the lining of the digestive tract, gases move across the delicate epithelial lining of the lungs, and wastes are excreted in the feces and urine. These chemical exchanges occur at specialized sites, yet they may affect every cell, tissue, and organ in a matter of moments. The cardiovascular system provides an efficient internal collection and distribution network.

All environmental interactions involve a certain amount of risk and wear-and-tear. Injuries, dangerous microorganisms, and chemical toxins are common, and damage to living tissues can threaten homeostasis and cause illness or even death. The lymphatic system defends the body against such threats. The vessels of the lymphatic system are connected to those of the cardiovascular system, and extracellular fluid is continuously moving from the bloodstream, into the tissues, then into the lymphatic vessels, and back to the bloodstream. For this reason the blood and lymphatic vessels are often said to be part of a combined *circulatory system*. As you will see in this unit, these two systems have tremendous clinical significance.

MAJOR SECTIONS INCLUDED WITHIN THIS UNIT:

The Cardiovascular System

The Lymphatic System

END-OF-UNIT CLINICAL PROBLEMS

The Cardiovascular System

The components of the cardiovascular system include the blood, heart, and blood vessels. Blood flows through a network of thousands of miles of vessels in the body, transporting nutrients, gases, wastes, hormones, and ions and redistributing the heat generated by active tissues. The exchange of materials between blood and peripheral tissues occurs across the thin walls of tiny capillaries that are situated between the arterial and venous systems. The total capillary surface area for exchange is truly enormous, averaging about 6300 square meters—about 50 percent larger than a football field.

Because the cardiovascular system plays a key role in supporting all other systems, disorders of this system will affect every cell in the body. One method of organizing the many potential disorders involving the cardiovascular system is by the nature of the primary problem—whether it affects the blood, the heart, or the vascular network. ▶ **Figure 52a,b** provides an introductory overview of major blood disorders and cardiovascular disorders; examples of the major categories shown are discussed in later sections of the *Applications Manual*.

The Physical Examination and the Cardiovascular System

Individuals with cardiovascular problems commonly seek medical attention. In many cases, they describe one or more of the following as chief complaints:

- **Weakness and fatigue.** These symptoms develop when the cardiovascular system can no longer meet tissue demands for oxygen and nutrients. They may occur because cardiac pumping function is impaired, as in *cardiomyopathy* (p. 132) or other forms of *heart failure* (pp. 136–137), or because the blood is unable to carry normal amounts of oxygen, as in the various forms of *anemia* (pp. 128–129). In the early stages of these conditions, the individual feels healthy at rest, but becomes weak and fatigued with any significant exertion because the cardiovascular system cannot keep pace with the rising tissue oxygen demands. In more advanced stages of illness, weakness and fatigue are problems that persist at rest.

- **Cardiac pain.** This pain is usually perceived as a deep pressure pain felt in the substernal region; it typically radiates down the left arm or up into the shoulder and neck. Cardiac pain has two major causes:

  1. Constant severe pain can result from inflammation of the pericardial sac and adjacent parietal pleura, a condition known as *pericarditis*. This *pericardial pain* can superficially resemble the pain experienced in a *myocardial infarction (MI)*, or heart attack. Pericardial pain differs from the pain of an MI in that (a) it may change with breathing or with changes in position, and is often less severe when sitting; (b) a fever may be present; and (c) the pain does not respond to the administration of drugs, such as *nitroglycerin*, that dilate coronary blood vessels.

  2. Cardiac pain also results from inadequate blood flow, or ischemia, to the myocardium. This type of pain is called

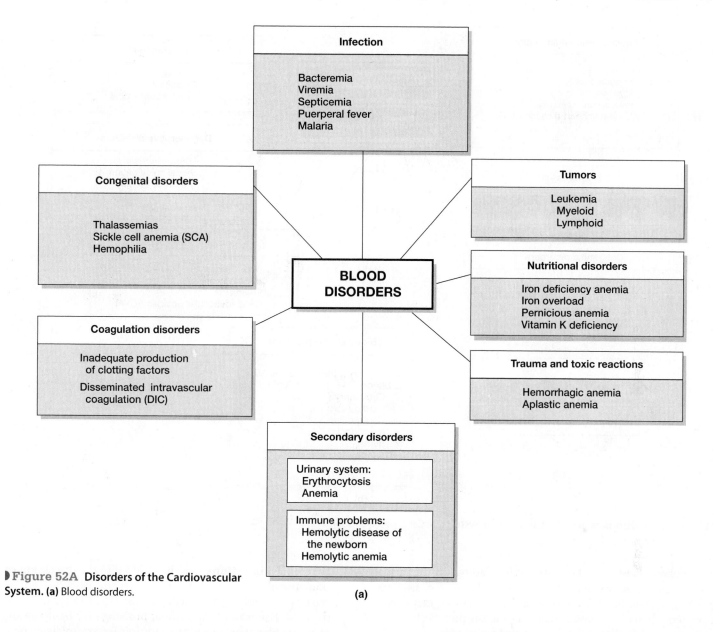

Infection

Bacteremia
Viremia
Septicemia
Puerperal fever
Malaria

Congenital disorders

Thalassemias
Sickle cell anemia (SCA)
Hemophilia

Tumors

Leukemia
Myeloid
Lymphoid

BLOOD DISORDERS

Nutritional disorders

Iron deficiency anemia
Iron overload
Pernicious anemia
Vitamin K deficiency

Coagulation disorders

Inadequate production
of clotting factors

Disseminated intravascular
coagulation (DIC)

Trauma and toxic reactions

Hemorrhagic anemia
Aplastic anemia

Secondary disorders

Urinary system:
Erythrocytosis
Anemia

Immune problems:
Hemolytic disease of
the newborn
Hemolytic anemia

▶ **Figure 52A Disorders of the Cardiovascular System. (a)** Blood disorders.

(a)

Cardiovascular System

myocardial ischemic pain. Ischemic pain occurs in *angina pectoris* and in an MI. Angina pectoris most commonly results from the narrowing of coronary blood vessels by atherosclerosis. The associated pain first appears during physical exertion, when myocardial oxygen demands increase, and the pain is relieved by rest and drugs such as nitroglycerin, which dilate coronary vessels and improve coronary blood flow. The pain associated with an MI is commonly felt as a heavy weight or a constriction of the chest. The pain of an MI is also distinctive because (a) it is not necessarily linked to exertion; (b) it is persistent (lasting longer than 15–20 minutes) and is not relieved by rest, nitroglycerin, or other coronary vasodilators; and (c) nausea, vomiting, and sweating may occur during the attack.

- **Palpitations.** Palpitations are a person's perception of an altered heart rate. The individual may complain of the heart "skipping a beat" or "racing." The most likely cause of palpitations is an abnormal pattern of cardiac activity known as an *arrhythmia.* The detection and analysis of arrhythmias are considered in a later section (pp. 134–135).

- **Pain on extremity exercise.** Individuals with advanced peripheral vascular disease have atherosclerosis with narrowing of peripheral arteries and may experience ischemic pain in the extremities during exercise. The pain, called claudication, may be so severe that the person is unwilling or unable to walk or perform other common activities. The underlying problem is the constriction or partial occlusion of major arteries, such as the external iliac and femoral arteries to the lower limb muscles, by plaque formation.

These are only a few of the many symptoms that can be caused by cardiovascular disorders. In addition, the individual may have characteristic observable signs of underlying cardiovascular problems. A partial listing of important signs includes the following:

- **Edema** is an increase of fluid in tissues that can occur when (a) the pumping efficiency of the heart is decreased, or (b) the plasma protein content of the blood is reduced, or (c) venous pressures are abnormally high. If the person can stand or sit the tissues of the limbs are most commonly affected. For example, after prolonged standing or sitting, gravity overpowers

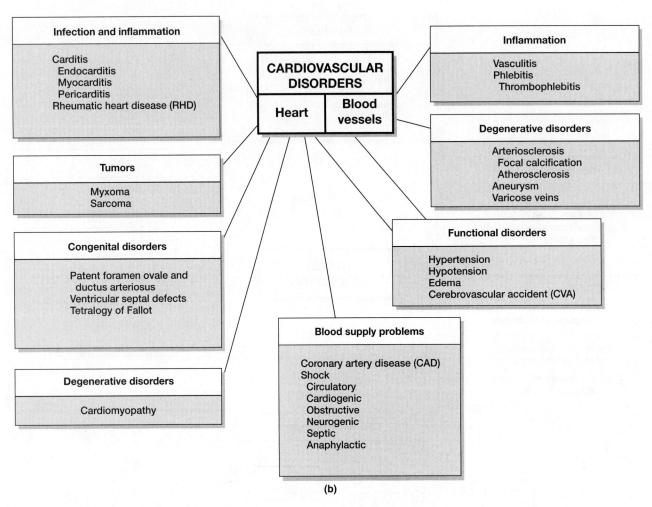

Figure 52B Disorders of the Cardiovascular System. (b) Cardiovascular disorders.

Cardiovascular System

lymphatic flow from the lower limbs, leading to swelling of the feet, ankles, and legs. If the person is bedridden, swelling of the eyelids, hands, and the area over the sacrum may occur. When edema is so severe that pressing on the affected area leaves an indentation, the sign is called *pitting edema.*

- Breathlessness, or *dyspnea,* occurs when cardiac output is inadequate for tissue oxygen demands. Dyspnea is one symptom of *pulmonary edema,* a buildup of fluid within the alveoli of the lungs. Pulmonary edema and dyspnea are typically associated with *congestive heart failure (CHF)* (pp. 136–138).

- **Varicose veins** are dilated superficial veins that are visible at the skin surface. This condition, which develops when venous valves malfunction, can be caused or exaggerated by increased systemic venous pressures. Varicose veins are considered further on page 139.

- There may be characteristic and distinctive changes in skin coloration. For example, *pallor* is the lack of normal red or pinkish color in the skin of a light-skinned person or the conjunctiva, nail beds, and oral mucosa of darker skinned people. Pallor accompanies many forms of anemia, but can also be the result of inadequate cardiac output, shock, or circulatory collapse. *Cyanosis* is the bluish color of the same areas that occurs when tissues are deficient in oxygen. Cyanosis generally results from cardiovascular or respiratory disorders.

- **Vascular skin lesions** were introduced in the discussion of skin disorders on page 44. Characteristic vascular lesions can occur in clotting disorders (pp. 131–132) and in *leukemia* (p. 129). For example, abnormal bruising may be the result of a disorder that affects the clotting system, platelet production, or the structure of blood vessels. *Petechiae,* which appear as purple spots on the skin, are seen in individuals with certain types of leukemia or other diseases associated with low platelet counts.

Diagnostic Procedures for Cardiovascular Disorders

Functional abnormalities of the heart and blood vessels can be detected through physical assessment and the recognition of characteristic signs and symptoms. In some cases, the initial detection of a cardiovascular disorder occurs during the physical examination:

1. When the vital signs are taken, the pulse is checked for strength, rate, and rhythm. Weak or irregular heartbeats are detectable.

2. Blood pressure is measured with a stethoscope and blood pressure cuff (sphygmomanometer). Unusually high or low readings can alert the examiner to potential problems with cardiac or vascular function. However, a diagnosis of

chronic hypertension (pp. 140–141) is usually not made on the basis of a single reading but after several readings over a period of time.

3. The heart sounds are assessed by listening with a stethoscope (auscultation):

- Cardiac rate and rhythm can be checked and arrhythmias detected.

- Abnormal heart sounds, or *murmurs*, may indicate problems with atrioventricular or semilunar valves. Murmurs are noted in relation to their location in the heart (as determined by the position of the stethoscope on the chest wall), the time of occurrence in the cardiac cycle, whether the sound is low pitched or high pitched, and whether variations in intensity are present.

- Usually nothing is heard during auscultation of normal blood vessels. *Bruits* are the sounds that result from turbulent blood flow, in many cases around an obstruction within an artery. Bruits are typically heard where large atherosclerotic plaques have formed.

The structural basis of heart and blood vessel problems are often determined by the use of imaging scans and x-rays and by monitoring electrical activity in the heart. For problems with the various components of blood, laboratory tests performed on blood samples provide information needed for diagnosis. **Table 27a–c**

Table 27A	Examples of Tests Used in Diagnosing Cardiovascular Disorders	
Diagnostic Procedure	**Method and Result**	**Representative Uses**
Electrocardiography (ECG/EKG)	Electrodes placed on the chest detect electrical activity of cardiac muscle during a cardiac cycle; information is transmitted to a monitor for visualization. Also, a paper record is produced.	Heart rate can be determined through study of the ECG; abnormal wave patterns may occur with cardiac irregularities such as myocardial infarction, chamber hypertrophy, or arrhythmias, such as premature ventricular contractions (PVCs), and defects in the conduction system.
Echocardiography	Standard ultrasound examination of the heart.	Detects structural abnormalities of the heart; useful in determination of valve function, chamber size, vessel size, and ejection fraction.
Transesophageal echocardiography	Use of an ultrasound probe attached to an endoscope and inserted into the esophagus.	Provides enhanced view of posterior and inferior heart chambers.
Phonocardiography	Heart sounds are monitored and graphically recorded.	Detects murmurs and abnormal heart sounds (a useful aid for teaching cardiac auscultation).
Exercise stress test	ECG, blood pressure, and heart rate are monitored during exercise on treadmill. Often teamed with before-and-after echocardiograms.	Coronary artery narrowing is suspected if ECG patterns or echocardiography results change during or after exercise.
Chest x-ray	X-rays penetrate body, recorded on film sheet with radiodense tissues in white-on-black negative image.	Determines abnormal shape or size of the heart and abnormalities of the aorta and great vessels; detects masses or fluid in lungs.
Cardiac catheterization (coronary arteriography)	For the study of the left side of the heart, a catheter is inserted into an artery; for the right side of the heart, the catheter is inserted into a vein. Arteries and veins in the arm, neck, or groin may be used. Contrast dye is injected to visualize vessels and chambers as x-rays are taken.	Detects blockages and spasms in coronary arteries; also helps evaluate congenital defects and ventricular hypertrophy and determine the severity of a valvular defect; used to monitor volumes and pressures within the chambers.
MRI of heart	Noninvasive images of the heart and coronary arteries; limited by heart motion and imaging speed.	Evaluates heart anatomy and function, detects coronary artery calcification, indicating atherosclerotic plaques and possible blockages.
Technesium scan	Radionuclides are injected into the bloodstream; accumulation occurs in perfused areas of the heart. A computer image is produced for examination.	Detects tissue areas with a reduced perfusion; determines blood flow through the coronary arteries; after an MI, it helps determine extent of damage from infarction; may detect exercise ischemia when used with stress test.
Pericardiocentesis	Needle aspiration of fluid from the pericardial sac for analysis; therapeutic for cardiac tamponade.	See "Pericardial fluid analysis," **Table 27b**.
Venous Doppler ultrasound	Transducer is placed over the vessel to be examined, and the echoes are analyzed by computer to provide information on blood velocity and flow direction.	Detects arterial or venous blockage or venous valve insufficiency; determines cardiac efficiency; monitors fetal circulation and blood flow in umbilical vessels.
Venography	Radiopaque dye is injected into a peripheral vein of a limb, and an x-ray study is performed to detect venous blockage; largely replaced by venous Doppler ultrasound.	Detects venous thrombosis (phlebitis, Deep Venous Thrombosis [DVT]).

Cardiovascular System

Table 27B	Laboratory Tests Useful in Diagnosing Cardiovascular Disorders	
Laboratory Test	**Normal Values in Blood Plasma or Serum**	**Significance of Abnormal Values**
Creatine kinase (CPK, CK)	Adult male: 50–290 U/L Adult female: 40–240 U/L	Prolonged elevated levels of CK indicate muscle damage; levels usually do not rise until 6–12 hours following myocardial infarction (MI)
Isoenzymes of CK (CK-MM) (CK-MB) (CK-BB)	Varies with method; the normal CK-MB is 0–6% of the total.	CK-MM helps diagnose muscle disease. CK-MB provides information about damaged cardiac muscle. Levels rise within 6–12 hours following MI. CK-BB helps diagnose brain infarct or stroke
Aspartate aminotransferase (AST)	Adults: 7–45 U/L	Elevated in congestive heart failure due to liver damage
Serum electrolytes		
Sodium	Adults: 135–145 mEq/L	Decreased levels occur with water retention caused by heart failure, hypotension; increased levels can occur with dehydration
Potassium	Adults: 3.5–5.0 mEq/L	Decreased levels occur when certain diuretics are taken; increased levels occur with kidney failure
Troponin T	Adults: 0–0.1 ng/mL	Cardiac-specific proteins not normally found in blood; preferred biochemical marker for early detection of MI
Serum lipoproteins	HDL: 29–77 mg/dL LDL: 60–130 mg/dL	Higher risk of coronary heart disease occurs when LDL values are above 100 mg/dL and HDL is less than 45 mg/dL
Serum cholesterol	Adults: 150–240 mg/dL (desirable level below 200 mg/dL)	When above 200 mg/dL, higher risk for atherosclerosis and coronary artery disease; elevated levels occur in familial hypercholesterolemia
Serum triglycerides	Adults: 10–160 mg/dL (older adults may have higher values)	Elevated with obesity, a high-fat diet, and poorly controlled diabetes mellitus and some AIDS treatments
Pericardial fluid analysis		Fluid is analyzed for appearance, number, and types of blood cells; protein and glucose levels; and the presence of infectious organisms

summarizes information about representative diagnostic procedures used to evaluate the health of the cardiovascular system.

▶ Disorders of the Blood

Changes in blood volume have direct effects on other body systems. Clinicians use the terms *hypovolemia* (hī-pō-vō-LĒ-m[e]-ah), *normovolemia* (nor-mō-vō-LĒ-m[e]-ah), and *hypervolemia* (hī-per-vō-LĒ-m[e]-ah) to refer to low, normal, and excessive blood volumes, respectively. Hypovolemia and hypervolemia are potentially dangerous, because variations in blood volume affect other components of the cardiovascular system. For example, an abnormally large blood volume can place a severe stress on the heart, which must keep the extra fluid circulating through the lungs and throughout the body. Hypovolemia, such as dehydration secondary to gastrointestinal problems or blood loss from trauma, is clinically much more common than hypervolemia. Short-term therapies to treat hypovolemia may include the intravenous use of an electrolyte solution or a **transfusion**—the administration of blood components to restore blood volume or to remedy a deficiency in blood composition. Whole blood, packed red blood cells, plasma, platelets, extracted proteins, or clotting factors may be transfused; these therapies will be considered in a later section.

Polycythemia

An elevated hematocrit with a normal blood volume constitutes **polycythemia** (po-lē-sī-THĒ-mē-uh). There are several types of polycythemia. We will consider *erythrocytosis* (e-rith-rō-sī-TŌ-sis), a polycythemia that affects only red blood cells (RBCs), later in this section. **Polycythemia vera** (PV, "true polycythemia") is an increase in the numbers of all types of blood cells resulting from excess bone marrow production, while the absolute blood volume is unchanged. As RBCs are by far the most numerous blood component, the hematocrit may double to reach 80–90, at which point the tissues become starved for oxygen because RBCs are blocking the smaller vessels. PV seldom strikes young people; most cases involve patients 60–80 years of age. Several treatment options are available, but none cures the condition. A somatic (not inherited) mutation in the *JAK2* gene of a hematopoietic stem cell is the major cause of PV.

Abnormal Hemoglobin

Several inherited disorders are characterized by the production of abnormal hemoglobin. Two of the best known are thalassemia and sickle cell anemia.

Table 27C	Representative Hematology Studies for Diagnosing Blood Disorders	
Laboratory Test	**Normal Values in Whole Blood, Plasma, or Serum**	**Significance of Abnormal Values**
Complete Blood Count		
RBC count	Adult males: 4.7–6.1 million/mm^3 Adult females: 4.0–5.0 million/mm^3	Increased RBC count, Hb, and Hct occur in polycythemia vera, congenital heart disease, and events that induce chronic hypoxia, such as moving to a high altitude, or lung disease
Hemoglobin (Hb, Hgb)	Adult males: 14–18 g/dL Adult females: 12–16 g/dL	Decreased RBC count, Hb, and Hct occur with hemorrhage and the various forms of anemia, including hemolytic anemia, iron deficiency anemia, vitamin B$_{12}$ deficiency, sickle cell, and thalassemia, or as a result of bone marrow failure or leukemia
Hematocrit (Hct)	Adult males: 42–52 Adult females: 37–47	See sections on RBC count and hemoglobin
RBC indices		
Mean corpuscular volume (MCV; measure of average volume of a single RBC)	Adults: 80–98 μm^3	Increased MCV and MCH occur in types of macrocytic anemia, including vitamin B$_{12}$ and folate deficiency
Mean corpuscular hemoglobin (MCH; measure of average amount of Hb per RBC)	Adults: 27–31 pg	Decreased MCV and MCH occur in types of microcytic anemia, such as iron deficiency anemia and thalassemia
Mean corpuscular hemoglobin concentration (MCHC)	Adults: 32–36% (derived by dividing the total Hb concentration number by the Hct value)	Decreased levels (hypochromic erythrocytes) suggest iron deficiency anemia or thalassemia
WBC count	Adults: 3500–10,000/mm^3	Increased levels occur with chronic and acute infections, tissue death (MI, burns), leukemia, parasitic diseases, and stress; decreased levels occur in aplastic and pernicious anemias, overwhelming bacterial infection (sepsis), viral infections, AIDS, and cancer chemotherapy
Differential WBC count	Neutrophils: 50–70%	Increased levels occur in acute bacterial infection, myelocytic leukemia, rheumatoid arthritis, and stress; decreased levels occur in aplastic and pernicious anemia, viral infections, radiation treatment, and with some medications
	Lymphocytes: 20–30%	Increased levels occur in lymphocytic leukemia, infectious mononucleosis, and viral infections; decreased levels occur in radiation treatment, AIDS, and corticosteroid therapy
	Monocytes: 2–8%	Increased levels occur in chronic inflammation, viral infections, and tuberculosis; decreased levels occur in aplastic anemia and corticosteroid therapy
	Eosinophils: 2–4%	Increased levels occur in allergies, parasitic infections, and some autoimmune disorders; decreased levels occur with steroid therapy
	Basophils: 0.5–1%	Increased levels occur in inflammatory processes and during healing; decreased levels occur in hypersensitivity reactions
Platelet count	Adults: 150,000–400,000/mm^3	Increased count can cause vascular thrombosis and occurs in polycythemia vera; decreased levels can result in spontaneous bleeding and occurs in different types of anemia, in some leukemias, and cancer chemotherapy
Clotting System Tests		
Bleeding time (amount of time for bleeding to stop after a small incision is made in skin)	3–7 minutes	Prolongation occurs in patients with decreased platelet count, anticoagulant therapy, aspirin ingestion, leukemia, or clotting factor deficiencies
Clotting factors assay (coagulation factors I, II, V, VIII, IX, X, XI, XII)	Measured for their hemostatic activity	Decreased activity of the coagulation factors will result in defective clot formation; deficiencies can be caused by liver disease or vitamin K deficiencies
Plasma fibrinogen (Factor I)	200–400 mg/dL	Elevated values can occur in inflammatory conditions and acute infections and also with medications such as birth control pills; decreased levels occur in liver disease, leukemia, and disseminated intravascular coagulation (DIC)

Cardiovascular System

(continued)

Table 27C	Representative Hematology Studies for Diagnosing Blood Disorders (*Continued*)	
Laboratory Test	**Normal Values in Whole Blood, Plasma, or Serum**	**Significance of Abnormal Values**
Partial thromboplastin time (PTT)	35–50 seconds for blood to clot	Prolonged by heparin
Plasma prothrombin time (PT)	Within 2 seconds of control (control should be 11–15 seconds)	Prolonged time to clotting can occur after trauma, MI, or infection
		DIC and liver disease
		Therapeutic use of the anticoagulant warfarin/coumadin is monitored by the PT ratio of tested blood to control results, called the INR (international normalized ratio)
INR	1.0	Normal value without anticoagulation therapy
Therapeutic INR range	2.0–3.5	Target value for anticoagulation therapy
Hemoglobin electrophoresis		
Hemoglobin A	Adults: within 95–98% of the total Hb	Normal adult form of hemoglobin, different forms in thalassemia
Hemoglobin F	<2% of the total Hb after age 2 (newborn has 50–80% HbF)	Elevated levels after six months of age suggest thalassemia
Hemoglobin S	0% of the total Hb	Present in sickle cell anemia and the sickle cell trait
Serum bilirubin	Adults: 0.1–1.2 mg/dL	Increased levels occur with hemolysis, liver disease (hepatitis), liver cancer, and gallstone obstruction of bile ducts
Erythrocyte sedimentation rate (ESR) (measure of rate at which erythrocytes settle in column of whole blood collected in non-clotting tube)	Adult males: 1–15 mm/h Adult females: 1–20 mm/h	Increased by disease processes that increase the protein concentration in the plasma, including inflammatory conditions, infections, and cancer; not useful in diagnosis of specific disorders but a rough gauge of the severity of active inflammation
Bone marrow aspiration biopsy (involves laboratory examination of shape and cell size of erythrocytes, leukocytes, and megakaryocytes)	For the evaluation of hematopoiesis or the presence of tumor cells or infection	Increased RBC precursors with polycythemia vera; increased WBC precursors with leukemia; radiation therapy or chemotherapy can cause a decrease in all cell populations

Thalassemia

The various forms of **thalassemia** (thal-ah-SĒ-mē-uh) result from an inability to produce adequate amounts of normal alpha or beta chains of hemoglobin. As a result, the rate of RBC production is slowed by ineffective erythropoiesis, and the mature RBCs are fragile and short lived. The scarcity of healthy RBCs reduces the oxygen-carrying capacity of the blood and leads to problems with the development and growth of systems throughout the body. Compensatory erythroid hyperplasia in the bone marrow, liver, and spleen also causes problems. Individuals with severe thalassemia must periodically undergo *transfusions*—the administration of blood components—to keep adequate numbers of RBCs in the bloodstream. The iron in the transfused RBCs remains in the body and may lead to iron overload, which can affect cardiac muscle performance as well as damage many other organ systems.

The thalassemias are categorized as an **alpha-thalassemia** or a **beta-thalassemia**, depending, respectively, on whether the alpha or beta hemoglobin chains are affected. Prenatal genetic testing is available for thalassemia. Individuals inherit two copies of alpha-chain genes from each parent, and alpha-thalassemia develops when one or more of these genes are missing or inactive. The severity of the symptoms varies with the number of normal alpha-chain genes that remain functional. For example,

an individual with three normal alpha-chain genes will not develop symptoms, but can be a carrier, passing the defect to the next generation. A child whose parents are both carriers is at risk to develop a more severe form of the disease:

- Individuals with two copies, rather than four, of the normal alpha-chain gene (one from each parent) have somewhat impaired hemoglobin synthesis. This condition is known as *alpha-thalassemia trait*. The RBCs are small and contain less than the normal quantity of hemoglobin. About 2 percent of African Americans and many Southeast Asians have alpha-thalassemia trait.

- Individuals with only one copy of the normal alpha-chain gene have very small (*microcytic*) RBCs that are relatively fragile.

- Most individuals with no functional copies of the normal alpha-chain gene die before birth or shortly after, because the hemoglobin that is synthesized cannot bind and transport oxygen normally. The incidence of fatal alpha-thalassemia is highest among Southeast Asians.

Each person inherits only one gene for the beta hemoglobin chain from each parent. These genes contain several possible mutations, and beta-thalassemia can take a variety of forms. If an individual does not receive a copy of a normal, functioning gene from either parent, the condition of **beta-thalassemia major** develops. Symptoms of this condition include severe anemia with microcytosis and a low hematocrit (below 20) and enlargement of the spleen, liver, heart, and areas of red bone marrow. Treatments for persons with severe symptoms include transfusions, splenectomy (to slow the rate of RBC recycling), and bone marrow transplantation. Gene replacement therapy of a normal beta globulin gene may someday be developed. Beta-thalassemia minor, or beta-thalassemia trait, seldom produces clinical symptoms. An individual with the condition has one normal gene for the beta hemoglobin chain, and the rate of hemoglobin synthesis is depressed by roughly 15 percent. This decrease does not affect the RBCs' functional abilities, however, so no treatment is necessary. Blood counts from individuals with alpha-thalassemia trait or beta-thalassemia trait are similar to those from individuals who have iron deficiency anemia, and there is a risk of inappropriate treatment with iron supplements.

Sickle Cell Anemia

Sickle cell anemia (SCA) results from a mutation affecting the amino acid sequence of the beta chains of the Hb molecule. The abnormal subunit is called *hemoglobin S*. When blood contains abundant oxygen, the Hb molecules and the RBCs that carry them appear normal. But when the defective hemoglobin gives up enough of its bound oxygen, the adjacent Hb molecules cluster into rods, and the cells become stiff and curved (**) Figure 53**). This "sickling" makes the RBCs fragile and easily damaged. Moreover, even though RBCs that have folded to squeeze into a narrow capillary deliver their oxygen to the surrounding tissue, the cells can become stuck as sickling occurs. A circulatory blockage results, and nearby tissues become starved for oxygen.

An individual with sickle cell anemia carries two copies of the abnormal gene—one from each parent. If only one copy is present, the individual has a *sickling trait*. One African American in 12 and one in 100 Hispanic Americans carries the sickling trait. These genes are also present in some persons of Mediterranean, Middle Eastern, and East Indian ancestry. Most people with the sickling trait do not have signs of the disease.

In an individual with the sickling trait, most of the hemoglobin is of the normal form, and the RBCs function normally. But

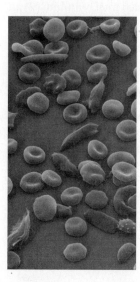

) Figure 53 "Sickling" in Red Blood Cells. At lower oxygen concentrations, red blood cells with hemoglobin S change shape, becoming more rigid and sharply curved, or "sickled."

the presence of the abnormal hemoglobin gives the individual some ability to resist the parasitic infection that causes **malaria**, a mosquito-borne illness. Malaria parasites enter the bloodstream when an individual is bitten by an infected mosquito. The microorganisms invade and reproduce within the RBCs. But when they enter the RBC of a person with the sickling trait, the cell responds by sickling. Either the sickling itself kills the parasite, or the sickling attracts the attention of a phagocyte that engulfs the RBC and kills the parasite. In either event, the individual better tolerates the parasitic infection, whereas individuals without the sickling trait are more likely to sicken and die of malaria. Genetic studies indicate that the sickling mutation has evolved at least five times at different locations in Africa and India—regions where malaria poses a serious health problem.

When sickled RBCs get stuck in small capillaries and obstruct blood flow, they cause pain and eventual damage to a variety of organs and systems, depending on the location and duration of the obstructions. In addition, the trapped RBCs die and break down, producing a characteristic hemolytic anemia. Only half of affected persons live past 50 years with most dying of infection or stroke. A normal RBC stays in circulation 120 days; sickled cells are gone in 10–20 days. Rapid RBC loss damages the spleen and exceeds the maximum rate of RBC production. Supplements of the vitamin folate (folic acid), which is needed in RBC production, help with this high turnover. Transfusions of normal blood can temporarily prevent additional complications, and treatment of affected infants and children with antibiotics reduces deaths due to infections. *Hydroxyurea* is an anticancer drug that stimulates the production of fetal hemoglobin, a slightly different form of hemoglobin normally produced during development of the fetus. The drug is effective, but has toxic side effects (not surprising in an anticancer drug). If a matched donor is available, a bone marrow transplant cures the disease. Mice genetically engineered to have human sickle cell anemia have been successfully treated by gene therapy to add normal human beta hemoglobin genes. This approach may someday be used to treat the disorder in humans.

Transfusions and Synthetic Blood

In a **transfusion**, blood components are provided to an individual whose blood volume has been reduced or whose blood is deficient in some way. The donated blood is obtained from screened donors under sterile conditions, tested for infectious agents, treated to prevent clotting and to stabilize the RBCs, and refrigerated. Transfusions of whole blood are most commonly used to restore the blood volume after massive blood loss. In an **exchange transfusion**, most of an individual's blood volume is drained off and simultaneously replaced with whole blood from another source. Most exchange transfusion patients are newborns with severe jaundice or immune-related anemias.

Chilled whole blood remains usable for about 3½ weeks. For longer storage or more efficient use, the blood must be *fractionated*. The RBCs are separated from the plasma, and may be used fresh for 42 days or treated with a special antifreeze solution, and then frozen. Plasma can be stored chilled, frozen, or freeze dried. This procedure permits the long-term storage of rare blood types that might not otherwise be available for emergency use. Frozen RBCs are usable for up to ten years, and plasma for one year.

Cardiovascular System

Fractionated blood has many uses. **Packed red blood cells** (PRBCs), with most of the plasma removed, are preferred for cases of anemia, in which the volume of blood may be close to normal but has below-normal oxygen-carrying capabilities. Plasma can be administered when massive fluid losses are occurring, such as after a severe burn. Plasma samples can be further fractionated to yield albumins, immunoglobulins, clotting factors, and plasma enzymes, each of which can be administered separately. White blood cells and platelets can also be preferentially collected, sorted, and stored for subsequent transfusion.

Some 9.5 million donors give over 16 million pints of blood, and 30 million units of blood components are transfused each year in the United States alone. Yet the demand for blood or blood components can exceed the supply. Moreover, many people are concerned about the risk of infection with hepatitis viruses or with HIV (the virus that causes AIDS) from contaminated blood. For these reasons, transfusion practices have changed during recent years. In general, fewer units of blood are now administered. Blood donors and collected blood are screened for 11 infectious diseases including hepatitis and HIV. There has also been an increase in **autologous transfusion**, in which blood is removed from an individual planning to have nonemergency surgery, stored, and later transfused back into the same individual when needed. In 2001, autologous blood was used in 4 percent of U.S. transfusions, and 2 percent of donated blood was discarded because of screening results.

Genetically engineered erythropoietin (EPO) can also be administered to help the patient's body restore its full complement of RBCs more quickly than it could without the hormone. Patients with kidney failure are often anemic and erythropoietin has reduced their need for transfusions. Moreover, new technology permits the reuse of blood "lost" during surgery. The blood is collected and filtered, the platelets are removed, and the remainder of the blood is reinfused into the patient.

Shortages of blood and anxieties over the safety of existing stockpiles persist. In addition, some people are unable or unwilling to accept transfusions for religious or other reasons. Thus, there has been widespread interest in the development of synthetic blood components. Genetic-engineering techniques are being used to address this problem. For example,

- It is now possible to produce one of the subunits of normal human hemoglobin using genetically modified bacteria. The isolated hemoglobin can be infused to increase oxygen transport and total blood volume. The hemoglobin molecules can be attached to inert carrier molecules that will prevent their filtration and loss at the kidneys. However, the differences between cell-free hemoglobin and red blood cells create other problems that still need to be solved.

- Hemoglobin can now be obtained from nonhuman sources, primarily cows. Technical progress has been rapid, but practical success has been limited. The possible risk of a transmissible prion disease, bovine spongiform encephalopathy (p. 26), is another concern. Genetically modified pigs and sheep may provide human hemoglobin and even human blood cells.

Whole-blood substitutes are highly experimental solutions still undergoing clinical evaluation. In addition to the osmotic agents in plasma expanders, these solutions contain small clusters of synthetic molecules built of carbon and fluorine atoms. The mixtures, known as **perfluorochemical (PFC) emulsions**, can carry roughly 70 percent of the oxygen of whole blood. Animals have been kept alive after an exchange transfusion that completely replaced their circulating blood with a PFC emulsion. Such an emulsion has the same advantages as other plasma expanders, plus the added benefit of transporting oxygen. Unfortunately, PFCs do not absorb oxygen as effectively as normal blood does. In addition, phagocytes appear to engulf the PFC clusters. These problems have limited the use of PFC emulsions in humans.

Bilirubin Tests and Jaundice

When hemoglobin is broken down, the heme units (minus the iron) are converted to bilirubin. Normal serum bilirubin concentrations range from 0.1 to 1.2 mg/dL. Of that amount, roughly 85 percent will be removed by the liver. Several clinical conditions are characterized by an increase in the total plasma bilirubin concentration. In such conditions, bilirubin diffuses into peripheral tissues, giving them a yellow color that is most apparent in the skin and over the sclera of the eyes. This combination of signs (yellow skin and eyes) is called **jaundice** (JAWN-dis).

Jaundice can have many causes, but blood tests that determine the concentration of different forms of bilirubin can provide useful diagnostic clues. For example, **hemolytic jaundice** results from the destruction of large numbers of RBCs. When this occurs, phagocytes release massive quantities of one form of *bilirubin* (*unconjugated*) into the blood. Because the liver cells continue to excrete a different form of bilirubin (*conjugated bilirubin*) in the bile, the blood concentration of conjugated bilirubin does not increase proportionally. A blood test from a person with hemolytic jaundice would reveal (1) an elevated total bilirubin, (2) high concentrations of unconjugated bilirubin, and (3) a conjugated bilirubin contribution of much less than 15 percent of the total bilirubin concentration.

These results are quite different from those seen in **obstructive jaundice**. In that condition, the ducts that remove bile from the liver are constricted or blocked. Liver cells cannot get rid of conjugated bilirubin, and large quantities diffuse into the blood. Diagnostic tests would then show (1) an elevated total bilirubin, (2) an unconjugated bilirubin contribution of much less than 85 percent of the total bilirubin concentration, and (3) high concentrations of conjugated bilirubin.

Iron Deficiencies and Excesses

If dietary supplies of iron are inadequate, hemoglobin production slows down and symptoms of *iron deficiency anemia* appear. This form of anemia can also be caused by any condition that produces a large or ongoing blood loss, because the iron in the lost blood cannot be recycled.

As the RBCs are replaced, iron reserves must be mobilized for use in the synthesis of new hemoglobin molecules. If those reserves are exhausted or if dietary sources are inadequate, iron deficiency results. In iron deficiency anemia, the RBCs cannot synthesize

Cardiovascular System

enough functional hemoglobin, and they are unusually small when they enter the bloodstream. The hematocrit declines, and the hemoglobin content and oxygen-carrying capacity of the blood are substantially reduced. Symptoms include weakness and fatigue.

Women are especially dependent on a normal dietary supply of iron, because their iron reserves are smaller than those of men. A healthy male's body has about 3.5 g of iron in the ionic form Fe^{2+}. Of that amount, 2.5 g is bound to the hemoglobin of circulating RBCs, and the rest is stored in the liver and bone marrow. In women, the total body iron content averages 2.4 g, with roughly 1.9 g incorporated into RBCs. Thus, a woman's iron reserves consist of only 0.5 g, half that of a typical man. Moreover, the monthly menstrual cycles of premenopausal adult women result in blood losses that further stress iron reserves. When the demand for iron increases out of proportion with dietary supplies, iron deficiency develops. At least 12 percent of menstruating women in the United States show signs of iron deficiency. Pregnancy also depletes iron reserves, because the woman must provide the iron needed to produce both maternal and fetal erythrocytes and replace blood lost at childbirth.

Good dietary sources of iron include liver, red meats, kidney beans, egg yolks, spinach, and carrots. Iron supplements can help prevent iron deficiency, but too much iron can be as dangerous as too little. Iron absorption across the digestive tract normally keeps pace with physiological demands. When the diet contains abnormally high concentrations of iron, or when hereditary factors increase the rate of absorption, the excess iron gets stored in peripheral tissues. This storage is called *iron loading*. Eventually, cells begin to malfunction as massive iron deposits accumulate in the cytoplasm. For example, iron deposits in pancreatic cells can lead to diabetes mellitus, and deposits in cardiac muscle cells lead to abnormal heart contractions and heart failure. (Evidence suggests that iron deposits in the heart caused by the overconsumption of red meats may contribute to heart disease.) Iron-loaded liver cells become nonfunctional, and cirrhosis or liver cancers may develop. Complications of iron loading may follow repeated transfusions of whole blood, because each unit of whole blood contains roughly 250 mg of iron. For example, as we noted previously, the severe forms of *thalassemia* result from a genetic inability to produce adequate amounts of one of the four chains of hemoglobin. Erythrocyte production and survival are reduced, and so is the blood's oxygen-carrying capacity. Most individuals with severe untreated thalassemia die in their twenties, but not because of the anemia. These patients are treated for severe anemia with frequent blood transfusions, which prolong life, but the excessive iron loading eventually leads to fatal heart problems. There are iron-chelating drugs that bind to iron in the body and are then excreted. Such treatment may prevent, delay, or reverse symptoms of iron overload.

Erythrocytosis and Blood Doping

In **erythrocytosis** (e-rith-rō-sī-TŌ-sis), the blood contains abnormally large numbers of RBCs. Erythrocytosis generally results from the increased release of erythropoietin (EPO) by tissues—especially the kidneys—when they are deprived of oxygen. After moving to high altitudes, people eventually develop erythrocytosis, because a lungful of air at that altitude contains fewer

oxygen molecules than it would at sea level, and the lower tissue oxygen levels that result stimulate EPO production. This increases the number of RBCs, compensating for the reduced amount of oxygen each RBC is transporting. The hematocrit of mountaineers training at altitudes of 10,000–12,000 feet may become as high as 65.

An individual whose heart or lungs are functioning inadequately may also develop erythrocytosis. For example, this condition is often seen in cases of heart failure and lung diseases, conditions we will discuss in later sections. Whether the blood fails to circulate efficiently or the lungs do not deliver enough oxygen to the blood, peripheral tissues remain poor in oxygen, despite the rising hematocrit. Having a higher concentration of RBCs increases the oxygen-carrying capacity of the blood, but it also makes the blood thicker and harder to push around the circulatory system. The workload on the heart increases, making a bad situation even worse.

The practice of **blood doping** has occurred among competitive athletes involved in endurance sports, such as cycling and cross-country skiing. The procedure entails removing whole blood from the athlete in the weeks before an event. The packed red cells are separated from the plasma and stored. By the time of the race, the competitor's bone marrow will have replaced the lost blood. Immediately before the event, the packed red cells are reinfused, increasing the hematocrit. The objective is to elevate the oxygen-carrying capacity of the blood and thereby increase endurance. The consequence is that the athlete's heart is placed under a tremendous strain and the thickened blood may clog small arteries, causing strokes, kidney damage, or heart attacks. The long-term effects are unknown, but the practice obviously carries a significant risk; as a performance-enhancing method it is on the World Anti-Doping Agency's prohibited list for competitive sports. Attempts to circumvent this rule by the use of administered EPO (now also prohibited) in 1992–1993 resulted in the tragic deaths of 18 European cyclists. During the 2002 Salt Lake City Winter Olympics, three skiers tested positive for darbepoetin, a compound similar to EPO, and several medals that they had won were rescinded. Even misguided recreational athletes have sought to use erythropoietin, and it has joined anabolic steroids (which increase muscle mass) on the illegal drug market.

A safer, milder erythrocytosis occurs in athletes who train at high altitudes. The U.S. Olympic training camp is located well above sea level in Colorado.

Blood Tests, RBCs, and Anemia

Several common blood tests are used to assess circulating RBCs (**Table 27c**): Automated blood-testing machines do the calculations and report the following measurements.

Reticulocyte Count

Reticulocytes are immature RBCs that are still synthesizing hemoglobin. Most reticulocytes remain in the bone marrow until they complete their maturation, but some enter the bloodstream. Reticulocytes normally account for about 0.8 percent of the erythrocyte population. Values above 1.5 percent or below 0.5 percent indicate that something is wrong with the rates of RBC survival or maturation.

Cardiovascular System

Hematocrit (Hct)

The hematocrit value is the percentage of whole blood occupied by red blood cells. The hematocrit of a normal adult averages 46 for men and 42 for women, with ranges of 42–52 for men and 37–47 for women.

Hemoglobin Concentration (Hb)

This test determines the amount of hemoglobin in the blood, expressed in grams per deciliter (g/dL). Normal ranges are 14–18 g/dL for males and 12–16 g/dL for females. The differences in hemoglobin concentration reflect the differences in hematocrit. For both genders, a normal, single RBC contains 27–33 picograms (pg) of hemoglobin.

RBC Count

Calculations of the RBC count—the number of RBCs per microliter of blood—are based on the hematocrit and hemoglobin content and can be used to get more information about the condition of the RBCs. Values typically reported in blood tests include:

- **Mean corpuscular volume (MCV)**, the average volume of an individual RBC, in cubic micrometers (μm^3). The MCV is calculated by dividing the hematocrit by the RBC count, using the formula

$$MCV = \frac{Hct}{RBC \; count \; (in \; millions)} \times 10$$

This is a "cookbook" method that takes advantage of the fact that the hematocrit closely approximates the relative volume of RBCs in any unit sample of whole blood. Normal values for the MCV range from 80 to 98. For a representative hematocrit of 46 and an RBC count of 5.2 million, the mean corpuscular volume is

$$MCV = \frac{46}{5.2} \times 10 = 88.5 \; \mu m^3$$

Cells of normal size are said to be **normocytic**, whereas larger-than-normal or smaller-than-normal RBCs are called **macrocytic** and **microcytic**, respectively.

- **Mean corpuscular hemoglobin (MCH)**, the average amount of hemoglobin in a single RBC, expressed in picograms. Normal values range from 27 to 33 pg. The MCH is calculated as

$$MCH = \frac{Hb}{RBC \; count \; (in \; millions)} \times 10$$

RBCs containing normal amounts of hemoglobin are said to be **normochromic**, whereas the names **hyperchromic** and **hypochromic** indicate higher-than-normal and lower-than-normal hemoglobin content, respectively.

Anemia

Anemia exists whenever the oxygen-carrying capacity of the blood is reduced, diminishing the delivery of oxygen to peripheral tissues. Such a reduction causes a variety of symptoms, including premature muscle fatigue, weakness, lethargy, and a lack of energy. Anemia may exist because the hematocrit is abnormally low or because the amount of hemoglobin in the RBCs is reduced. Standard laboratory tests can be used to differentiate among the various forms of anemia on the basis of the number, size, shape, and hemoglobin content of RBCs. As an example, **Table 28** shows how this information can be used to distinguish among four major types of anemia:

1. **Hemorrhagic anemia** results from severe blood loss. Erythrocytes are of normal size; each contains a normal amount of hemoglobin, and reticulocytes are present in normal concentrations, at least initially before the homeostatic mechanisms increase blood production. Blood tests would therefore show a low hematocrit and low hemoglobin, but the MCV, MCH, and reticulocyte counts would be normal.

2. In **aplastic** (ā-PLAS-tik) **anemia**, the bone marrow fails to produce new RBCs. Presumed causes include radiation, toxic chemicals, and immunologic or infectious diseases, but in most cases the precise cause is unknown. The 1986 nuclear accident in Chernobyl (in the former USSR) caused a number of cases of aplastic anemia. The condition is fatal unless surviving stem cells repopulate the marrow or a transplant of hematologic stem cells is performed. In aplastic anemia, the circulating RBCs are normal in all respects, but because new RBCs are not being produced, the RBC count, Hct, Hb, and reticulocyte count are low.

3. In **iron deficiency anemia**, normal hemoglobin synthesis cannot occur because iron reserves are inadequate (pp. 126–127). Because developing RBCs cannot make functional hemoglobin, they are unusually small. A blood test shows a low hematocrit, low hemoglobin content, low MCV, and low MCHC, but generally a normal reticulocyte count. An estimated 2 billion people worldwide have iron deficiency anemia.

4. In **pernicious** (per-NISH-us) **anemia**, normal RBC maturation ceases because the supply of vitamin B_{12} is inadequate. Erythrocyte production declines, and the RBCs are abnormally

Table 28	RBC Tests and Anemias				
Type of Anemia	**Hct**	**Hb**	**Reticulocyte Count**	**MCV**	**MCHC**
Hemorrhagic (acute)	Low	Low	Normal or high	Normal	Normal
Aplastic	Low	Low	Very low	Normal	Normal
Iron deficiency	Low	Low	Normal or low	Low	Low
Pernicious (B_{12} deficiency)	Low	Low	Very low	High	Normal or low
Hemolytic	Low	Low	Very high (three times normal)	Normal or low	Normal

Cardiovascular System

large and may develop a variety of bizarre shapes. Blood tests from a person with pernicious anemia indicate a low hematocrit with a high MCV and a normal or low reticulocyte count. A similar macrocytic anemia occurs with deficiency of another B vitamin, folate.

5. In **hemolytic anemia**, RBCs are breaking down in the bloodstream. The individual RBCs are generally normal in size and hemoglobin content, although in some cases RBC fragments are present. The hematocrit and hemoglobin concentration are low. In chronic cases, reticulocyte counts are high as reticulocytes enter the bloodstream prematurely in response to the anemia.

The Leukemias

Leukemias (*leukos*, white) result from the proliferation of abnormal, nonfunctioning white blood cells. Either myeloid cells (granulocytes or other cells of the bone marrow) or lymphoid cells (lymphocytes) are involved. Leukemias are further classified as *acute* (short and severe) or *chronic* (prolonged). Acute leukemias are linked to radiation exposure, hereditary susceptibility, benzene exposure (an industrial chemical also found in cigarette smoke), or unknown causes. Chronic leukemias are related to acquired chromosomal abnormalities in the cancerous cells or immune system malfunctions. The first symptoms appear as immature and abnormal white blood cells (WBCs) enter the bloodstream. As the number of white blood cells increases, they travel through the circulation, invading tissues and organs throughout the body.

These leukemic cells are extremely active, although nonfunctional, and they require abnormally large amounts of energy. As in other cancers, described elsewhere in this manual (pp. 36–40), leukemic cells multiply and replace normal cells, especially in the bone marrow. Red blood cell, normal WBC, and platelet formation decline, with resulting anemia, infection, and impaired blood clotting. Untreated leukemias are invariably fatal. Survival in untreated acute leukemia averages about three months; individuals with chronic leukemia may survive for years.

Effective treatments exist for some forms of leukemia, but not for others. For example, acute lymphoid leukemia (ALL) in children has an 85–90 percent 5-year survival rate, and most children with ALL are probably cured. For childhood acute myeloid leukemia (AML), the 5-year survival is 60–70 percent. Treatments are being developed that show promise in combating specific forms of leukemia. For example, the administration of *gamma-interferon*, a hormone of the immune system, has been highly effective in treating hairy cell leukemia. Chronic myeloid leukemia (CML) has responded dramatically to the tyrosine kinase inhibitor imatinib mesylate.

One option for treating acute leukemias is to perform a stem cell transplant, harvested from blood or bone marrow. Most bone marrow/stem cell transplants involve aggressive "ablative" therapy with radiation and chemotherapy, which is meant to destroy the patient's cancer and also destroys the patient's bone marrow stem cells. This result then requires the replacement of functioning stem cells for the patient's survival. Following the ablative therapy the individual receives an infusion of healthy bone marrow with its stem cells, which repopulate the blood and marrow tissues. Transplants using stem cells collected by filtering the donors' blood are possible, reducing the need for bone marrow collection.

If the hematologic stem cells are extracted from another person (a **heterologous transplant** or *allograft*), care must be taken to ensure that the blood types and tissue types are compatible. This reduces the risk that the new lymphocytes may attack the patient's tissues (graft-versus-host disease, or GVHD), with potentially fatal results. The best results are obtained when the donor is a close relative. Fetal blood with fetal hematologic stem cells obtained from the umbilical cord after birth appears to be more compatible with unrelated recipients, and stored cord blood banks exist in Europe and North America. While the number of stem cells required for successful transplantation increases with body size, cord blood transplantation has been successful in both children and adults. In older patients too fragile to tolerate aggressive therapy, non-ablative transplant therapy has been tried. Reduced amounts of chemo and radiation therapy are used, and an allograft of stem cells is performed. Any graft-versus-host disease that occurs is hoped to attack any remaining leukemia cells, with serious GVHD avoided by immunosuppressive treatment.

In an **autologous marrow transplant**, bone marrow is removed from the patient, cleansed of cancer cells, and reintroduced after radiation or chemotherapy treatment. Although this method produces fewer complications, the preparation and cleansing of the marrow are technically difficult, and a recurrence of leukemia is more likely.

Whatever the source (bone marrow, peripheral blood, or umbilical cord blood), successful transplants require a large number of hematologic stem cells, and may require ongoing immunosuppression for less-well-matched recipient/donor pairs.

The Clotting System: A Closer Look

▶**Figure 54a**, the clotting cascade, presents the interactions of the intrinsic and extrinsic pathways and the roles of the coagulation factors. ▶**Figure 54b** presents the fibrinolytic system, which is activated as thrombin forms. This linked pathway provides a mechanism for the gradual removal of a blood clot during tissue repair.

Testing the Clotting System

Several clinical tests check the efficiency of the clotting system (**Table 27c**).

Bleeding Time

This test measures the time it takes a small bleeding skin wound to clot and cease bleeding. It primarily tests platelet function. There are variations on the procedure, with normal values ranging from three to seven minutes. Aspirin prolongs bleeding time by affecting platelet function and suppressing the extrinsic pathway.

Coagulation Time

In this test, a sample of whole blood is observed under controlled conditions until a visible clot has formed. Normal values range from 3 to 15 minutes. The test has several potential sources of error and so is not very accurate. Nevertheless, it is the simplest test that can be performed on a blood sample.

Cardiovascular System

▶ **Figure 54** The Clotting System.

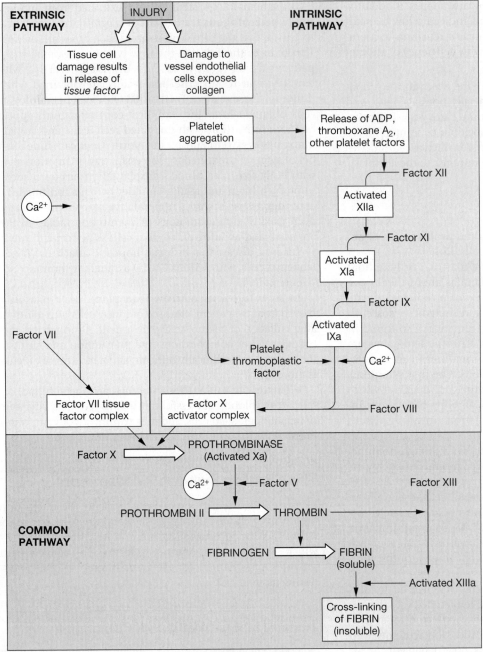

(a) The Clotting Cascade

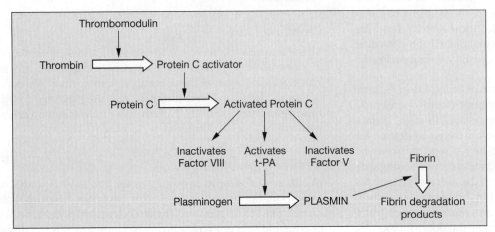

(b) The Fibrinolytic System

Partial Thromboplastin Time (PTT)

In this test, a plasma sample is mixed with chemicals that mimic the effects of activated platelets. Plasma is obtained by adding citrate ions to a blood sample. This ties up calcium ions and prevents clotting. Calcium ions are then added to the plasma, and the clotting time is recorded. Clotting normally occurs in 35–50 seconds if the enzymes and clotting factors of the intrinsic pathway are present in normal concentrations. The PTT is prolonged by heparin and is used to monitor its therapeutic use.

Prothrombin Time

This test checks the performance of the extrinsic pathway. The procedure is similar to that used in the PTT test, but the clotting process is triggered by exposure to a combination of tissue thromboplastin and calcium ions. Clotting normally occurs in 13–17 seconds. The **prothrombin time** (PT) is prolonged by the drug coumadin, and this anticoagulant therapy is monitored by repeat PT testing. For better standardization of test results, most laboratories report PT results that have been adjusted to the International Normalized Ratio (INR) for patients taking anticoagulant drugs. The desired INR varies with the reason for anticoagulation therapy.

Abnormal Hemostasis

Excessive or Abnormal Blood Clotting

If the clotting response is inadequately controlled, blood clots will begin to form in the bloodstream rather than at the site of an injury. These blood clots may not stick to the wall of the vessel, but continue to drift around until either plasmin digests them or they become lodged in a small blood vessel. A drifting blood clot is a type of **embolus** (EM-bō-lus; *embolos*, plug), an abnormal mass within the bloodstream. An embolus that becomes stuck in a blood vessel blocks circulation to the area downstream, killing the affected tissues. The sudden blockage is called an **embolism**, and the area of tissue damage caused by the circulatory interruption is one form of an **infarct**. Infarctions in the brain are known as *strokes*; infarctions in the heart are called *myocardial infarctions*, or *heart attacks*.

An embolus in the arterial system can get stuck in capillaries in the brain, causing an embolic stroke. An embolus in the venous system will probably become lodged in one of the capillaries of the lungs, resulting in *pulmonary embolism*.

A **thrombus** (*thrombos*, clot), or blood clot attached to a vessel wall, begins to form when platelets stick to the wall of an intact blood vessel. Often, the platelets are attracted to areas called *plaques*, where endothelial and smooth muscle cells contain large quantities of lipids and may be inflamed. (The mechanism of plaque formation will be discussed in a later section.) The thrombus enlarges, projecting into the lumen of the vessel and reducing its diameter. Smaller vessels (usually arteries such as the coronary arteries) may be completely blocked, creating an infarct, or a large chunk of the clot may break off, creating an equally dangerous embolus.

To treat an acute myocardial infarction (heart attack) caused by these circulatory blockages, clinicians may attempt systemic intravenous thrombolysis (*lysis*, dissolution) with t-PA ("tissue plasminogen activator") or Streptokinase. Streptokinase is relatively inexpensive but has more side effects, and t-PA is more often used in the United States. If available within 2 hours, emergency Percutaneous Coronary Interventions (PCI) with catheters and/or stents (a metal mesh tube that can hold a vessel open) introduced into the affected vessels can restore blood flow. Important anticoagulant drugs include:

- Heparin, which activates antithrombin-III.

- **Coumadin** (COO-ma-din) or *warfarin* (WAR-fa-rin), and **dicumarol** (dī-KOO-ma-rol), which depress the synthesis of several clotting factors by blocking the action of vitamin K.

- **Streptokinase** (strep-tō-KĪ-nās) and recombinant DNA–synthesized **tissue plasminogen activator (t-PA)**, which are enzymes that convert plasminogen to plasmin.

- Aspirin, which (1) inactivates platelet enzymes involved in the production of thromboxanes and prostaglandins and (2) inhibits the production of prostacyclin by endothelial cells. Daily ingestion of small quantities of aspirin reduces the sensitivity of the clotting process. This method has been proven effective in preventing heart attacks in people with significant coronary artery disease.

For some blood tests, it is necessary to prevent clotting within a collected blood sample to avoid changes in plasma composition. Blood samples can be stabilized temporarily by the addition of heparin or *EDTA* (*e*thylene*d*iamine*t*etroacetic *a*cid) to the collection tube. EDTA removes Ca^{2+} from plasma, effectively preventing clotting. In units of whole blood held for extended periods in a blood bank, *citratephosphate dextrose* (CPD) is typically added. Like EDTA, CPD ties up plasma Ca^{2+}.

Inadequate Blood Clotting

Hemophilia (hē-mō-FĒL-ē-uh) is an inherited disorder with inadequate production of one of the clotting factors, most commonly Factor VIII or IX. Single gene mutations in the *F8* and *F9* genes on the X chromosome cause hemophilia. The condition affects between 1 in 4000 and 1 in 16,000 people worldwide, 90 percent of whom are males. As in most X-linked diseases, affected females have milder problems. The severity of the condition depends on the degree of underproduction. In severe cases, extensive internal bleeding accompanies the slightest mechanical stress; hemorrhages occur spontaneously at joints and around muscles.

Transfusions of clotting factors, either concentrated from pooled plasma from blood donors or manufactured from recombinant gene splicing techniques (available for Factors VIII and IX), can reduce or control the symptoms of hemophilia. Pooled plasma must be sterilized in several ways to reduce the risk of blood-borne infections such as HIV or hepatitis. Interestingly, hemophiliacs who developed liver failure from hepatitis and survived liver transplantation were cured of hemophilia, since the clotting factors are synthesized in the liver. Gene therapy, once developed and perfected, may theoretically be curative.

The condition known as **von Willebrand disease** is the most common inherited coagulation disorder, affecting both sexes, and even occurs in dogs. The *von Willebrand factor (vWF)* is a plasma protein that binds and stabilizes Factor VIII. The symptoms and severity of the bleeding vary widely. Many individuals

Cardiovascular System

with mild forms of von Willebrand disease remain unaware of any bleeding problems until they have an accident or undergo surgery. Because a manufactured form of vWF is not yet available (one for dogs has been made and is effective), treatment consists of the administration of pooled serum Factor VIII, because pooled Factor VIII also contains normal vWF. In some forms of this disease in which normal vWF is produced but plasma levels are abnormally low, bleeding can be controlled by the intermittent use of an injection or nasal spray containing a synthetic form of antidiuretic hormone (ADH). The ADH stimulates the release of Factor VIII and vWF from endothelial cells. It is also effective in mild cases of hemophilia A.

In *disseminated intravascular coagulation* (DIC), also called *consumption coagulopathy*, bacterial toxins or factors released by damaged tissues activate thrombin, which then converts fibrinogen to fibrin within the circulating blood. Much of the fibrin is removed by phagocytes or dissolved by plasmin, but tiny clots may block small vessels and damage the associated tissues. If the liver cannot keep pace with the increased demand for fibrinogen, coagulation factors are depleted and uncontrolled bleeding may occur. DIC is one of the complicating factors of septicemia, a dangerous infection of the bloodstream that spreads bacteria and bacterial toxins throughout the body.

▶ Disorders of the Heart

Infection and Inflammation of the Heart

Many different microorganisms may infect heart tissue, leading to serious cardiac abnormalities. **Carditis** (kar-DĪ-tis) is a general term for inflammation of the heart. Clinical conditions resulting from cardiac infection are usually identified by the primary site of the infection. For example, infections that affect the endocardium produce symptoms of **endocarditis**, a condition that damages primarily the chordae tendineae and heart valves. Bacteria from infection in other parts of the body that spread through the blood cause bacterial endocarditis. The most severe complications of endocarditis result from the formation of blood clots on the damaged surfaces. These clots subsequently break free, entering the bloodstream as drifting emboli that may cause strokes, heart attacks, or kidney failure. The destruction of heart valves by the infection may lead to valve leakage, heart failure, and death.

Myocarditis, inflammation of the heart muscle, can be caused by bacteria, viruses, protozoans, or fungal pathogens that either attack the myocardium directly or produce toxins that damage the myocardium. The microorganisms implicated include those responsible for many of the conditions discussed in previous sections, such as diphtheria, syphilis, polio, tuberculosis, and malaria. The membranes of infected heart muscle cells become facilitated, and the heart rate may rise dramatically. Over time, abnormal contractions may appear and the heart muscle weakens; these problems may eventually prove fatal.

If the pericardium becomes inflamed or infected, fluid may accumulate around the heart (*cardiac tamponade*) or the elasticity of the pericardium may be reduced (*constrictive pericarditis*). In both conditions, the expansion and filling of the heart is restricted and cardiac output is reduced. Treatment involves draining the excess fluid or cutting a window in the pericardial sac.

The Cardiomyopathies

The **cardiomyopathies** (kar-dē-ō-mī-OP-a-thēz) include an assortment of diseases with a common symptom: the progressive, often irreversible degeneration of the myocardium. Cardiac muscle cells are damaged and replaced by fibrous tissue, and the muscular walls of the heart become thin and weak. As muscle tone declines, the ventricular chambers enlarge greatly. When the remaining cells cannot develop enough force to maintain cardiac output, symptoms of heart failure develop.

Chronic alcoholism and coronary artery disease are probably the most common causes of cardiomyopathy in the United States. Infectious agents, especially viruses, also cause cardiomyopathy. In addition, diseases affecting neuromuscular performance, such as muscular dystrophy (p. 66), damage cardiac muscle cells, as can starvation or chronic variations in the extracellular concentrations of calcium or potassium ions.

Several forms of cardiomyopathy are inherited. **Hypertrophic cardiomyopathy** (HCM) is an inherited disorder in which the walls of the left ventricle and the interventricular septum thicken to the point that the ventricle has difficulty pumping blood. Most people with HCM do not become aware of it until relatively late in life. However, HCM can also cause a fatal arrhythmia and has been implicated in the sudden deaths of several young athletes. The implantation of an electronic cardiac pacemaker has proved to be beneficial in controlling HCM-related arrhythmias.

Finally, there are a significant number of cases of *idiopathic cardiomyopathy*, a term used when the primary cause cannot be determined.

Heart Transplants and Assist Devices

Individuals with severe cardiomyopathy may be considered as candidates for a heart transplant. This surgery involves the removal of the weakened heart and its replacement with a heart taken from a suitable donor. To survive the surgery, the recipient must be in otherwise satisfactory health. In the United States in 2012, 2378 heart transplants were done. Over 42 percent of transplant recipients were age 50–64. More than 3500 patients, ranging in age from infancy to over 65, were waiting for a heart transplant. After successful transplantation, the 1-year survival rate is 89 percent and the 5-year survival rate is 74 percent. These rates are quite good, especially considering that most of these patients were expected to die if the transplant had not been performed.

Many individuals with cardiomyopathy who are initially selected for heart transplant surgery succumb to the disease before a suitable donor becomes available. Several total artificial hearts which are surgically sewn to the atria and replace the ventricles have been tried. One model, currently named SynCardia, a temporary Total Artificial Heart, was used in the first artificial heart transplant in 1982. It requires an external power source. The AbioCor artificial heart is totally implantable and does not need an attached external power source. A continuing problem with artificial hearts is that blood clots tend to form on the mechanical valves. When the clots break free, they become drifting emboli that plug peripheral vessels, producing strokes, kidney failure, and other complications. In addition, infections involving the mechanical connections of the heart sometimes lead to sepsis and death. These units have been successfully used as a "bridge to transplant" to maintain transplant candidates who are awaiting the arrival of a donor

Cardiovascular System

organ. Some new machines are called *left ventricular assist devices* (LVADs). As the name implies, they assist, rather than replace, the damaged heart. In one study, the survival rate was 75 percent for patients relying on such a device for 3½ months. The devices are also being called on to reduce the workload on weak hearts, giving them time to heal, or even to supplement the heart indefinitely, rather than acting solely as a bridge to heart transplantation. However, all these devices usually extend life for months, not years, beyond what is expected in end-stage heart disease.

As public-health knowledge and education about the risk factors for heart disease have improved, and more effective treatments for hypertension, high cholesterol, and heart failure have been developed, death rates from ischemic coronary artery disease have declined significantly. Between 1980 and 2010, the rates fell from 345 to 122 per 100,000 population. It has been estimated that from the mid-1960s to 2006 there were over 630,000 fewer deaths from coronary heart disease than would have been expected had mortality rates stayed the same.

RHD and Valvular Stenosis

Rheumatic (roo-MA-tik) **fever** is an inflammatory autoimmune condition that can develop after untreated infection by streptococcal bacteria ("strep throat"). Rheumatic fever typically affects children of 5–15 years of age; symptoms include high fever, joint pain and stiffness, and sometimes a distinctive full-body rash. Obvious symptoms generally persist for less than six weeks, although severe cases may linger for six months or more. The longer the duration of the inflammation, the more likely it is that carditis will develop. The carditis that does develop in 50–60 percent of individuals may escape detection, as scar tissue gradually forms in the myocardium and the heart valves. Valve condition deteriorates over time, and valve problems serious enough to affect cardiac function may not appear until 10–20 years after the initial infection.

During the interim, the affected valves become thickened and may also calcify to some degree. This thickening narrows the opening guarded by the valves, producing a condition called **valvular stenosis** (ste-NŌ-sis; *stenos*, narrow). The resulting clinical disorder is known as **rheumatic heart disease (RHD)**. The thickened cusps stiffen in a partially closed position, but the valves do not completely block the blood flow, because the edges of the cusps are rough and irregular. Regurgitation (reverse blood flow through the valve) may occur, and much of the blood pumped out of the heart may leak/flow back in. The abnormal valves are also much more susceptible to bacterial infection, a type of endocarditis (p. 132). Fortunately, with the detection and prompt antibiotic treatment of strep infections, the number of cases of RHD has declined dramatically since the 1940s.

Mitral stenosis and **aortic stenosis** are the most common forms of RHD. About 40 percent of patients with RHD develop mitral stenosis, and two-thirds of them are women. The reason for the correlation between gender and mitral stenosis is unknown. In mitral stenosis, blood enters the left ventricle at a slower-than-normal rate; in mitral regurgitation, when the ventricle contracts, blood flows back into the left atrium, as well as forward through the aortic valve (which may itself be stenotic and leaky) into the aortic trunk. The right and left ventricles discharge identical amounts of blood with each beat, so, as the output of the left ventricle declines, blood "backs up" in the pulmonary circuit. Venous pressures then rise in the pulmonary circuit, and the right ventricle must develop greater pressures to force blood into the pulmonary trunk. In severe cases of mitral stenosis, both left and right ventricles have to work much harder to maintain adequate circulation. The heart weakens and peripheral tissues begin to suffer oxygen and nutrient deprivation from the reduced cardiac output. (We discuss this condition, called congestive heart failure or CHF in more detail in a later section.)

Valvular heart disease can be a birth defect or caused by other heart diseases. Aortic stenosis accounts for 25 percent of valvular disease. Stenosis frequently develops in persons born with bicuspid aortic valves, 80 percent of whom are males. If rheumatic fever is the cause, the mitral valve is usually affected as well. Symptoms of aortic stenosis are initially less severe than those of mitral stenosis. Although the left ventricle enlarges and works harder, normal circulatory function can be maintained for years. Clinical problems develop only when the opening narrows enough to prevent adequate blood flow. Symptoms then resemble those of mitral stenosis.

One reasonably successful treatment for severe stenosis involves the replacement of the damaged valve. ▶ Figure 55a shows a stenotic heart valve; two possible replacements are a valve from a pig (▶ Figure 55b) and a synthetic valve (▶ Figure 55c), one of

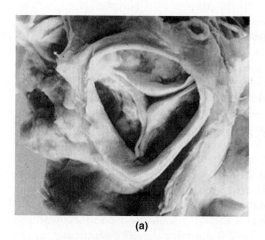
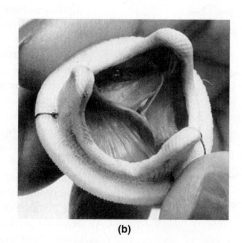

(a)   (b)   (c)

▶ **Figure 55 Artificial Heart Valves. (a)** A stenotic semilunar valve; note the irregular, stiff cusps. **(b)** Intact Bioprosthetic heart valve, which uses the valve from a pig's heart. **(c)** Medtronic Hall prosthetic heart valve.

Cardiovascular System

a number of designs that have been employed. Pig valves do not require anticoagulant therapy, but may wear out and begin leaking after roughly 10 years of service. The plastic or stainless-steel components of the artificial valve are more durable, but they activate the clotting system of the recipient, leading to inflammation, clot formation, and other complications. Synthetic-valve recipients must take anticoagulant drugs to prevent strokes and other disorders caused by the formation of intracardiac blood clots and emboli. Valve replacement operations are quite successful, with about 95 percent of the surgical patients surviving for three years or more and 70 percent surviving more than nine years.

Diagnosing Abnormal Heartbeats

Damage to the conduction pathways caused by mechanical distortion, ischemia, infection, or inflammation can affect the normal rhythm of the heart. The resulting condition is called a **heart block**, or **conduction deficit**. ▶Figure 56a shows the electrocardiogram of a normal heart, and this figure also shows heart blocks of varying severity. In a **first-degree heart block** (▶Figure 56b), the AV node and the proximal portion of the AV

(a) Normal

(b) First-degree heart block (long P–R interval)

Skipped ventricular beat

2:1 block (ventricles follow every other atrial beat)

3:1 block (ventricles follow every third atrial beat)

(c) Second-degree blocks

(d) Complete block (third-degree block)
(atrial beats occur regularly, ventricular beats occur at slower, unrelated pace)

▶**Figure 56 Heart Blocks.**

bundle slow the passage of impulses that are heading for the ventricular myocardium. As a result, a pause appears between the atrial and ventricular contractions. Although a delay exists, the regular rhythm of the heart continues, and each atrial beat is followed by a ventricular contraction.

If the delay lasts long enough, the nodal cells will still be repolarizing from the previous beat when the next impulse arrives from the SA node. The arriving impulse will then be ignored, the ventricles will not be stimulated, and the normal "atria–ventricles, atria–ventricles" pattern will disappear. This condition is a **second-degree heart block** (▶Figure 56c). A mild second-degree block may produce only an occasional skipped beat, but with more substantial delays, the ventricles will follow every second atrial beat. The resulting pattern of "atria, atria–ventricles, atria, atria–ventricles" is known as a **two-to-one (2:1) block**. Three-to-one or even four-to-one blocks are also encountered.

In a **third-degree heart block**, or **complete heart block**, the conducting pathway stops functioning (▶Figure 56d). The atria and ventricles continue to beat, but their activities are no longer synchronized. The atria follow the pace set by the SA node, beating 70–80 times per minute, and the ventricles follow the commands of the AV node, beating at a rate of 40–60 beats per minute. A temporary heart block can be induced by stimulating the vagus nerve. In addition to slowing the rate of impulse generation by the SA node, such stimulation inhibits the AV nodal cells to the point at which they cannot respond to normal stimulation. Comments such as "my heart stopped" or "my heart skipped a beat" generally refer to this phenomenon. The pause typically lasts just a few seconds. Longer delays end when a conducting cell, normally one of the Purkinje fibers, depolarizes to threshold. This phenomenon is called **ventricular escape**, because the ventricles are escaping from the control of the SA node. Ventricular escape can be a lifesaving event if the conduction system is damaged. Even without instructions from the SA or AV nodes, the ventricles will continue to pump blood at a slow (40 to 50 beats per minute), but steady, rate.

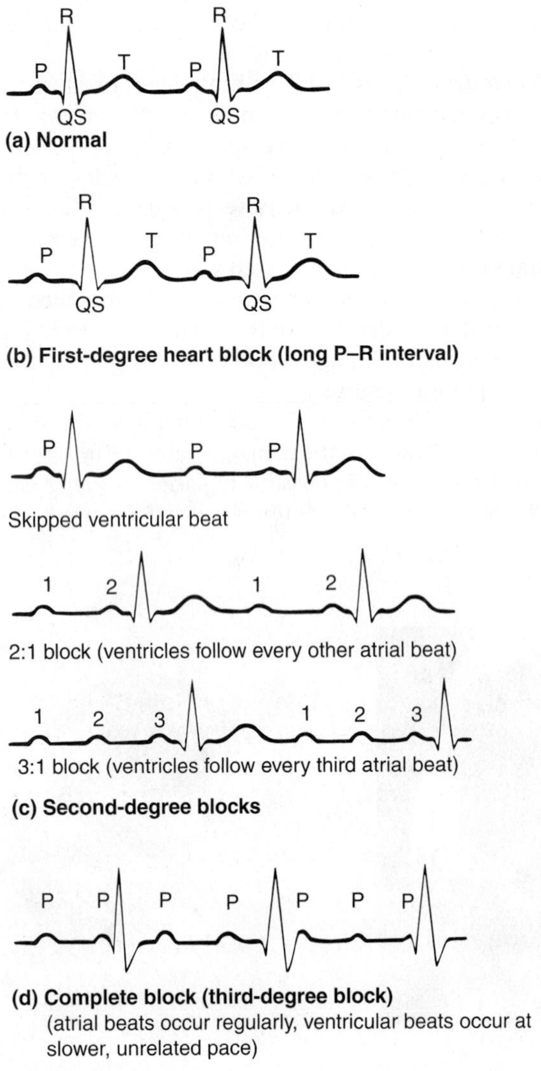

Tachycardia and Fibrillation

Additional important examples of arrhythmias are shown in ▶Figure 57. ▶Figure 57a shows a normal heart rhythm. **Premature atrial contractions (PACs)**, indicated in ▶Figure 57b, often occur in healthy individuals. In a PAC, the normal atrial rhythm is momentarily interrupted by a "surprise" atrial contraction. Stress, caffeine, and various drugs may increase the incidence of PAC, presumably by increasing the permeabilities of the SA pacemakers. The impulse spreads along the conduction pathway, and a normal ventricular contraction follows the atrial beat. In **paroxysmal** (par-ok-SIZ-mal) **atrial tachycardia**, or **PAT**, a premature atrial contraction triggers a flurry of atrial activity (▶Figure 57c). The ventricles are still able to keep pace, and the heart rate jumps to about 180 beats per minute. In **atrial flutter**, the atria contract in a coordinated manner, but the contractions occur very frequently. Affecting at least 2.6 million Americans, **atrial fibrillation** (fi-bri-LĀ-shun) (AF) is the most common type of arrhythmia. In AF the impulses move over the atrial surface at rates of perhaps 500 beats per minute (▶Figure 57d). The atrial wall quivers instead of producing the normal organized contraction that helps fill the ventricles. The ventricular rate in atrial flutter or atrial fibrillation cannot follow the atrial rate

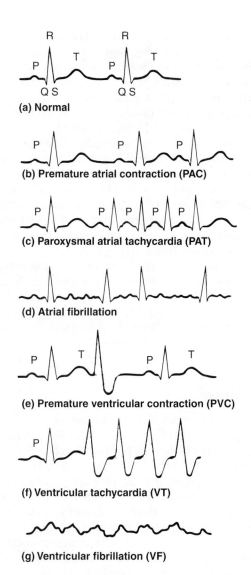

Figure 57 Cardiac Arrhythmias.

increased by exposure to epinephrine, to other stimulatory drugs, or to ionic changes that depolarize cardiac muscle plasma membranes. Similar factors may be responsible for periods of **ventricular tachycardia** (defined as four or more PVCs without intervening normal beats), also known as **VT** or *V-tach* (Figure 57f). Multiple PVCs and VT often precede the most serious arrhythmia, **ventricular fibrillation (VF)** (Figure 57g). The resulting condition, also known as **cardiac arrest**, is rapidly fatal, because the heart quivers instead of pumping blood. During ventricular fibrillation, the cardiac muscle cells are overly sensitive to stimulation, and the impulses are traveling from cell to cell, around and around the ventricular walls. A normal rhythm of coordinated pumping contractions cannot become established, because the ventricular muscle cells are stimulating one another at such a rapid rate. The problem is sustained by a rise in free intracellular calcium ion concentrations, due to a massive stimulation of alpha and beta receptors following sympathetic activation.

Treating Problems with Pacemaker Function

Symptoms of severe **bradycardia** (below 50 beats per minute) include weakness, fatigue, fainting, and confusion. Drug therapies are seldom helpful, but artificial pacemakers can be used with considerable success. Wires are run to the atria, the ventricles, or both, depending on the nature of the problem, and the unit delivers small electrical pulses to stimulate the myocardium. Internal pacemakers are surgically implanted, batteries and all. These units last seven to eight years or more before another operation is required to change the battery. External pacemakers are used for temporary emergencies, such as immediately after cardiac surgery. Only the wires are implanted, and an external control box is worn on a belt.

More than 418,000 artificial pacemakers were implanted in the United States in 2009 (Figure 58). The simplest provide constant stimulation to the ventricles at rates of 70–80 impulses

but usually beats faster than normal. Despite the fact that the atria are now essentially nonfunctional, if the ventricular rate is not too fast the condition may go unnoticed, especially in older individuals who lead sedentary lives. In chronic atrial fibrillation blood flowing slowly through weakly quivering atria may form clots near the atrial walls. Pieces of a clot may break off, creating emboli, and increase the risk of stroke. As a result, most people diagnosed with this condition are placed on anticoagulant therapy. PACs, PAT, atrial flutter, and even atrial fibrillation (except for the stroke risk) are not considered very dangerous, unless they are prolonged or associated with some more serious indications of cardiac damage, such as coronary artery disease, congestive heart failure, and/or valve problems.

In contrast, ventricular arrhythmias can be fatal. Because the conduction system functions in one direction only, from atria to ventricle, a ventricular arrhythmia is not linked to atrial activities. **Premature ventricular contractions (PVCs)** occur when a Purkinje cell or ventricular myocardial cell depolarizes to threshold and triggers a premature contraction (Figure 57e). Single PVCs are common and not dangerous. The cell responsible is called an *ectopic pacemaker*. The frequency of PVCs can be

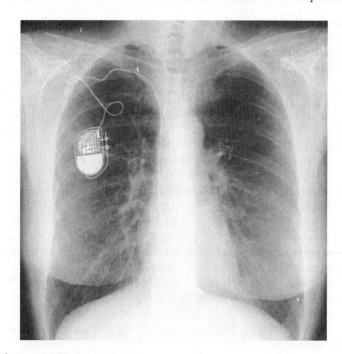

Figure 58 An Artificial Pacemaker.

per minute. More sophisticated pacemakers stimulate the atria and ventricles in sequence and may vary their rates to adjust to changing circulatory demands, such as during exercise. Others are able to monitor cardiac activity and respond whenever the heart begins to function abnormally. A **defibrillator** is a device that attempts to eliminate ventricular fibrillation and restore normal cardiac rhythm. An external defibrillator has two electrodes that are placed in contact with the chest, and a powerful electrical shock is administered. The electrical stimulus depolarizes the entire myocardium simultaneously. With luck, after repolarization, the SA node will be the first area of the heart to reach threshold. Thus, the primary goal of defibrillation is not just to stop the fibrillation, but to give the ventricles a chance to respond to normal SA commands.

Early defibrillation can result in dramatic recovery of an unconscious cardiac-arrest victim. **Automatic external defibrillators (AEDs)** are easily used portable machines that can detect lethal ventricular rhythms in persons who have collapsed and administer a defibrillating shock. These devices are increasingly being placed in or at planes, airports, sports fields, and shopping centers to be available for use in emergencies by the public much as fire extinguishers are. Implantable cardioverter defibrillators detect abnormal ventricular rhythms and administer a shock to restore normal rhythm. These devices can prevent sudden cardiac death in people who have survived an episode of ventricular tachycardia and/or ventricular fibrillation. Over 114,000 of the cardioverters were implanted in 2009.

Tachycardia, usually defined as a heart rate of over 100 beats per minute, increases the workload on the heart. Cardiac performance suffers at very high heart rates, because the ventricles do not have enough time to refill with blood before the next contraction occurs. Chronic or acute incidents of tachycardia may be controlled by drugs that affect the permeability of pacemaker membranes or block the effects of sympathetic stimulation.

Examining the Heart

Many techniques can be used to examine the structure and performance of the heart. No single diagnostic procedure provides the complete picture, so the tests used will vary with the suspected nature of the problem. A standard chest x-ray will show the basic size, shape, and orientation of the heart. An electrocardiogram (ECG or EKG) records the electrical activity of the heart, providing basic information on heart rhythm and heart size, and detecting signs of cardiac damage. Additional details require more specialized procedures to enhance the clarity of the images.

Coronary arteriography (ar-tē-rē-OG-ra-fē) is often used to look for abnormalities in the coronary circulation. In this procedure, a catheter is inserted into the femoral artery in the leg or the radial artery at the wrist and is maneuvered back along the arterial passageways until its tip reaches the aortic sinus. A radiopaque dye can then be released at the openings of the coronary arteries, and its distribution can be followed in a series of high-speed x-rays. The images obtained are called **coronary angiograms** (❱ Figure 59a). For direct analyses of cardiac performance, the monitoring of chamber pressures, or the collection of blood samples, a catheter may be introduced past the aortic or tricuspid valves and into the ventricles and atria. The catheters enter via a femoral artery and the aorta, or from the venous system by way of a femoral vein and the inferior vena cava. As technology improves, even the smaller

vessels of the arm are being used for heart catheterization. Because the heart is constantly moving, ordinary ultrasound, computerized tomography (CT), and magnetic resonance imaging (MRI) scans create blurred images. Special instruments and computers, however, that can generate images at high speeds can be used with these techniques to develop three-dimensional still or moving pictures of the heart as it beats (❱ Figure 59b–c). The resulting images are dramatic, but the cost, complexity of the equipment, and radiation exposure in the case of coronary CT exams have so far limited their use. MRI scans can detect calcifications in coronary arteries, which are associated with coronary artery disease and may be detectable before symptoms appear. The value of scanning for early coronary heart disease versus the cost and complications of further diagnosis and treatment in asymptomatic persons detected by these scans is controversial.

Ultrasound analysis of the heart, called **echocardiography** (ek-ō-kar-dē-OG-ra-fē) (❱ Figure 59b), provides images that lack the clarity of CT or MRI scans, but the equipment is relatively inexpensive and portable, making it an important and common diagnostic tool. Recent advances in data processing have made the images suitable for following details of cardiac contractions, including valve function and blood flow dynamics.

Heart Failure

Heart failure exists when the cardiac output is insufficient to meet the circulatory demand. The initial symptoms of heart failure depend on whether the problem involves the left ventricle, the right ventricle, or both. However, over time, these differences are reduced or eliminated; for example, the major cause of right ventricular failure is left ventricular failure. ❱ Figure 60 is a flowchart that shows the basic causes and effects of heart failure and indicates potential therapies.

Suppose that the left ventricle cannot maintain normal cardiac output due to damage to the ventricular muscle or high arterial pressures (hypertension, pp. 140–141). In effect, the left ventricle can no longer keep pace with the right ventricle, and blood backs up into the pulmonary circuit. This venous congestion is responsible for the term **congestive heart failure (CHF)**. The right ventricle now works harder, elevating pulmonary arterial pressures and forcing blood through the lungs and into the weakened left ventricle.

At the capillaries of the lungs, arterial and venous pressures are now elevated. The elevated pressure pushes additional fluid out of the capillaries and into the interstitial fluids, most notably at the lungs. The buildup of fluid and compression of the airways reduce the effectiveness of gas exchange, leading to hypoxia and shortness of breath, typically the first obvious sign of congestive heart failure. The fluid buildup begins at a pulmonary postcapillary pressure of about 25 mm Hg. At a capillary pressure of about 30 mm Hg, fluid not only enters the tissues of the lungs, but crosses the alveolar walls and begins to fill the air spaces. This condition is called **pulmonary edema**. Abnormal sounds *(moist rales)* can be heard at the base of each lung on auscultation.

Over time, the less muscular right ventricle may become unable to generate enough pressure to force blood through the pulmonary circuit. Venous congestion now occurs in the systemic circuit, the ankles may become distended with fluid ("ankle edema"), the liver may become swollen with blood, and cardiac

Cardiovascular System

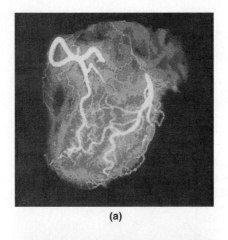

(a)

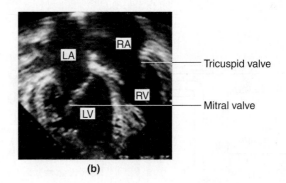

RA

LA

RV

LV

—— Tricuspid valve

—— Mitral valve

(b)

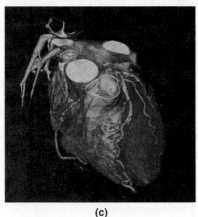

(c)

▶ **Figure 59 Monitoring the Heart. (a)** A coronary angiogram. **(b)** An echocardiogram. **(c)** Reconstructed three-dimensional CT scan of the heart.

Cardiovascular System

output declines further. When the reduction in systemic pressures lowers blood flow to the kidneys, renin and erythropoietin are released. The result is elevated blood volume as salt and water retention at the kidneys increase and RBC production accelerates. This rise in blood volume actually complicates the situation, as it tends to increase venous congestion and cause widespread edema.

The increased volume of blood in the venous system leads to a distension of the veins, making superficial veins more prominent. When the heart contracts, the rise in pressure at the right atrium produces a pressure pulse in the large veins. This venous pulse can be seen most easily over the right external jugular vein.

The treatment of CHF in the recent past involved bed rest to reduce the cardiac workload, reduced dietary salt, and diuretics—medicines that promote fluid loss by increasing salt and water loss at the kidneys. Digoxin, a medicine derived from digitalis (itself extracted from the foxglove plant), reduces symptoms by increasing the force of cardiac muscle contractions and helps control some heart rhythm abnormalities that contribute to CHF. The outmoded practice of "bloodletting" or "bleeding," which, along with foxglove, was common medical treatment in the 17th and 18th centuries, did cause a temporary reduction in blood volume and thereby reduced symptoms of CHF. However, the relief was only temporary, and the practice had many disadvantages, such as producing anemia.

Recent studies have shown that new medicines in combination with older remedies reduce mortality from CHF in selected groups of seriously ill patients. Drugs that lower peripheral vascular resistance (vasodilators), such as the ACE inhibitors and nitroglycerin, reduce the workload of the left ventricle and improve cardiac output. Spironolactone, a mild diuretic that is an aldosterone antagonist, and beta-blockers that decrease sustained adrenergic activation both reduce the cardiac workload and beta-blockers also alleviate some arrhythmias. The treatment of severe heart failure may involve all of these drugs and modalities up to and including heart transplantation. Obviously, however, prevention is preferable.

Abnormal Conditions Affecting Cardiac Output

Various drugs, abnormal variations in ion concentrations, and changes in body temperature can alter the basic rhythm of contraction established by the SA node. Earlier we noted that several drugs, including caffeine and nicotine, have a stimulatory effect on excitable membranes in the nervous system (p. 101). These drugs also cause an increase in heart rate. Caffeine acts directly on the conducting system and increases the rate of depolarization at the SA node. Nicotine acts indirectly by stimulating the activity of sympathetic neurons that innervate the heart.

Disorders affecting either ion concentrations or body temperature can have direct effects on cardiac output by changing the stroke volume, the heart rate, or both. Abnormal ion concentrations can change both the contractility of the heart, by

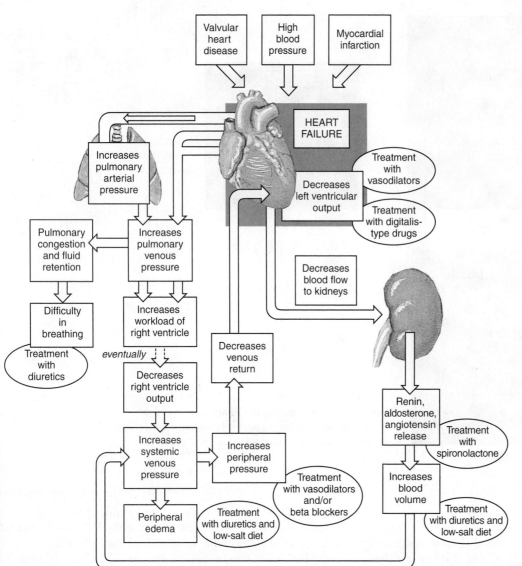

▶ **Figure 60 Heart Failure.**

Hyperkalemia and Hypokalemia

Changes in the extracellular K^+ concentration affect the transmembrane potential and the rates of depolarization and repolarization. In **hyperkalemia** (hī-per-ka-LĒ-mē-uh), in which K^+ concentrations are high, the muscle cells depolarize, and repolarization is inhibited. Cardiac contractions become weak and irregular; in severe cases, the heart eventually stops in diastole after ventricular fibrillation, a state known as *asystole*. In **hypokalemia** (hī-pō-ka-LĒ-mē-uh), in which K^+ concentrations are abnormally low, the membranes of cardiac muscle cells hyperpolarize and become less responsive to stimulation. Meanwhile, the hyperpolarization of nodal cells leads to a reduction in the heart rate. Blood pressure falls, and, in severe cases, the heart eventually stops in systole. Severe hyperkalemia and hypokalemia are life threatening, and require immediate corrective action.

Hypercalcemia and Hypocalcemia

In **hypercalcemia** (hī-per-kal-SĒ-mē-uh), a condition in which the extracellular concentration of Ca^{2+} is elevated, cardiac muscle cells become extremely excitable. Their contractions become powerful and prolonged. In extreme cases, the heart goes into an extended state of contraction that is generally fatal. In **hypocalcemia** (hī-pō-kal-SĒ-mē-uh), in which the Ca^{2+} concentration is abnormally low, the contractions become very weak and may cease altogether.

Abnormal Body Temperatures

Temperature changes affect metabolic operations throughout the body. For example, a reduction in temperature slows the rate of depolarization at the SA node, lowers the heart rate, and reduces the strength of cardiac contractions. (In open-heart surgery, the exposed heart may be deliberately chilled until it stops beating.) An elevated body temperature accelerates the heart rate and the contractile force. That is one reason your heart may seem to race and pound when you have a fever.

affecting the cardiac muscle cells, and the heart rate, by affecting the SA nodal cells. The most obvious and clinically important examples of problems with ion concentrations involve K^+ and Ca^{2+}.

Disorders Affecting Blood Vessels and Circulation

Aneurysms

An **aneurysm** (AN-ū-rizm) is a bulge in the weakened wall of a blood vessel (generally an artery) or a heart chamber (usually in the left ventricle). The bulge resembles a bubble in the wall of a tire—and like a bad tire, the affected artery may suffer a catastrophic "blowout," or rupture. The most dangerous aneurysms are those involving arteries of the brain, where they cause strokes, and of the aorta, where a rupture can cause fatal bleeding in a matter of minutes.

Aneurysms are thought to be caused by chronic high blood pressure, although any inflammation, trauma, or infection, such as syphilis, that weakens vessel walls can lead to an aneurysm. In addition, at least some aortic aneurysms have been linked to inherited disorders, such as *Marfan's syndrome*, which weakens connective tissues in vessel walls, and polycystic kidney disease which also increases risk for aneurysms.

It is not known whether other genetic factors are involved in the development of other types of aneurysms. Fibrous scar tissue replacing muscle tissue damaged in a heart attack may result in a left ventricular aneurysm. The damaged area doesn't contract and, if large enough, reduces cardiac output. In addition, the area is a site at which a blood clot may form, posing a risk of embolism.

Most aneurysms form gradually as vessel walls become less elastic. When a weak point develops, the arterial pressures distort the wall, creating an aneurysm. Unfortunately, because many aneurysms are painless, they are likely to go undetected until they leak or rupture.

When aneurysms are detected by ultrasound or other scanning procedures, the risk of rupture can sometimes be estimated on the basis of their size. For example, an aortic aneurysm larger than 6 cm has a 50 percent chance of rupturing within 10 years. Depending on size and location, aneurysm treatment includes reducing hypertension if present, watchful waiting to see if they enlarge, and when the risk of rupture exceeds the risk of repair, surgery. A large aneurysm in an accessible area, such as the abdomen, may be surgically removed and the vessel repaired. ▶ **Figure 61** shows a large aortic aneurysm and the steps involved in its surgical repair with a synthetic patch. Endovascular repair, the insertion of a tube graft into the artery with guide wires and catheters, does not require a large abdominal incision. This procedure is less invasive and may be appropriate for many patients.

Problems with Venous Valve Function

One of the consequences of aging is an increase in the fragility of connective tissues throughout the body. Blood vessels are no exception: With age, the walls of veins begin to sag. This change generally affects the superficial veins of the legs first, because at these locations gravity opposes blood flow. The situation is aggravated by a lack of exercise and by an occupation requiring long hours of standing or sitting. When there is little muscular activity in the leg to help keep the blood moving, venous blood pools on the proximal (heart) side of each valve. As the venous walls are distorted, the valves become less effective, and gravity can then pull blood back toward the capillaries. This pulling further impedes normal blood flow, and the veins become distended. The sagging, swollen vessels are called **varicose** (VAR-i-kōs) **veins.** Though superficial varicose veins are relatively

Cardiovascular System

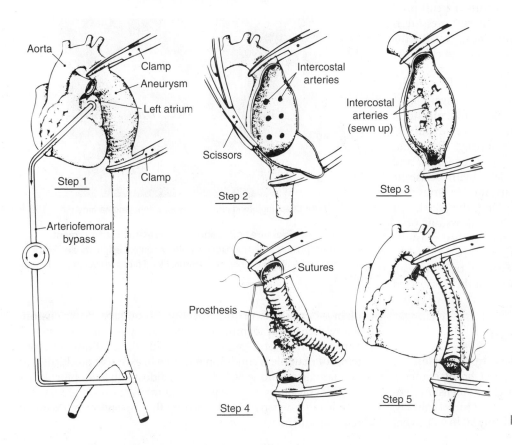

Step 1 — Aorta, Clamp, Aneurysm, Left atrium, Clamp, Arteriofemoral bypass

Step 2 — Intercostal arteries, Scissors

Step 3 — Intercostal arteries (sewn up)

Step 4 — Sutures, Prosthesis

Step 5

▶ **Figure 61 Repair of an Aneurysm.**

harmless, they may be painful, lead to chronic lower leg sores, and are unsightly. Surgical procedures are sometimes used to remove or constrict them.

Checking the Pulse and Blood Pressure

You can feel your pulse at any of the large or medium-sized arteries. The usual procedure involves using your fingertips to flatten an artery against a relatively solid mass, preferably a bone. When the vessel is compressed, you feel your pulse as an episodic pressure against your fingertips. The inside of the wrist is commonly used, because the *radial artery* can easily be pressed against the distal portion of the radius (▶ Figure 62a). Other accessible arteries include the *external carotid, brachial, temporal, facial, femoral, popliteal, posterior tibial,* and *dorsalis pedis arteries.* Firm pressure exerted at some of these locations, called **pressure points**, can reduce arterial bleeding distal to the site.

Blood pressure not only forces blood through the circulatory system, but also pushes outward against the walls of the containing vessels, just as air pushes against the walls of an inflated balloon. As a result, we can measure blood pressure indirectly by determining how forcefully the blood presses against the vessel walls.

The instrument used to measure blood pressure is called a **sphygmomanometer** (sfig-mō-ma-NOM-e-ter; *sphygmos,* pulse + *manometer,* device for measuring pressure). The liquid heavy-metal mercury (Hg) enclosed in a glass tube was used in sphygmomanometers but as it is neurotoxic (see p. 78), carefully calibrated mechanical devices are now used. An inflatable cuff is placed around the arm in such a position that its inflation compresses the brachial artery (▶ Figure 62b). A stethoscope is placed over the artery distal to the cuff, and the cuff is then inflated. A tube connects the cuff to a pressure gauge that reports the cuff pressure (in mm Hg). Inflation continues until the cuff pressure is roughly 30 mm Hg above the pressure sufficient to collapse the brachial artery completely, stop the flow of blood, and eliminate the sound of the pulse.

The air is then slowly let out of the cuff. When the pressure in the cuff falls below systolic pressure, blood can again enter the artery. At first, blood enters only at peak systolic pressures, and the stethoscope picks up the sound of blood pulsing through the artery. As the pressure falls further, the sound changes, because the artery is remaining open for longer and longer periods. When the cuff pressure falls below diastolic pressure, blood flow becomes continuous and the sound of the pulse becomes muffled or disappears. Thus, the pressure at which the pulse appears corresponds to the peak systolic pressure; when the pulse fades, the pressure has reached diastolic levels. The distinctive sounds heard during this test, called *sounds of Korotkoff* (sometimes spelled *Korotkov* or *Korotkow*), are produced by turbulence as blood flows past the constricted portion of the artery.

Hypertension and Hypotension

Elevated blood pressure is called **primary hypertension**, or *essential hypertension,* if no obvious cause can be determined. Known risk factors include a family history of hypertension, gender (males are at higher risk), high plasma cholesterol, obesity, chronic stresses, and cigarette smoking. Probably 90 to 95 percent

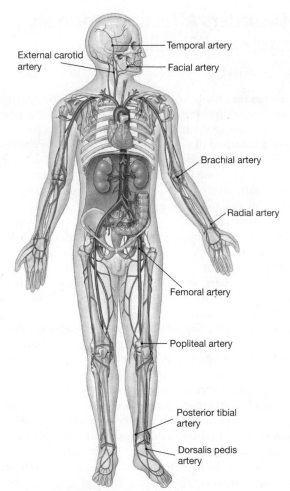

(a) Location of pressure points

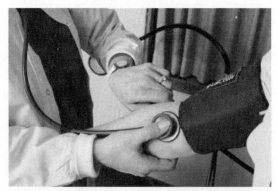

(b) Checking blood pressure using a sphygmomanometer

▶ **Figure 62 Pressure Points and Blood Pressure Measurement.**
(a) Pressure points used to monitor the pulse or control peripheral bleeding.
(b) Using a sphygmomanometer to measure blood pressure over brachial artery.

of hypertension is in this category. **Secondary hypertension**, where there is an identifiable cause, can appear as the result of abnormal hormonal production outside the cardiovascular system. For example, a condition resulting in excessive production of antidiuretic hormone (ADH), renin, aldosterone, or epinephrine typically produces hypertension, and many forms of kidney disease lead to hypertension caused by fluid retention or excessive renin production.

Hypertension significantly increases the workload on the heart, and the left ventricle gradually enlarges. The increased muscle mass requires a greater oxygen demand. When the coronary circulation cannot keep pace, symptoms of coronary ischemia or decreased coronary output may appear.

Increased arterial pressures also place a physical stress on the walls of blood vessels throughout the body. This stress promotes or accelerates the development of arteriosclerosis and increases the risks of aneurysms, heart attacks, and strokes. Vessels supplying the retinas of the eyes are often affected, and hemorrhages and associated circulatory changes can produce disturbances in vision. Because a routine physical includes the examination of these vessels, retinal changes may provide the first evidence that hypertension is affecting peripheral circulation.

One of the most difficult aspects of hypertension is that in most cases no obvious symptoms are present. As a result, clinical problems may not appear until the condition has reached the crisis stage. There is therefore considerable interest in the early detection and prompt treatment of hypertension.

Treatment consists of a combination of lifestyle changes and physiological therapies. Quitting smoking, getting regular exercise, and restricting one's dietary intake of salt, fats, and calories will improve peripheral circulation, prevent increases in blood volume and total body weight, and reduce plasma cholesterol levels. These strategies may be sufficient to prevent, delay onset, or control hypertension if it has been detected before significant cardiovascular damage has occurred. Most therapies involve antihypertensive drugs, such as diuretics, calcium channel blockers, and ACE inhibitors, singly or in combination. Diuretics promote the loss of water and sodium ions at the kidneys, lowering blood volume, and calcium channel blockers cause vasodilation to further reduce blood pressure. **ACE (angiotensin-converting enzyme) inhibitors**, such as *captopril*, cause vasodilation and lower blood pressure by preventing the conversion of angiotensin I to angiotensin II.

In **hypotension**, blood pressure declines and peripheral systems begin to suffer from oxygen and nutrient deprivation. One clinically important form of hypotension can develop with decreased blood volume or after antihypertensive drugs have been administered. Problems may appear when the individual changes position from lying down to sitting or from sitting to standing. Each time you sit up or stand, blood pressure in your carotid sinus drops because your heart must suddenly counteract gravity to push blood up to your brain. The fall in pressure triggers the carotid reflex, and blood pressure returns to normal. But if blood volume is low or the carotid response is blunted by beta-blockers or other antihypertensive drugs, blood pressure at the brain may fall so low that the individual becomes weak, dizzy, disoriented, or unconscious. This condition is known as **orthostatic hypotension** (or-thō-STAT-ik; *orthos*, straight + *statikos*, causing to stand), or simply **orthostasis** (or-thō-STĀ-sis). You may have experienced a brief episode of orthostasis when you stood suddenly after reclining for an extended period. The carotid reflex typically slows with age, so older people must sit and stand more carefully than in their younger years in order to avoid orthostatic hypotension. Diabetic neuropathy involving the autonomic nervous system may cause severe orthostatic hypotension.

Edema

Edema (e-DĒ-muh) is an abnormal accumulation of interstitial fluid. Edema has many causes, and we will encounter specific examples in other sections. The underlying problem in all types of edema is a disturbance in the normal balance between hydrostatic and osmotic forces at the capillary level. For instance:

- When a capillary is damaged, plasma proteins can cross the capillary wall and enter the interstitial fluid. The resulting elevation of the interstitial colloid osmotic pressure (ICOP) reduces the rate of capillary reabsorption and produces a localized edema. This is why swelling usually occurs at a bruise.

- In starvation, the liver cannot synthesize enough plasma proteins to maintain normal concentrations in the blood. The blood colloid osmotic pressure (BCOP) declines, and fluids begin moving from the blood into peripheral tissues. In children, fluid accumulates in the abdominopelvic cavity, producing the swollen bellies typical of starvation victims. A reduction in BCOP is also seen after severe burns and in several types of liver or kidney diseases.

- In the U.S. population, most serious cases of edema result from increases in venous pressure or total circulatory pressure. The increase may result from heart problems (such as heart failure), venous blood clots that elevate venous pressures, or other circulatory abnormalities. The net result is an increase in capillary hydrostatic pressure (CHP) that accelerates fluid movement into the tissues.

Edema can also result from problems with other organ systems, such as a blockage of lymphatic vessels or impaired kidney function:

- If the lymphatic vessels in a region become blocked, the volume of interstitial fluid will continue to rise, and the interstitial hydrostatic pressure (IHP) will gradually increase until capillary filtration ceases. In *filariasis* (pp. 147–148), a condition we will discuss in a later section, parasites can block lymphatic vessels and cause a massive regional edema known as *elephantiasis*.

- If the kidneys malfunction and leak protein into the urine and/or retain sodium and water due to decreased function, blood volume may rise and BCOP decline. This situation ultimately leads to elevated CHP and peripheral edema.

The Causes and Treatment of Cerebrovascular Disease

Symptoms of cerebrovascular disease appear when atherosclerosis reduces the circulatory supply to the brain. If the blood flow to a portion of the brain is completely shut off, a *cerebrovascular accident (CVA)*, or *stroke*, occurs. Roughly 88 percent of strokes are ischemic from **cerebral thrombosis** (clot formation at a plaque) or **cerebral embolism** (intravascular drifting blood clots, fatty masses, or air bubbles). **Cerebral hemorrhages** (rupture of a blood vessel, often at the site of an aneurysm) cause most other strokes. The observed symptoms and their severity vary with the vessel involved and the location of the blockage.

If the circulatory blockage disappears in a matter of minutes, the effects are temporary and the condition is called a **transient ischemic attack (TIA)**. A TIA typically indicates

Cardiovascular System

that cerebrovascular disease exists, so preventive measures can be taken to forestall more serious incidents. For example, taking aspirin each day slows blood clot formation in patients who experience TIAs and thereby reduces the risks of cerebral thrombosis and cerebral embolism.

If the blockage persists for a longer period, neurons die. Stroke symptoms are initially exaggerated by swelling and distortion of the injured neural tissues; if the individual survives (129,000 people died of cerebrovascular disease in 2010), in many cases brain function gradually improves. Early emergency treatment can reduce brain damage. The warning signs of a stroke are:

- Sudden numbness or weakness of the face, arm, or leg (especially on one side of the body)
- Sudden confusion, trouble speaking or understanding speech
- Sudden trouble seeing in one or both eyes
- Sudden trouble walking, dizziness, loss of balance or coordination
- Sudden severe headache with no known cause

Brain damage from ischemic strokes caused by cerebral thrombosis or embolism is reduced when treated in the first three hours after a stroke. First, a rapid head CT scan is done to distinguish between an ischemic or a hemorrhagic stroke (where anticoagulation would be harmful). Then, thrombolysis by the blood clot–dissolving medicine t-PA (which is also used in early treatment of heart attacks) is given. Often, antiplatelet agents (aspirin) or other anticoagulants are given to reduce the risk of future strokes.

Carotid artery stenosis (narrowing) from atherosclerotic plaques causes 5 to 10 percent of ischemic strokes. Preventive surgery significantly reduces the risk of future strokes. In this procedure, plaques within the carotid artery(ies) are removed when they have reduced the luminal diameter by 60 to 70 percent.

The very best solution is to prevent or restrict plaque formation by controlling the risk factors involved, which include smoking, untreated hypertension, diabetes, and elevated cholesterol levels in the blood. As public-health knowledge and education about risk factors for strokes have spread, and as treatments for hypertension and high cholesterol have improved, age-adjusted death rates from strokes have declined from 180 per 100,000 population in 1950 to 39.1 per 100,000 in 2010.

Shock

Shock is an acute circulatory crisis marked by low blood pressure (hypotension) and inadequate peripheral blood flow. Severe and potentially fatal signs and symptoms develop as vital tissues become starved for oxygen and nutrients. Common cardiovascular causes of shock are (1) a drop in cardiac output after bleeding or other fluid losses, (2) damage to the heart, (3) external pressure on the heart, and (4) extensive peripheral vasodilation.

Circulatory Shock

Symptoms of **circulatory shock** appear after fluid losses of about 30 percent of total blood volume. The cause can be bleeding or fluid losses to the environment, as in dehydration or after severe burns. All cases of circulatory shock share the same six basic signs and symptoms:

1. Hypotension, with systolic pressures below 90 mm Hg.

2. Pale, cool, and moist ("clammy") skin. The skin is pale and cool due to peripheral vasoconstriction; the moisture reflects the sympathetic activation of sweat glands.

3. Frequent confusion and disorientation, due to a drop in blood pressure at the brain.

4. A rise in heart rate and a rapid, weak pulse.

5. Decreased urine formation, because the reduced blood flow to the kidneys slows or stops urine production.

6. A drop in blood pH (*acidosis*), due to lactic acid generation in oxygen-deprived tissues.

In mild forms of circulatory shock, homeostatic adjustments can cope with the situation. During this period, peripheral blood flow is reduced, but remains adequate to meet tissue demands.

If the blood loss is greater, these adjustments cannot restore homeostasis. When blood volume declines by more than 35 percent, blood pressure remains abnormally low, venous return is reduced, and cardiac output is inadequate despite sustained vasoconstriction and the mobilization of the venous reserve. This leads to several potentially fatal positive feedback loops that are diagrammed in ▶ **Figure 63**. Positive feedback loop 1 begins when the low cardiac output damages the myocardium. This damage leads to a further reduction in cardiac output. Reduced CO accelerates oxygen starvation in peripheral tissues, and the resulting chemical changes promote intravascular clotting that further restricts blood flow (feedback loop 2) and increased capillary permeability that further reduces blood volume (feedback loop 3). These changes reduce both blood pressure and venous return. If unbroken, these feedback loops will eventually depress the mean arterial blood pressure (MAP) below 50 mm Hg (▶ **Figure 63b**). At this point carotid sinus baroreceptors trigger a massive activation of the vasomotor centers. In effect, the goal now is to preserve blood flow to the brain at any cost. The sympathetic output causes a sustained and maximal vasoconstriction. This reflex, called the *central ischemic response*, reduces peripheral circulation to an absolute minimum, but it elevates blood pressure to about 70 mm Hg and at least temporarily improves blood flow to cerebral vessels. However, unless prompt treatment is provided, blood pressure will again decline. As blood flow to the brain decreases, the individual becomes disoriented and confused (feedback loop 4).

The stage known as irreversible shock begins when oxygen starvation in the heart, liver, kidneys, and CNS has caused such extensive tissue damage that death will occur even *with* medical treatment. The end comes when conditions in tissues throughout the body become so abnormal that arteriolar smooth muscles and precapillary sphincters become unable to contract, despite the commands of the vasomotor centers. The result is a widespread peripheral vasodilation and an immediate and fatal decline in blood pressure. This event is called *circulatory collapse*. The blood pressure in many tissues then falls so low that the capillaries collapse like deflating balloons, and blood flow ceases.

Other Types of Shock

Circulatory shock caused by low blood volume from traumatic blood loss or severe dehydration is relatively common, but other

Cardiovascular System

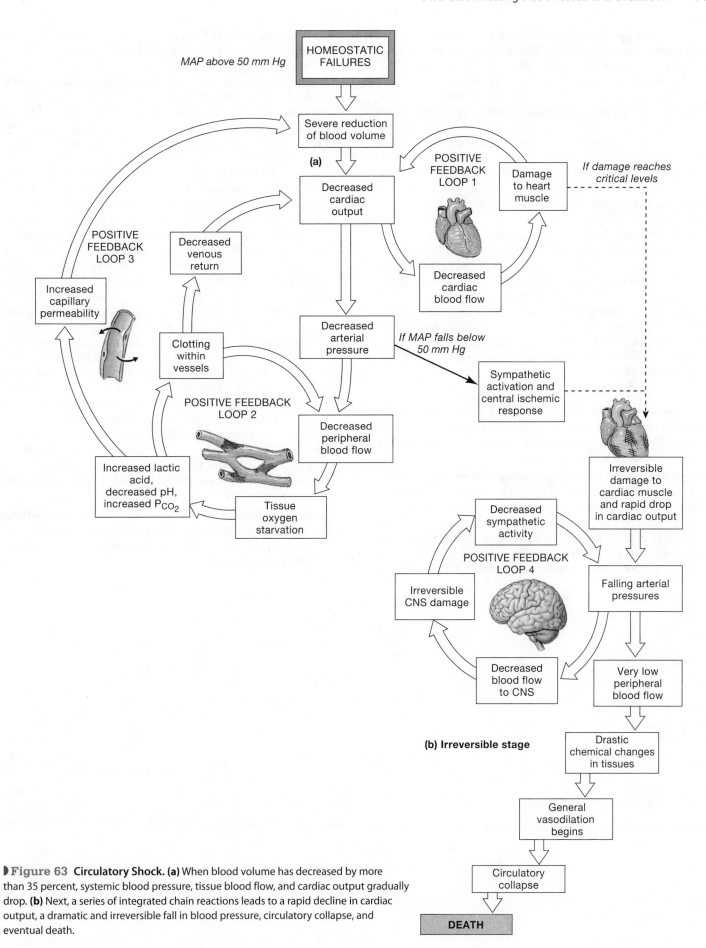

Figure 63 Circulatory Shock. (a) When blood volume has decreased by more than 35 percent, systemic blood pressure, tissue blood flow, and cardiac output gradually drop. **(b)** Next, a series of integrated chain reactions leads to a rapid decline in cardiac output, a dramatic and irreversible fall in blood pressure, circulatory collapse, and eventual death.

Cardiovascular System

forms of shock can develop when the blood volume is normal. **Cardiogenic** (kar-dē-ō-JEN-ik) **shock** occurs when the heart becomes unable to maintain a normal cardiac output. The most common cause is failure of the left heart ventricle as a result of coronary artery disease and a myocardial infarction. Between 5 and 10 percent of patients surviving a heart attack must be treated for cardiogenic shock. Restoring blocked coronary circulation (caused by a coronary artery thrombus) with thrombolytic drugs such as *tissue plasminogen activator* (t-PA) to dissolve the thrombus or PCI to mechanically remove it can improve ventricular function, thereby relieving the peripheral symptoms. Cardiogenic shock can also be the result of valvular heart disease, cardiomyopathy, or ventricular arrhythmias.

In **obstructive shock**, ventricular output is reduced because tissues or fluids are restricting the expansion and contraction of the heart. For example, a buildup of fluid in the pericardial cavity (cardiac tamponade; see "Infection and Inflammation of the Heart," p. 132) can compress the heart and limit ventricular filling.

Distributive shock is the result of a widespread, uncontrolled vasodilation. The blood pressure falls dramatically, leading to a reduction in blood flow and the onset of shock. Three important examples are neurogenic shock, septic shock, and anaphylactic shock:

1. **Neurogenic** (noo-rō-JEN-ik) **shock** can be caused by general or spinal anesthesia and by trauma or inflammation of the brain stem or spinal cord. The underlying problem is damage to the vasomotor center or to the sympathetic tracts or nerves, leading to a loss of vasomotor tone.

2. **Septic shock** results from the massive release of endotoxins—poisons derived from the cell walls of bacteria during a systemic infection. In some persons an overwhelming immune response to the infection occurs. Immune chemicals released into the blood to combat the infection trigger widespread inflammation. These compounds cause a vasodilation of precapillary sphincters throughout the body, where intravascular blood may clot resulting in drops in peripheral resistance, blood pressure, and blood flow, which damage the body's organs by depriving them of nutrients and oxygen. Symptoms of septic shock generally resemble those of other types of shock, but the skin is flushed and the individual often has a high fever. For this reason, septic shock is also known as "warm shock."

   One form of septic shock, called **toxic shock syndrome (TSS)**, results from toxins released by an infection by the bacterium *Staphylococcus aureus*. Symptoms include high fever, sore throat, vomiting, diarrhea, and a generalized rash. As the condition progresses, shock, respiratory distress, and kidney or liver failure can develop; 5 to 15 percent of cases prove fatal. These symptoms result from the entry of bacteria or bacterial toxins into the bloodstream. Toxic shock cases were sporadically reported until 1978, when it appeared in a group of children and was recognized as a syndrome. Between 1978 and 1980 TSS was diagnosed in an increasing number of young women. Although other sources of infection are possible, infection most often occurs during menstruation, and the chances of infection are increased with the use of super-absorbent tampons. When this relationship was discovered, the brands involved were taken off the market, and the incidence has declined steadily since 1980. Only 55 percent of current TSS cases are associated with menstruation. However, TSS continues to occur at a low but significant rate in men, women, and children after abrasion, burn injuries, or other wounds. The treatment for TSS includes the administration of fluids, the removal of the focus of infection (such as the removal of a tampon or the cleansing of a wound), and antibiotic therapy. A similar syndrome involving streptococcal bacteria occurs and is associated with necrotizing fasciitis (see p. 151).

3. Extensive peripheral vasodilation also occurs in **anaphylactic** (an-a-fi-LAK-tik) **shock,** a dangerous allergic reaction. This type of shock, which is caused by a misdirected immune response, can be life-threatening.

▶ The Lymphatic System and Immunity

The lymphatic system consists of the fluid *lymph*, a network of *lymphatic vessels*, specialized cells called *lymphocytes*, and an array of *lymphoid tissues* and *lymphoid organs* throughout the body. This system has three major functions: (1) to protect the body through the *immune response*, (2) to transport fluid from interstitial spaces to the bloodstream, and (3) to help distribute hormones, nutrients, and wastes.

The immune response produced by activated lymphocytes is responsible for the detection and destruction of pathogenic microorganisms, wayward cells, or toxic substances that can cause illness. For example, viruses, bacteria, and tumor cells are usually recognized and eliminated by cells of the lymphatic system. Immunity is the specific resistance to disease, and all the cells and tissues involved with the production of immunity are considered to be part of an *immune system*. Note that while the lymphatic system is an anatomically distinct system, the immune system is a physiological system that includes the lymphatic system, as well as components of the integumentary, cardiovascular, respiratory, digestive, and other systems.

History and Physical Examination and the Lymphatic System

Individuals with lymphatic system disorders experience a variety of signs and symptoms. The pattern observed depends on whether the problem affects the immune functions or the circulatory functions of the system (▶ **Figure 64**). Important symptoms and signs include the following (some will be discussed in more detail in a later section):

- **Recurrent infections** occur for a variety of reasons including multiple exposures, anatomical susceptibility, and inadequate immune response. Tonsillitis (p. 148) and *adenoiditis* are common recurrent infections in children. Serious infections throughout the body are common among persons with immunodeficiency disorders such as AIDS (acquired immunodeficiency disease) (pp. 155–157) or *severe combined immunodeficiency disease (SCID)*. When the immune response is inadequate, the individual may not be able to overcome even a minor infection. For these people, respiratory system infections are very common, and recurring, chronic gastrointestinal infections can produce chronic diarrhea. The pathogens involved often do not affect persons with a normal immune response. Infections are also a problem for individuals who take medications that suppress the immune response. Examples of immunosuppressive drugs include anti-inflammatories such as the *corticosteroids* (e.g., *Prednisone*) as well as more specialized drugs such as *methotrexate* and *cyclosporine*.

- Infections may cause regional or systemic *lymph node enlargement*. Lymph nodes also enlarge in cancers of the lymphatic system, such as *lymphoma* (p. 149), or when primary tumors in other tissues have metastasized to regional lymph nodes. Whether regional lymph nodes are involved is important in the diagnosis and treatment of many cancers. The onset and duration of swelling; the size, texture, and mobility of the nodes; the number of affected nodes; and the degree of

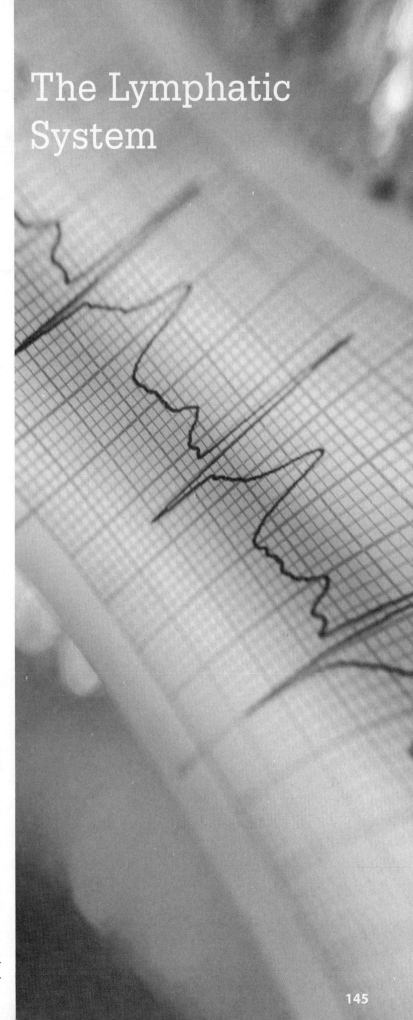

The Lymphatic
System

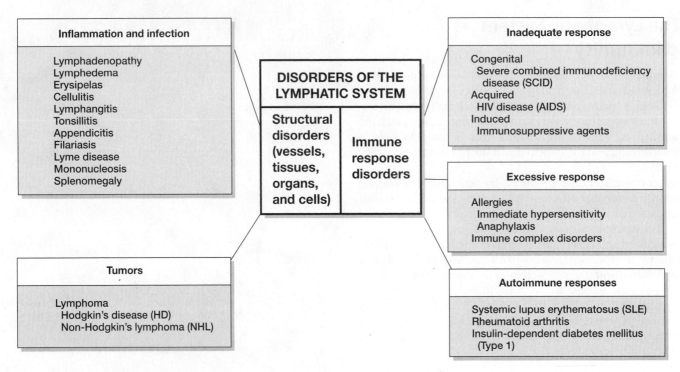

▶Figure 64 Disorders of the Lymphatic System.

tenderness all assist in clinical diagnosis. For example, nodes containing cancer cells tend to be large or enlarging, hard, fixed in place, and nontender. On palpation, these nodes feel like dense masses rather than individual lymph nodes. In contrast, infected lymph nodes tend to be rounded, freely mobile, and very tender.

- **Lymphangitis** consists of *erythematous* (red) streaks on the skin that may develop with an inflammation of superficial lymphatic vessels. Lymphangitis often occurs in the limbs, with reddened streaks that start at an infection site and progress toward the trunk. Before the linkage to the lymphatic system was known, this sign was called "blood poisoning."

- **Splenomegaly** (p. 150) is an enlargement of the spleen that can result from acute infections such as *endocarditis* (p. 132) and *mononucleosis* (p. 150), chronic infections such as *malaria* (p. 125), or cancers such as *leukemia* (p. 129). The spleen can be examined through palpation or percussion (p. 7) to detect splenic enlargement. In percussion, an enlarged spleen produces a distinctive dull sound in the left upper quadrant of the abdomen. The patient's history may also reveal important clues. For example, an individual with an enlarged spleen may report a feeling of fullness after eating a small meal, probably because the enlarged spleen limits gastric expansion.

- Prolonged *weakness* and *fatigue* typically accompany immunodeficiency disorders, *Hodgkin's disease* and other *lymphomas* (p. 149), and *mononucleosis* (p. 150).

- **Skin lesions**, such as hives (urticaria), or contact dermatitis (pp. 46, 158), can develop during allergic reactions. Immune responses to a variety of allergens, including animal dander, pollen, dust, medications, and some foods, may cause such lesions.

- **Respiratory problems**, including rhinitis (inflamed nasal mucosa) and wheezing (caused by bronchospasm), may accompany the allergic response to allergens such as pollen, hay, dust from many sources, and mildew. *Bronchospasm* occurs when the bronchial smooth muscles contract and constrict the airways making breathing difficult. Bronchospasm, which defines an asthmatic attack and often accompanies severe allergic reactions, may be a response to the appearance of antigens or irritants within the respiratory passageways, or may accompany more widespread exposure to antigens.

When lymphatic circulatory functions are impaired, the most common sign is *lymphedema*, a tissue swelling caused by the buildup of interstitial fluid. Lymphedema can result from trauma and scarring of lymphatic vessels and nodes or from a lymphatic blockage due to a tumor or infections. Lymphedema can also be due to congenital malformations.

Diagnostic procedures and laboratory tests used to detect disorders of the lymphatic system are detailed in **Tables 29 and 30** (pp. 147, 148).

Disorders of the Lymphatic System

Disorders of the lymphatic system that affect the immune response can be sorted into three general categories, as diagrammed in ▶**Figure 64**:

1. **Disorders resulting from an insufficient immune response.** This category includes immunodeficiency disorders,

Table 29   Representative Laboratory Tests of Blood for Disorders of the Lymphatic System and Immunity

Laboratory Test	Significance of Normal Values	Abnormal Values
Complete Blood Count	See laboratory tests for blood disorders, Table 27c	
WBC count	See laboratory tests for blood disorders, Table 27c	
Differential WBC count	See laboratory tests for blood disorders, Table 27c	
Immunoglobulin electrophoresis (IgA, IgG, IgM)	Adults: IgA: 85–330 mg/dL IgG: 565–1765 mg/dL IgM: 45–250 mg/dL IgD and IgE: values should be minimal	Increased levels of IgG occur with chronic infections; IgA levels increase with chronic infections, especially involving mucosa; IgE levels increase with allergic reactions and skin sensitivities; IgM levels are highest early in the immune response; IgM does not cross the placenta
Antinuclear antibody (ANA)	No antinuclear antibodies detected	Positive test occurs in up to 95% of persons diagnosed with SLE; false positive (usually low levels) can occur with rheumatoid arthritis and other autoimmune disorders, and in the elderly
Anti-DNA antibody test	Low levels or none	Positive test with high levels of antibodies occurs in 40–80% of persons with SLE
Total complement assay	Total complement: 41–90 hemolytic units C_3: 55–120 mg/dL C_4: 20–50 mg/dL	Total complement and C_3 are decreased in SLE and glomerulonephritis and increased in rheumatic fever, rheumatoid arthritis, and certain types of malignancies
Rheumatoid factor test	Negative	Positive test in 80% of cases of rheumatoid arthritis, but results may also be positive in SLE and myositis, other inflammatory conditions, false positives similar to ANA
HIV/AIDS Serology Enzyme-linked immunosorbent assay (ELISA)	Negative	Positive test indicates detection of antibodies against HIV. Tests given in the early stages of infection may yield a negative result; positive results may not develop for two months. HIV-positive status is assigned after two different tests are positive for the antibodies and the Western blot is positive
Western blot	Negative	Positive test indicates detection of antibodies to specific viral proteins

such as AIDS and SCID, the immature immune system of infants, and the decreased immune response in the elderly and those on immunosuppression medication. Individuals with depressed immune defenses can develop life-threatening diseases caused by microorganisms that are harmless to other individuals.

2. **Disorders resulting from an excessive immune response.** Conditions such as allergies and *immune complex disorders* can result from an immune response that is out of proportion to the size of the stimulus.

3. **Disorders resulting from an inappropriate immune response.** Autoimmune disorders result when normal tissues are mistakenly attacked by T cells or the antibodies produced by activated B cells. For instance, in *thrombocytopenic purpura*, the body forms antibodies against its own platelets. We will consider representative disorders from each of these categories.

Lymphedema

Lymphedema is an accumulation of fluid within a tissue due to inadequate lymphatic drainage. Temporary lymphedema can result from tight clothing or focal inactivity. Chronic lymphedema can result from the formation of scar tissue owing to repeated bacterial infections or from surgery that cuts or removes lymphatic vessels and nodes. Axillary or inguinal lymph node removal (sometimes required during surgery for breast cancer or malignant melanoma) or radiation treatment can lead to lymphedema of the adjacent limb (**Figure 65**). Lymphedema can also result from parasitic infections. For example, in **filariasis** (fil-a-RĪ-a-sis), immature stages of a parasitic roundworm (generally *Wuchereria bancrofti*) are transmitted by mosquitoes or black flies. The adult worms accumulate within lymphatic vessels and lymph nodes. Repeated scarring of the passageways eventually blocks lymphatic drainage, producing extreme lymphedema with permanent distension of tissues. The

Table 30	Representative Diagnostic Tests for Disorders of the Lymphatic System and Immunity	
Diagnostic Procedure	**Method and Results**	**Representative Uses**
Skin Tests		
Hypersensitivity response	Antigens are extracted in a sterile, diluted form; examples include animal dander, pollen, certain foods, medications (especially penicillin) that cause hypersensitivities, house dust mites, and insect venom	Identifies specific antigens that may cause hypersensitivity reaction (allergy)
Prick test	A small amount of antigen is applied as a prick to the skin. Positive test: Erythema, hardening, and swelling appear around puncture area; usually, the area affected is measured and must be of a certain size to qualify for a positive test	As above; also a screening test for tuberculosis exposure, but not as accurate as intradermal test
Intradermal test	Antigen is injected into the skin to form a 1–2 mm bleb. Positive test: Within 15 minutes to two days, a firm papule is produced that is larger than 5 mm in diameter and larger than those seen in the control group	As above
Patch test	A patch is impregnated with the antigen and applied to the surface of the skin. Positive test: The patch provokes an allergic focal rash, usually over several hours to several days	As above (Nickel allergy may be inadvertently tested for by wearing jewelry or watches containing nickel.)
Tuberculin skin test	Injection of tuberculin protein into the dermis. Positive test: Red, hardened area >10 mm wide appears around injection site 48–72 hours later	Indicates presence of antibodies to the organism that causes tuberculosis (infection may be active or dormant). Borderline tests may be repeated within a week to increase accuracy
Other Tests		
Nuclear, CT, PET, or ultrasound	Scan images created by radiation from radioisotopes, x-rays, or ultrasound reveal position, shape, and size of spleen, liver, or lymphoid tissues	Detects abscesses, tumors of the liver, spleen, lymphatic system, and infarcts of liver or splenic tissue
Biopsy of lymphoid tissue	Surgical excision of suspicious lymphoid tissue for pathological examination	Determines potential malignancy and staging of cancers in progress. Usually lymph nodes adjacent to tumors
Lymphangiography	Dye is injected into a lymphatic vessel in the distal portion of a limb and travels to nodes; x-rays are taken as the dye accumulates in the lymph nodes	Identifies and determines the stages of Hodgkin's disease and other lymphomas and detects the cause of lymphedema. Technically difficult to do, rarely done

Lymphatic System

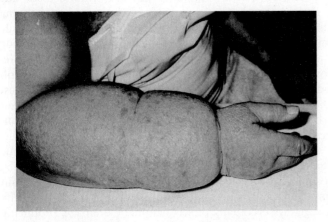

▶ Figure 65 **Lymphedema.**

limbs or external genitalia typically become grossly distended; this parasite-induced lymphedema is called **elephantiasis** (el-e-fan-TĪ-a-sis).

Therapy for chronic lymphedema consists of treating infections with antibiotics and (when possible) reducing the swelling. One treatment involves the application of elastic wrappings that squeeze the tissue. This external compression elevates the hydrostatic pressure of the interstitial fluids and opposes the entry of additional fluid from the capillaries.

Infected Lymphoid Nodules

Lymphoid nodules can be overwhelmed by a pathogenic invasion. The result is a localized infection accompanied by regional swelling and discomfort. The tonsils are a first line of defense against infection of the pharyngeal walls.

An individual with **tonsillitis** has infected tonsils. Symptoms include a sore throat, high fever, and often leukocytosis (an abnormally high white blood cell count). The affected tonsils (normally, the pharyngeal tonsils) become swollen and inflamed, sometimes enlarging enough to partially block the entrance to the trachea. Breathing then becomes difficult or, in severe cases, impossible. If the infection proceeds, abscesses may develop within the tonsillar or peritonsillar tissues. Bacteria may enter the bloodstream by passing through the lymphatic capillaries and vessels to the venous system.

In the early stages, antibiotics may control the infection, but once abscesses have formed, the best treatment involves surgical drainage of the abscesses and **tonsillectomy**, the removal of the tonsil. Tonsillectomy was once highly recommended prior to the development of antibiotics, in order to prevent recurring tonsillar infections. The procedure does reduce the incidence and severity of subsequent infections, but questions have arisen about the overall cost to the individual, especially now that antibiotics are available to treat severe infections.

Appendicitis is infection and erosion of the epithelial lining of the appendix, a blind pouch off the intestinal tract near the start of the colon; its walls are dominated by lymphoid tissues. Several factors may be responsible for the initial ulceration—notably, bacterial or viral pathogens. Bacteria that normally inhabit the lumen of the large intestine then cross the epithelium and enter the underlying tissues. Inflammation involving the appendiceal lymphoid nodules occurs, and the opening between the appendix and the rest of the intestinal tract may become constricted and blocked. Mucus secretion and pus accumulate, and the organ becomes increasingly distended. Eventually, the appendix may rupture, or *perforate*. If it does, bacteria will be released into the warm, dark, moist confines of the peritoneal space, where they can cause a life-threatening peritonitis. The most effective treatment for appendicitis is surgical removal of the appendix—an appendectomy.

Lymphomas

Lymphatic malignancies involve T cells and B cells and occur in both circulating phases (such as the leukemias, introduced on page 129) and solid phases. **Lymphomas** are an important group of solid-phase malignancies. These malignant tumors consist of cancerous lymphocytes or lymphocytic stem cells. About 79,000 cases of lymphoma are diagnosed in the United States each year. There are many types of lymphoma. One form, called **Hodgkin's disease (HD)**, accounts for roughly 11.4 percent of all lymphoma cases. Hodgkin's disease most commonly strikes individuals at ages 15–35 years or those over age 50. The reason for this pattern of incidence is unknown. Although the cause of the disease is uncertain and lymphoma is not contagious, previous infection with several viruses increases the risk of developing Hodgkin's lymphoma.

Other types of lymphoma are usually grouped together under the heading of **non-Hodgkin's lymphoma (NHL)**. These lymphomas are extremely diverse. More than 85 percent of NHL cases are associated with chromosomal abnormalities of the tumor cells, typically involving *translocations*, in which sections of chromosomes have been swapped from one chromosome to another. The shifting of genes from one chromosome to another interferes with the normal regulatory mechanisms, and the cells grow uncontrollably (become cancerous). The nature of the cancer depends on which of the many types of lymphocyte are affected. A combination of inherited and environmental factors may be responsible for specific translocations. For example, one form involving B lymphocytes, called **Burkitt's lymphoma**, develops only after genes from chromosome 8 have been

translocated to chromosome 14. (There are at least three variations.) Burkitt's lymphoma mostly affects male children in Africa and New Guinea who have been infected with the *Epstein–Barr virus (EBV)*. This highly variable virus is also responsible for infectious mononucleosis (described below).

The EBV infects B cells, but under normal circumstances most infected cells are destroyed by the immune system. After recovery, perhaps one in a million memory B cells remain as infected reservoirs of the virus. This virus affects many people, and childhood exposure generally produces lasting immunity. Children who develop Burkitt's lymphoma may have a genetic susceptibility to EBV infection. In addition, the presence of another illness, such as malaria, can weaken the immune system enough that sporadically occurring lymphoma cells are not recognized, and therefore not destroyed.

The first symptom usually associated with solid lymphomas is a painless and often overlooked enlargement of lymph nodes. The involved nodes have a firm, rubbery texture. Until patients develop secondary symptoms, such as recurrent fevers, night sweats, gastrointestinal or respiratory problems, or weight loss, they may be unaware of any underlying lymph node changes.

In planning treatment, clinicians consider the histological structure of the nodes and the stage of the disease. In a biopsy, the node is described as *nodular* or *diffuse*. A nodular node retains a semblance of normal structure, with follicles and germinal centers. In a diffuse node, the interior of the node has changed and the follicle structure has broken down. In general, nodular lymphomas progress more slowly than do the diffuse forms, which tend to be more aggressive. Conversely, nodular lymphomas are more resistant to treatment and are more likely to recur even after remission has been achieved.

The most important factor influencing which treatment is selected is the stage of the disease. Table 31 presents a simplified classification of lymphomas in terms of the stage. When the condition is diagnosed early (stage I or stage II), localized therapies may be effective. For example, the cancerous node(s) may be surgically removed and the region(s) irradiated to kill residual cancer cells. Success rates are very high when a lymphoma is detected in an early stage. For Hodgkin's disease, localized radiation can produce remission that lasts 10 years or more in more than 90 percent of patients. The treatment of localized NHL is somewhat less effective. The 5-year remission rates average 60–80 percent for all types; success rates are higher for nodular forms than for diffuse forms.

Although these results are encouraging, few lymphomas are diagnosed in the early stages. For example, only 10–15 percent of NHL patients are diagnosed at stage I or stage II. For lymphomas at stages III and IV, most treatments involve chemotherapy. Combination chemotherapy, in which two or more drugs are administered simultaneously, is the most effective treatment. For Hodgkin's disease, a four-drug combination with the acronym **MOPP** (nitrogen Mustard, Oncovin [*vincristine*], Prednisone, and Procarbazine) produces lasting remission in 80 percent of patients. Other chemotherapeutic and/or local radiation regimens are also used successfully. For relapsed cases of NHL, several specifically designed monoclonal antibodies that target lymphoma cells are available. One of these antibodies, called *rituximab*, is lethal to lymphoma cells, whereas other antibodies carry radioactive molecules or toxins that kill the targeted cells.

Lymphatic System

Table 31	Cancer Stages in Lymphoma
Stage I	Involvement of a single node or region (or of a single extranodal site)
	Typical treatment: surgical removal or localized irradiation (or both); in slowly progressing forms of non-Hodgkin's lymphoma, treatment may be postponed indefinitely
Stage II	Involvement of nodes in two or more regions (or of an extranodal site and nodes in one or more regions) on the same side of the diaphragm
	Typical treatment: surgical removal and localized irradiation that includes an extended area around the cancer site (the *extended field*)
Stage III	Involvement of lymph node regions on both sides of the diaphragm. This is a large category that is subdivided on the basis of the organs or regions involved. For example, in stage III, the spleen contains cancer cells
	Typical treatment: combination chemotherapy, possibly including monoclonal antibody with or without radiation; radiation treatment may involve irradiating all thoracic and abdominal nodes, plus the spleen (*total axial nodal irradiation*, or *TANI*)
Stage IV	Widespread involvement of extranodal tissues above and below the diaphragm; involvement of bone marrow
	Treatment: highly variable, depending on the circumstances. Combination chemotherapy is always used and may be combined with irradiation. The "last resort" treatment involves massive chemotherapy and whole-body irradiation followed by a bone marrow transplant

A **bone marrow** or **hematologic stem cell transplant** is a treatment option for acute, late-stage lymphoma. When suitable donor marrow stem cells are available, the patient receives whole-body irradiation, chemotherapy, or some combination of the two sufficient to kill tumor cells throughout the body. Unfortunately, this treatment also destroys normal bone marrow stem cells. Donor stem cells are then infused. Within two weeks, the donor cells colonize the bone marrow and begin producing red blood cells, granulocytes, monocytes, megakaryocytes, and lymphocytes.

Potential complications of stem cell transplantation include the risk of infection and bleeding while the donor marrow is becoming established. The immune cells of the donor may also attack the tissues of the recipient, a response called **graft-versus-host disease (GVHD)**. For a person with stage I or stage II lymphoma without bone marrow involvement, bone marrow can be removed and stored (frozen) for over 10 years. If other treatment options fail or if the person comes out of remission at a later date, an autologous marrow transplant can be performed. This procedure eliminates both the need for finding a matched donor and the risk of GVHD.

Disorders of the Spleen

The spleen responds like a lymph node to infection, inflammation, or invasion by cancer cells. The enlargement that follows is called **splenomegaly** (splen-ō-MEG-a-lē; *megas*, large); the spleen can rupture under these conditions. One relatively common condition that causes temporary splenomegaly is **mononucleosis**. This condition, also known as the "kissing disease," or "glandular fever" results from acute infection by the Epstein–Barr virus. In addition to enlargement of the spleen, symptoms of mononucleosis may include fever, sore throat with tonsillitis, widespread swelling of lymph nodes, increased numbers of atypical lymphocytes in the blood, and, after recovery, the presence of circulating antibodies to the virus. The condition typically affects young adults (ages 15–25 years) in the spring or fall. Treatment is symptomatic as no drugs are effective against this virus, but the illness usually resolves in two to four weeks. The most dangerous aspect of the disease is the risk of rupturing the temporarily enlarged spleen, which becomes fragile. Patients are cautioned against heavy exercise or

contact sports. If the spleen does rupture, severe hemorrhaging can occur. Death may follow unless a transfusion and an immediate splenectomy are performed.

An individual whose spleen is missing or nonfunctional has **hyposplenism** (hī-pō-SPLĒN-izm). Persons with sickle cell anemia often develop hyposplenism from recurrent splenic infarcts. Hyposplenism usually does not pose a serious problem. Hyposplenic individuals, however, are more prone to some bacterial infections, including infection by *Streptococcus pneumoniae*, than are individuals with normal spleens, so immunization against the infections caused by the pneumoniae, meningococcal, and Haemophilus influenzae type b bacteria are recommended. In **hypersplenism** where an underlying cause can't be identified or treated, the spleen may become overactive. Increased phagocytic activities lead to anemia (low RBC count), leukopenia (low WBC count), and thrombocytopenia (low platelet count). Splenectomy is the only known cure for hypersplenism.

The Complement System: A Closer Look

The complement system (▶ Figure 66) is an innate fixed defense system that we are all born with and that attacks many different pathogens. Variations of the complement system exist in some invertebrates (animals without backbones) and all vertebrates. In humans it has 11 main serum proteins. These proteins, when activated, contribute to inflammation and can either damage bacteria directly or serve as attractants for the white blood cells of the more elaborate and adaptive immune system. If these "complementary" (to the immune system) actions are deficient, the affected individual will have a much greater risk of bacterial infections. The proteins C1q, C1r, and C1s circulate as a single complex designated as C1. The proteins C2–C9 are individual proteins.

Complications of Inflammation

When bacteria invade the dermis of the skin, the production of cellular debris and toxins reinforces and exaggerates the inflammation process. **Pus** is an accumulation of debris, fluid, dead and dying cells, and necrotic tissue components. Pus commonly forms at an infection site in the dermis; an **abscess** is an accumulation

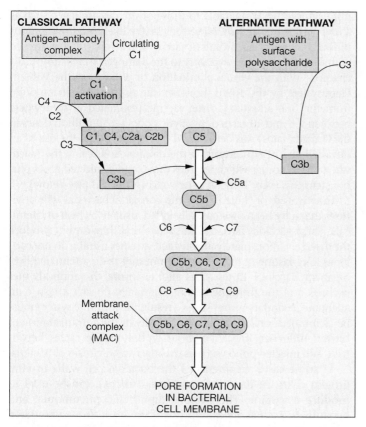

CLASSICAL PATHWAY   ALTERNATIVE PATHWAY

▶Figure 66 **The Complement Pathways.**

of pus in an enclosed tissue space. In the skin, an abscess can form as pus builds up inside the fibrin clot that surrounds an injury site. If the cellular defenses succeed in destroying the invaders, the pus will be either absorbed or surrounded by a fibrous capsule; this capsule is one type of **cyst**. (Cysts can also form in the absence of infection.)

Erysipelas (er-i-SIP-e-las; *erythros*, red + *pella*, skin) is a widespread inflammation of the dermis caused by bacterial infection. If the inflammation spreads into the subcutaneous layer and deeper tissues, the condition is called **cellulitis** (sel-ū-LĪ-tis). Erysipelas and cellulitis develop when bacterial invaders break through the fibrin wall. The bacteria involved produce large quantities of two enzymes: *hyaluronidase*, which liquifies the ground substance, and *fibrinolysin*, which breaks down fibrin and prevents clot formation. These are serious conditions that require prompt antibiotic therapy. One rare, but very aggressive, form of group A streptococcal bacteria, often mixed with other bacteria, may invade the subcutaneous tissues, destroying muscle, fat, and skin. This infection, called *necrotizing fasciitis*, is most common in persons with suppressed immune systems and chronic diseases. Surgical removal of the affected tissue plus aggressive antibiotic treatment may be lifesaving.

Transplants and Immunosuppressive Drugs

Organ transplantation is a treatment option for patients with severe disorders of the kidneys, liver, heart, lungs, or pancreas. Finding a suitable donor is the first major problem. Unless the donor can live without the donated organ, recently deceased

donors are required. After surgery has been performed, the major problem is **graft rejection**. In graft rejection, T cells are activated by contact with MHC proteins on plasma membranes in the donated tissues. The cytotoxic T cells that develop then attack and destroy the foreign cells. Graft rejection is much less likely when the major histocompatibility complex (MHC) proteins of the donor and the recipient are identical. Historically, the first successful organ transplant involved a kidney donated by the recipient's identical twin; the kidneys are paired, and the donor can survive with one kidney. The greater the difference in MHC structure between donor and recipient, the greater is the likelihood that the graft will be rejected. The process of tissue typing assesses the degree of similarity between the MHC complexes of two individuals. For this process, lymphocytes are collected and examined; they can easily be obtained from the blood and bear both Class I and Class II MHC molecules. There are many different MHC proteins, and finding a perfect match among nonidentical siblings, other relatives, or nonrelated persons can be difficult. Full siblings have only a 25 percent chance of being complete matches. For patients with a diverse ethnic background, finding a matched unrelated donor is almost impossible. With more people being tested and put on the international bone marrow donor registry, there is more hope for these patients. For example, in 1996, an adopted Korean American soldier found a suitable donor in Korea, and more recently a Chinese American child from Hawaii found a donor in Taiwan. Unlike solid organ donors, bone marrow donors do not lose an irreplaceable organ. The bone marrow stem cells left in the donor soon replace the donated bone marrow, and the donor's health is unaffected.

Significant improvements in transplant success can be made by **immunosuppression**, a reduction in the sensitivity of the immune system, of the recipient. Until recently, the drugs used to produce immunosuppression did not selectively target the immune response. For example, **prednisone** (PRED-ni-sōn), a corticosteroid, was used because it has anti-inflammatory effects that reduce the number of circulating leukocytes and depresses the immune response. However, corticosteroid use also causes undesirable side effects in systems other than the immune system. Another drug, **azathioprine** (a-za-THĪ-o-prĭn), is more powerful, but has greater associated risks. It reduces the rates of cellular growth and replication throughout the body. When this drug is used, hematopoiesis slows dramatically, and undesirable side effects may develop in the reproductive, nervous, and integumentary systems.

An understanding of the communication among T cells, macrophages, and B cells has now led to the development (in the 1980s and '90s) of drugs with more selective immunosuppressive effects. **Cyclosporin A (CsA)** suppresses all aspects of the immune response, primarily by inhibiting helper T cells while leaving suppressor T cells relatively unaffected. In the early 1980s, before the use of cyclosporin, the five-year survival rate for liver transplants was below 20 percent. In the mid-1990s, the survival rate was approximately 70 percent, and newer drugs, such as *tacrolimus* (FK 506), which inhibits lymphocytes by the same mechanism as cyclosporin but is much more powerful, have further improved survival. Genetically engineered antibodies have been developed that target specific T and B lymphocytes; these drugs provide more precise control of key phases of the immune response. The goal is individually tailored drug treatment combinations to limit the complications of immunosuppression.

An obvious problem posed by the use of any immunosuppressive drug is that the recipient becomes more susceptible than normal to viral, bacterial, and fungal infections. Immunosuppression at the lowest effective dose usually continues indefinitely after a transplant is performed. The transplant recipient must be monitored closely and treated with antibiotics at the first signs of infection.

A more subtle risk of immunosuppression is the reduction in immune surveillance by natural killer (NK) cells. Transplant patients are up to 100 times more likely to develop cancer than are others in their age group. Lymphoma-type cancers are the most common; some of these cancers appear to be linked to post-transplant infection with the Epstein–Barr virus.

Despite the risks, organ transplants are effective. Unfortunately, each day in the United States, many people die while awaiting an organ transplant, and dozens are added to the transplant waiting list. In 2013, 118,212 persons were awaiting organ transplants in the United States. However, in 2012, the most recent year for which data are available, U.S. surgeons performed 28,051 transplant surgeries, an annual number basically unchanged since 2004. The breakdown was: 16,485 kidneys, 6256 livers, 2378 hearts, 1754 lungs, 29 heart-lung combinations, 801 kidney-pancreas combinations, 242 pancreases, and 106 intestines. There is a serious disparity between the demand for organ transplantation and the number of surgeries performed. The underlying problem is that suitable organs are in short supply, due to a lack of organ donors.

Removing parts of a living donor's lung or liver for transplant has been successful, but donor deaths have occurred. Using genetic engineering and cloning techniques, pigs that carry human MHC proteins are being developed, in the hope that pig organs might be used for human transplants. Human stem cell cloning could lead to the production of new organs for transplants as well. These scenarios are complicated and controversial, involving societal, ethical, financial, and technical issues.

Immunization

Immunization is the controlled activation of the immune system by the administration of antigens or by the administration of antibodies that can combat an existing infection. In **active immunization**, a primary response to a particular pathogen is intentionally stimulated before an individual encounters the pathogen in the environment. The result is lasting immunity against that pathogen. If or when an immunized person is exposed to the pathogen, a secondary immune response produces antibodies that prevent clinical illness. Immunization is accomplished by the administration of a **vaccine**—a preparation of antigens derived from a specific pathogen. A vaccine can be given orally, nasally, or by intramuscular or subcutaneous injection. Most vaccines consist of the pathogen in whole or in part—either dead, living but weakened, or inactivated.

Before live bacteria or viruses are administered, they are weakened, or **attenuated** (a-TEN-ū-ā-ted), to lessen or eliminate the chance that a serious infection will develop from exposure to the vaccine. The rubella, mumps, measles, smallpox, chicken pox, yellow fever, tuberculosis, oral typhoid, and oral polio vaccines are examples of vaccines that use live attenuated organisms. Despite attenuation, the administration of live microorganisms may produce mild symptoms comparable to those of the disease itself, such as a low-grade fever or rash. Nevertheless, the likelihood that serious illness will develop as a result of vaccination is very small compared with the risks posed by exposure to the pathogen *without* prior vaccination. With the virtual elimination of polio from the Western Hemisphere by the 1990s, however, the risk of contracting polio from the oral attenuated-virus vaccine (estimated at somewhere between 2.5 and 10 cases of paralytic polio per 10 million doses in the United States) was determined to be greater than the risk of an unvaccinated person acquiring the disease. As a result, the inactivated polio vaccine, which requires three or four injected doses plus booster shots every 10 years, is now recommended (see below).

Inactivated, or "killed," vaccines consist of bacterial cells or viruses that have been mechanically killed, usually by heat or chemicals. These vaccines have the advantage that they cannot produce the disease. Unfortunately, inactivated vaccines usually do not confer as long-lasting or as strong an immunity as do attenuated live-organism vaccines. In the years after exposure, the antibody titer declines and the immune system eventually doesn't produce an adequate secondary response. As a result, the immune system must be "reminded" of the antigen periodically by the administration of *boosters*. Influenza, injected typhoid, typhus, plague, rabies, hepatitis A, and injected polio vaccines use inactivated viruses or bacteria.

In some cases, fragments of the bacterial cell walls or viral protein coats, or their toxic products (toxins), can be used to produce a vaccine. The tetanus, diphtheria, pneumonia, and hepatitis B vaccines are examples. Information about attenuated and inactivated vaccines is presented in **Table 32a**, and the impact of vaccines on disease in the United States is detailed in **Table 32b**.

Gene-splicing techniques can now be used to incorporate antigenic compounds from pathogens into the cell walls of harmless bacteria. Once purified, these antigens are used in vaccines; the immune system responds by producing antibodies and memory B cells that are effective against the pathogen.

Years are required to develop safe, effective vaccines. The measles vaccine took 9 years, polio 46 years, and typhoid over 100 years. As vaccines are neither perfectly safe nor perfectly effective, vaccination is about weighing risks and benefits. Malaria vaccine research has been unsuccessful for 129 years. Many current vaccines are newer improvements of earlier vaccines and the diseases they aim to prevent are still with us in our interconnected world. Vaccines provide both individual and community protection from diseases. Most of the vaccine-preventable diseases are spread from person to person. As a result of "herd" immunity, unvaccinated persons get some protection but are at much higher risk of disease once exposed and are much more likely to infect others as well. Some health care employers prohibit unvaccinated employees from patient contact for this reason.

The CDC has a detailed discussion of vaccines and immunizations at www.cdc.gov. Those considering immunizations for themselves and family should research both the vaccines AND the diseases.

Passive immunization is used in immune-compromised persons or if the individual has already been exposed to a dangerous pathogen or toxin and there is not enough time for active immunization to take effect. In passive immunization, the patient receives a dose of antibodies that will attack the pathogen and overcome the infection, even without the help of the host's own immune system. Passive immunization provides only short-term

Table 32A — Immunizations Currently Available

Vaccine-Preventable Diseases	Type of Immunity Provided	Type of Vaccine	Remarks
Viruses			
Poliovirus	Active / Active	Live, attenuated / Inactivated	Oral / Boosters every 10 years
Chicken pox	Active	Live, attenuated	A related vaccine can prevent Shingles/Herpes zoster
Rubella	Active / Passive	Live, attenuated / Human antibodies (pooled)	
Mumps	Active	Live, attenuated	
Measles (rubeola)	Active / Passive	Live, attenuated / Human antibodies (pooled)	May need second booster
Hepatitis A	Active / Passive	Inactivated / Human antibodies (pooled)	May need second booster
Hepatitis B	Active / Passive	Inactivated / Human antibodies (pooled)	May need periodic boosters; antigens may also be produced by gene-splicing technologies
Smallpox	Active	Live, attenuated (related) virus	Boosters every 10 years (no longer required, as disease has been eradicated, but bioterrorism concerns may change situation)
Yellow fever	Active	Live, attenuated	Boosters every 10 years
Herpes varicella/ zoster (Shingles)	Passive	Human antibodies (pooled)	Important for immunosuppressed patients exposed to varicella infection
Influenza A and B	Active	Inactivated	Required annually because of waning immunity and shifting strains of viruses
Rabies	Passive / Passive / Active	Human antibodies / Horse antibodies / Inactivated	Produced by gene-splicing technologies / Rarely used because of allergic reasons / Boosters required
Human papilloma viruses (HPV)	Active	Recombinant, genetically engineered protein vaccine	Prevents infection with 2 or 4 strains of HPV virus that cause most cases of cervical cancer and genital warts, as well as some oral and penile cancers; recommended for females and males ages 9 to 26
Rotavirus	Active	Live, attenuated	Oral, prevents infant diarrhea
Bacteria			
Meningococcal	Active	Cell wall antigen	Prevents meningitis in young adults
Pertussis (whooping cough)	Active	Cell wall antigen	Most often combined with tetanus and diphtheria vaccines
Anthrax	Active	Cell antigen of a virulent strain	Not available to general public
Typhoid	Active	Inactivated or live, attenuated	Boosters every 2, 3, or 5 years, depending on the type of vaccine
Tuberculosis	Active	Live, attenuated	Variable effectiveness
Plague	Active	Inactivated or live, attenuated	Boosters every 1–2 years
Tetanus	Active / Passive	Toxins used as antigens / Human antibodies (pooled)	Boosters every 5–10 years
Diphtheria	Active / Passive	Toxins used as antigens / Horse antibodies	Boosters every 10 years
Haemophilus influenza (HIB)	Active	Inactivated	Prevents early childhood meningitis
Pertussis	Active	Antigens only	Primarily for those under 6 years; booster after age 11
Streptococcal pneumonia	Active	Antigenic components of bacterial cell walls	Booster every 10 years
Botulism	Passive	Horse antibodies	
Rickettsia: typhus	Active	Inactivated	Boosters yearly if high risk of exposure
Lyme disease	Active	Inactivated	No longer available in the United States
Other			
Snake bite	Passive	Horse antibodies	
Spider bite	Passive	Horse antibodies	
Venomous fish spine	Passive	Horse antibodies	

Lymphatic System

Table 32B	Comparison of Twentieth Century Annual Morbidity* and Current Morbidity of Vaccine-Preventable Diseases of Children in the United States		
Disease	**Twentieth Century Annual Morbidity**	**2000[†]**	**Percentage Decrease**
Smallpox	48,164	0	100
Diphtheria	175,885	4	99.99
Measles	503,282	81	99.98
Mumps	152,209	323	99.80
Pertussis	147,271	6755	95.40
Polio (paralytic)	16,316	0	100
Rubella	47,745	152	99.70
Congenital rubella syndrome	823	7	99.10
Tetanus	1314	26	98.00
Haemophilus influenzae type b and unknown (children <5 years)	20,000	167	99.10

* Typical average during the 3 years before vaccine licensure.

† Provisional data.

Table 32b adapted from Malone, K. M. and Hinman, A. R. (2007). Vaccination Mandates: The public health imperative and individual rights. In Law in Public Health Practice, Goodman R. A., Hoffman R. E., Lopez W., Matthews G. W., Rothstein M., and Foster K., eds. Oxford Scholarship Online DOI 10.1093/acprof:oso/9780195301489.001.00001

resistance to infection, because the antibodies are eventually removed from circulation and are not replaced.

The antibodies provided during passive immunization have traditionally been acquired by collecting and combining antibodies from the sera of many presumably immune individuals. These *pooled sera* are used to obtain sufficient quantities of antibodies, but the procedure is very expensive. In addition, improper treatment of the sera carries the risk of accidental transmission of an infectious agent, such as a hepatitis virus or the virus responsible for AIDS, if inadvertently, sera from an infected individual is used. Antibodies can also be obtained from the blood of a domesticated animal (typically a horse) exposed to the same antigen. Unfortunately, recipients frequently suffer allergic reactions to horse serum proteins.

At present, antibody preparations are available to prevent or treat hepatitis A, hepatitis B, herpes zoster (or varicella–chicken pox), diphtheria, tetanus, rabies, measles, rubella, botulism, and the venoms of certain fish, snakes, and spiders. Gene-splicing genetic engineering technology can now be used to reproduce pure antibody preparations that are free of antigenic or viral contaminants; this technology should eventually eliminate the need for pooled plasma. Passive antibodies against rabies are produced in this way.

Passive immunity occurs naturally during fetal development. During pregnancy, maternal IgG antibodies cross the placental barriers and enter the fetal circulation, where they persist and partially protect the infant for three to six months after birth.

Lyme Disease

Because of the size and contents of medically related textbooks, it is tempting to think that most illnesses are known and understood. But clinical, epidemiologic, and laboratory sleuthing can still reveal a "new" disease. In November 1975, the town of Lyme, Connecticut, experienced an epidemic of adult arthritis and juvenile arthritis. Between June and September, 59 cases were reported, 100 times the statistical average for a town of its size. Symptoms were unusually severe: Victims experienced chronic fever and a prominent rash that began as a red bull's-eye centered around what appeared to be an insect bite (▶ Figure 67). Joint degeneration and nervous system problems often occurred somewhat later. It took almost two years to track down the cause of this condition, now known as **Lyme disease**.

Lyme disease is caused by the bacterium *Borrelia burgdorferi*, which normally lives in white-footed mice. The disease is transmitted to humans and other mammals by the bite of a tick that harbors that bacterium. Deer, which can carry infected adult ticks without becoming ill, have helped spread the ticks through populated areas. The high rate of infection among children reflects the fact that they play outdoors during the summer in fields where deer may also be found. Children are thus more likely to encounter—and be bitten by—infected ticks. After 1975, the Lyme disease problem became regional and then national in scope. The incidence in the United States is now about 35,000 cases per year, and Lyme disease has also been reported in Europe and northern Asia. Prevention involves avoiding ticks. Early diagnosis and antibiotic treatment cures most cases.

Although some of the joint destruction results from the deposition of immune complexes by a mechanism comparable to that

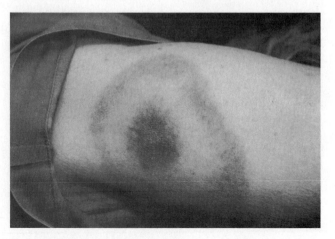

▶ Figure 67 Lyme Disease.

responsible for rheumatoid arthritis, many of the symptoms (fever, pain, skin rash) develop in response to the release of interleukin-1 (IL-1) by activated macrophages. The cell walls of *B. burgdorferi* contain lipid–carbohydrate complexes that stimulate the secretion of IL-1 in large quantities. By stimulating the body's specific and nonspecific defense mechanisms, IL-1 exaggerates the inflammation, rash, fever, pain, and joint degeneration associated with the primary infection. A vaccine was developed, not widely used, and is no longer available. If chronic arthritic and neurological symptoms develop (which may resemble fibromyalgia and chronic fatigue syndrome), treatment is less satisfactory.

AIDS

Acquired immunodeficiency syndrome (AIDS), or *late-stage HIV disease*, develops after infection by several strains of an RNA virus known as the *human immunodeficiency virus (HIV)*. Most people with HIV in the United States are infected with HIV-1; HIV-2 infections are most common in West Africa.

Symptoms of HIV Disease

The initial infection may produce a flulike illness, with fever and swollen lymph nodes, a few weeks after exposure to the virus. Exposure generally triggers the production of antibodies, which appear in the blood within two to twelve weeks. Antibody and HIV viral load tests can be used to diagnose infection. Further symptoms may not appear for 5–10 years or more. During this period, the viral content of the blood varies, but the viruses are at work within lymphoid tissues, especially in the lymph nodes. There, they selectively infect and kill helper T cells, also called CD4 or T4 cells. A steady decline in the number of CD4 T cells ensues and, for reasons as yet unknown, a decline in the number of dendritic cells in lymphoid tissues as well.

The Centers for Disease Control and Prevention (CDC) in Atlanta, which monitor infectious diseases, now recognize three categories of HIV/AIDS disease on the basis of CD4 T cell counts:

1. In *Category 1*, the CD4 T cell count is depressed (normal is over 800 per mm^3) but above 350 per mm^3, and these cells account for at least 29 percent of the total circulating lymphocytes.

2. In *Category 2*, the CD4 T cell count is between 200 and 350 per mm^3, and these cells account for 14–28 percent of the total circulating lymphocytes.

3. In *Category 3*, the CD4 T cell count is below 200 per mm^3, and these cells account for 14 percent or less of the total circulating lymphocytes. This category is **late-stage, advanced HIV disease**, or AIDS. AIDS is diagnosed if the CD4 T cell count is less than 350 per mm^3 and infection with an AIDS-defining condition occurs.

Symptoms commonly do not appear until the CD4 T cell concentration falls below 350 per mm^3 (Category 2). The symptoms that first appear are typically mild, consisting of lymphadenopathy, chronic nonfatal infections, diarrhea, and weight loss. By contrast a person with AIDS develops a variety of life-threatening infections and disorders. There are over 26 recognized AIDS-defining conditions, and the average life expectancy after diagnosis (without specific anti-HIV treatment) is two years. The life expectancy is short because the reduction in helper T cell activity impairs the immune response. This effect is magnified, because suppressor T cells are relatively unaffected by the virus. Over time, circulating antibody levels decline, cellular immunity is reduced, and the body is left without defenses against a wide variety of bacterial, viral, fungal, and protozoan invaders.

This vulnerability is what makes AIDS dangerous. The effects of HIV on the immune system are not by themselves life threatening, but the infections that result when the immune system is weakened certainly are. With the suppression of immune function, ordinarily harmless pathogens can initiate lethal infections known as *opportunistic infections*. In fact, the most common and dangerous pathogens for an individual with AIDS are microorganisms that seldom cause illnesses in individuals with normal immune systems. Persons with advanced HIV/AIDS are especially prone to lung infections (pneumonia) caused by the fungus *Pneumocystis carinii*.

In addition to pathogenic invasions, immune surveillance is depressed and the risk of cancer increases—as in transplant patients, who have to take immunosuppressive drugs (pp. 151–152). One of the most common cancers in persons with AIDS is *Kaposi's sarcoma*, a cancerous tumor of blood vessels in skin or mucous membranes. This condition is associated with concurrent herpes type 8 viral infection. Women with AIDS are at increased risk of cervical cancer, which is associated with one type of human papillomavirus.

Incidence

In 2010, there were about 1.1 million people living with HIV/AIDS in the United States and 15,500 deaths. The CDC statistics show that the incidence (the number of new cases in a given year) of HIV/AIDS in the nation has been stable at an estimated 50,000 new cases diagnosed a year from 2008 to 2011. Because HIV can remain in the body for 2 to 15 years without producing clinical symptoms, the number of individuals infected is higher than the number of reported cases of AIDS. The total number of U.S. deaths attributed to this disorder since its discovery in 1982 exceeds 635,000.

The World Health Organization (WHO) estimates that in 2012 as many as 35.3 million people worldwide were living with HIV/AIDS, with 2.3 million new infections, and 1.6 million deaths over the course of the year (❯**Figure 68**). AIDS has caused over 35 million deaths worldwide since 1982.

Persons with undiagnosed cases of HIV infection are most likely to spread the disease. Widespread testing of pregnant women and treatment of those found to be infected has significantly reduced neonatal HIV cases.

Modes of Infection

Infection with HIV occurs through intimate contact with the body fluids of an infected individual. Although many body fluids, including saliva, tears, and breast milk, carry the virus, the major routes of adult infection involve contact with infected blood, semen, or vaginal secretions. Four major transmission routes have been identified:

1. Sexual transmission

2. Sharing needles or syringes with someone infected with HIV

3. Prenatal, during birth, and breast-feeding with fetal or infant exposure

4. Receipt of HIV-infected blood or tissue products

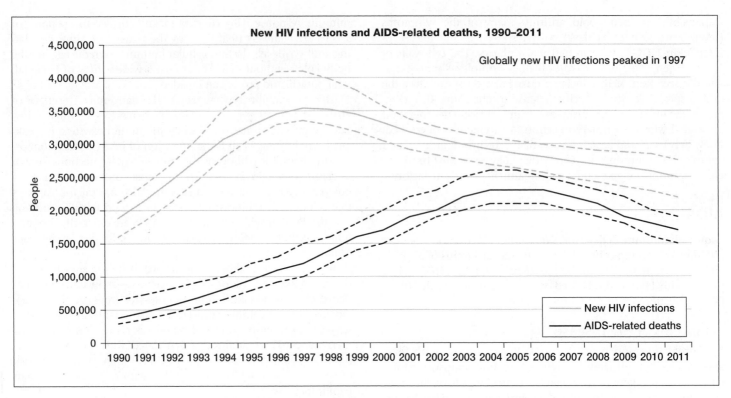

New HIV infections and AIDS-related deaths, 1990–2011

Globally new HIV infections peaked in 1997

— New HIV infections
— AIDS-related deaths

▶ **Figure 68** **World Health Organization and United Nations Program on 2012 HIV/AIDS Estimates.**
Based on World Health Organization and United Nations Program 2012.

Sexual Transmission

In the United States, AIDS was first noted in male patients who had male-to-male sexual contact, but over time more women have become infected. Worldwide, 80 percent of all infections have resulted from heterosexual intercourse.

Whether homosexual or heterosexual contact is involved, a sex partner whose epithelial defenses are weakened is at increased risk of infection. This factor accounts for the relatively higher rate of transmission by anal intercourse, which tends to damage the delicate lining of the anorectal canal. The presence of genital ulcers from other sexually transmitted diseases, such as syphilis, chancroid, or herpes, also increases the risk of HIV transmission.

Injecting Drug Use

In the United States, over 20 percent of new HIV infections of both sexes are associated with injecting drug use. Although only small quantities of blood are inadvertently transferred when needles are shared, this practice injects HIV directly into the body. All injections of HIV-infected fluid whether into a vein, a muscle, or the skin, risk infection. It is a highly effective way to transmit the infection. Injecting drug users who are diabetic, and so have access to sterile needles, have a lower rate of HIV infection.

Receipt of Blood or Tissue Products

Early in the AIDS epidemic, about 3 percent of persons with AIDS were infected with HIV after they received a transfusion of infected blood products or an organ transplant from an infected individual. Most of these infections occurred before 1985 when tests were developed to screen blood and tissue donations. With careful screening of blood and blood products,

the rate of new transmission by this route is now essentially zero in the United States.

Prenatal Exposure

A pregnant, HIV-positive woman will infect her child either prenatally, at birth, or from breast-feeding in 35 percent of pregnancies. Treatment with anti-AIDS medicine around the time of delivery has reduced the transmission rate to 17 percent or lower. Avoidance of breast-feeding by infected mothers also protects the infant.

Prevention of AIDS

The best defense against AIDS is to avoid exposure to HIV. The most obvious precaution is to avoid sexual contact with infected individuals. All forms of sexual intercourse with an infected individual carry the risk of transmitting the virus. Recent studies have shown that circumcised men were up to 60 percent less likely to become infected than uncircumcised men. The use of synthetic (latex) condoms is recommended for those who do not know the HIV status of a partner. (Condoms that are not made of synthetic materials are effective in preventing pregnancy, but do not block the passage of viruses.) Although condom use does not provide absolute protection, it reduces the risk of infection dramatically. Contrary to popular belief, significant risks are associated with oral sex, as well as other forms of sexual contact.

Blood and blood products are carefully screened for the presence of HIV-1. A relatively simple **ELISA** (enzyme-linked immunosorbent assay) blood test exists for the detection of HIV-1 antibodies; a positive reaction indicates previous exposure to the

Lymphatic System

virus. There can be false positives, so infection is confirmed by more sensitive Western Blot tests. Rarely, there can be false negatives to the ELISA test, and donated blood is screened by even more sensitive tests that can detect HIV viral particles, reducing the risk of infection by transfusion or through the use of blood products from pooled sera. Pooled sera can also be treated with detergents and chemicals that kill the virus, or heat-treated by exposure to temperatures sufficient to kill the virus, but too low to denature blood proteins permanently. A rapid (20-minute) saliva test for HIV was approved by the FDA for use in early 2004.

Because the incubation period is variable, a positive test for HIV infection does not mean that the individual has AIDS. It does mean, however, that the individual is likely to develop AIDS sometime in the future and is a carrier capable of infecting others. In terms of the spread of the disease, the most dangerous individuals are those who seem perfectly healthy and have no idea that they are carrying the virus. The CDC has recommended increased routine screening for HIV in higher risk populations.

Despite intensive efforts, a vaccine has yet to be developed that will provide immunity from HIV infection.

Treatment

There is no cure for AIDS. However, the length of survival for a person with the disease has been steadily increasing, because (1) new drugs are available that slow the progression of AIDS, and (2) improved antibiotic therapies have helped overcome infections that would otherwise prove fatal. This combination is extending lives as the search for more effective treatment continues. However, overcoming a bacterial infection in an individual with AIDS might require antibiotic doses up to 10 times greater than those used to fight infections in HIV-free individuals. Moreover, once the infection has been overcome, the patient may have to continue taking that drug for the rest of his or her life. As a result, some AIDS patients take dozens of pills a day just to prevent recurrent infections.

In 1995, treatment regimens involving a combination of drugs that attack different stages of HIV virus replication became available. When used in high-income countries, such "highly active anti-retroviral therapy," or HAART, reduced the excess death rate in HIV-infected persons by 85 percent. In many cases, viral loads became undetectable, and patients got up from their (expected) deathbeds to attend informational meetings on "how to prepare for the future you didn't think you'd have." Hopes of a complete cure were dashed when viable virus particles were detected in the lymph nodes of people undergoing this protocol. The initial difficulties of complying with stringent treatment schedules, side effects, and the significant cost (around $10,000 per patient per year for medication alone) were barriers to successful therapy. Moreover, viral mutation has caused relapses in some patients, as well as the development of drug-resistant viral strains. On the positive side, combination pills and newer drugs are simplifying treatment regimens, and an increase in access to lower cost generic HAART drugs in lower income countries starting in 2003 has produced similar results. As of 2011, the World Health Organization estimates that the number of AIDS-related deaths worldwide was 1.7 million, a decline of 24 percent from 2005.

Genetic Engineering and Gene Therapy

Our understanding of disease mechanisms has been profoundly influenced by recent advances in genetic engineering. The information and technical capabilities to copy, add to or subtract from, move, and otherwise manipulate DNA and RNA and their protein products developed since the 1980s are now used in clinical diagnosis and treatment. Using laboratory bacteria, animals, or in some cases human tissue, this trend is sure to continue, and several lines of research have the potential for broad application. These projects involve a mixture of genetic engineering, computer analysis, and protein biochemistry. Researchers can identify the individual B cells responsible for producing a given antibody. A specific B cell can then be isolated in a petri dish and fused with a cultured cancer cell. This technique produces a **hybridoma** (hī-bri-DŌ-ma), a cancer cell that manufactures a single type of antibody. Because hybridomas, like other cancer cells, undergo rapid mitotic divisions, culturing the original hybridoma cell soon produces an entire population, or **clone**, of genetically identical cells. A cloned hybridoma will produce large quantities of a **monoclonal** (mo-nō-KLŌ-nal; *mono*, one) **antibody (mAb)**. Because they are produced from genetically identical cells, monoclonal antibodies consist of identical molecules, which can be separated and purified.

One important use for this technology has been the development of antibody tests for the clinical analysis of body fluids. Labeled antibodies can be used to detect small quantities of specific antigens in a sample of plasma or other body fluids. For example, a popular home pregnancy test relies on monoclonal antibodies that detect small amounts of a placental hormone, *human chorionic gonadotropin (hCG)*, in the urine of a pregnant woman. Other monoclonal antibodies are used in standard blood-screening tests for sexually transmitted diseases and urine tests for ovulation.

Monoclonal antibodies can also be used to provide passive immunity to disease. Passive immunizations that include monoclonal antibodies cause fewer of the unpleasant side effects associated with antibodies from pooled human or animal sera, because the product does not contain plasma proteins, viruses, or other contaminants. The antibodies can be made to order by exposing a population of B cells to a particular antigen. mAb antibodies directed at lymphoma tumors were first in clinical use. In 2013, there were 40 FDA approved mAb drugs available to treat various cancers, the autoimmune disorders rheumatoid arthritis and inflammatory bowel disease, and to prevent organ transplant rejection. These medicines are very expensive to make and can have serious toxicities including severe or fatal allergic reactions.

Genetic engineering can be used to promote immunity in other ways as well. One interesting approach involves gene-splicing techniques. The genes that code for an antigenic protein of a viral or bacterial pathogen are identified, isolated, and inserted into a harmless bacterium that can be cultured in the laboratory. The clone that eventually develops will produce large quantities of pure antigen that can then be used to stimulate a primary immune response. Vaccines against hepatitis were developed in this manner, and a similar strategy may be successful in developing vaccines for malaria or AIDS.

A more controversial experimental technique involves taking a pathogen and adding or removing genes to make it harmless. The modified pathogen can then be used to produce active immunity without the risk of severe illness. Fears that the engineered organism could mutate or regain its pathogenic properties have so far limited the use of this approach.

Hybridomas that manufacture other products of the immune system can also be produced. Interferons are not effective against all viruses, but can control certain forms of virus-induced cancers and hepatitis. Interleukins may also prove useful in increasing the intensity of the immune response.

Fetal Infections

Fetal infections are rare, because the developing fetus acquires passive immunity from IgG antibodies produced by the mother that cross to the fetus through the placenta. These defenses don't work, however, if the mother has had no prior infection or the maternal antibodies are unable to cope with a bacterial or viral infection. The fetus may then begin producing its own IgM antibodies. Blood drawn from a newborn infant or taken from the umbilical cord of a developing fetus can be tested for the presence of IgM antibodies. This procedure provides concrete evidence of congenital infection, because maternal IgM antibodies are large and cannot cross the placenta. For example, a newborn infant with congenital syphilis will have IgM antibodies that target the pathogenic bacterium involved (*Treponema pallidum*). Fetal or neonatal (newborn) blood can also be tested for antibodies against the rubella (German measles) virus or other pathogens.

In the case of congenital syphilis, antibiotic treatment of the pregnant woman also treats the fetus and can prevent or reduce fetal damage. In the absence of antibiotic treatment, fetal syphilis can cause liver and bone damage, hemorrhaging, and a susceptibility to secondary infections in the newborn infant. There is no satisfactory treatment for congenital rubella infection, which can cause deafness and other significant developmental abnormalities. For that reason, rubella immunization has been recommended for all young children (to slow the spread of the disease in the population) and for women of childbearing age. The vaccination, which contains live, attenuated viruses, must be administered before pregnancy to prevent maternal infection and possible fetal damage during a subsequent pregnancy.

Delayed Hypersensitivity and Skin Tests

Delayed hypersensitivity begins with the sensitization of cytotoxic T cells. At the initial exposure, macrophages called

antigen-presenting cells (APCs) present antigens to T cells. On subsequent exposures, these T cells respond by releasing cytokines, which stimulate macrophage activity and produce a massive inflammatory response in the immediate area (❱ Figure 69). Examples of delayed hypersensitivity include the many types of contact dermatitis, such as poison ivy.

Skin tests can be used to check for delayed hypersensitivity. The antigen is administered by shallow injection, and the site is inspected minutes to days later. Most skin tests inject antigens taken from bacteria, fungi, viruses, or parasites. If the individual has previously been exposed to the antigen and has developed antibodies to it, the injection site will become red, swollen, and invaded by macrophages, T cells, and neutrophils within two to four days. These signs are considered a positive test, which indicates previous exposure, but does not necessarily indicate the current presence of the disease. The skin test for tuberculosis, the *ppd*, is the most commonly used test of this sort. Allergy skin testing for plant and animal sensitivities involves different parts of the immune system and produces a more immediate and shorter response.

Immune Complex Disorders

Under normal circumstances, immune complexes (antigen–antibody linkages) are promptly eliminated by phagocytosis. But when an antigen appears suddenly in high concentrations, the local phagocytic population may not be able to cope with the situation. The immune complex may then enlarge further, eventually forming insoluble granules that are deposited in affected areas. The presence of these complexes triggers the extensive activation of complement, leading to inflammation and tissue damage at the site (❱ Figure 70). This condition is known as an **immune complex disorder**. The process of immune complex formation is further enhanced by neutrophils, which release enzymes that attack the inflamed cells and tissues. The most serious immune complex disorders involve deposits within blood vessels and in the filtration membranes of the kidneys.

Systemic Lupus Erythematosus

Systemic lupus erythematosus (LOO-pus e-rith-ē-ma-TŌ-sis), or **SLE**, appears to result from a generalized breakdown in the antigen recognition mechanism. An individual with severe SLE manufactures *autoantibodies* against the body's own nucleic acids

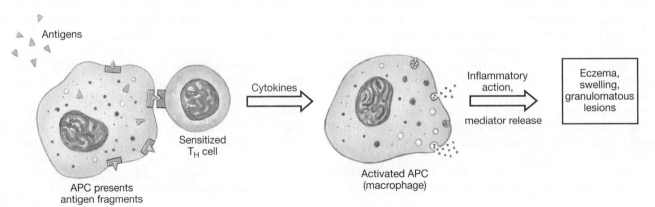

❱ Figure 69 **Delayed Hypersensitivity.**

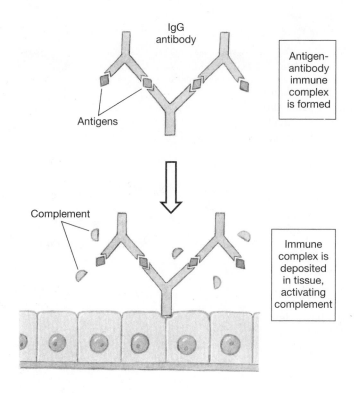

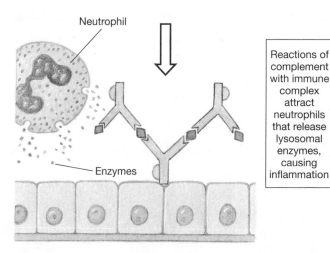

Figure 70 Immune Complex Hypersensitivity.

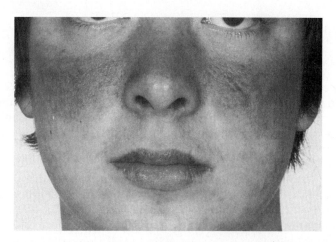

Figure 71 Butterfly Rash of Systemic Lupus Erythematosus.

States varies from 1.8 to 7.6 cases per 100,000 population. SLE has no known cure, but 90 percent of persons with SLE survive 10 years after diagnosis. Treatment consists of controlling the symptoms and if necessary depressing the immune response through the administration of specialized drugs or corticosteroids.

Autoimmune Disorders

Autoimmune disorders result when plasma cells begin producing antibodies that bind to normal "self" antigens throughout the body. **Table 33** lists representative autoimmune disorders and the targets of the antibodies responsible for these conditions.

(the ANA tests for antinuclear antibodies), ribosomes, clotting factors, blood cells, platelets, and lymphocytes. The immune complexes form deposits in peripheral tissues, producing anemia, kidney damage, arthritis, and vascular inflammation. CNS function deteriorates if the blood flow through damaged cranial or spinal vessels slows or stops.

The most visible sign of this condition is the presence of a butterfly-shaped rash centered over the bridge of the nose (**Figure 71**). For some people, *discoid lupus* affects mainly the skin and is not widespread in the body like systemic lupus or drug-induced lupus; drug-induced lupus usually resolves if the offending drug is stopped. Systemic lupus affects women nine times as often as it affects men, and the incidence in the United

Table 33	Autoimmune Disorders
Disorder	**Antibody Target**
Psoriasis	Epidermis of skin
Vitiligo	Melanocytes of skin
Rheumatoid arthritis	Connective tissues at joints
Myasthenia gravis	Synaptic ACh receptors
Multiple sclerosis	Myelin sheaths of axons
Addison's disease	Adrenal cortex
Graves disease	Thyroid follicles
Hypoparathyroidism	Chief cells of parathyroid
Thyroiditis	Thyroid-binding globulin
Type I diabetes mellitus	Beta cells of pancreatic islets
Rheumatic fever	Myocardium and heart valves
Systemic lupus erythematosus	DNA, cytoskeletal proteins, other tissue components
Thrombocytopenic purpura	Platelets
Pernicious anemia	Parietal cells of stomach
Chronic hepatitis	Hepatocytes of liver

Lymphatic System

1. Noticing that her teeth have tartar buildup, Melanie visits her dentist for a cleaning. The dentist reminds Melanie that she has not been in for the last five years and recommends that she update her health history form. Melanie, however, forgets to inform the dentist that as a child she had heart surgery and has had residual abnormalities. The dentist cleans her teeth and remarks about how the tartar on her teeth has gone below the gum line, causing bleeding during the procedure. He advises Melanie to floss and keep her teeth clean so that bacteria in her mouth won't produce more plaque. Several weeks later, Melanie becomes very ill and has the following symptoms for several days: night sweats, fever and chills, joint pain, and tachycardia. She is admitted to the hospital for further testing. What might be happening?

2. Bill's father, Emilio, says he has chest pain and, worried about a heart attack, they rush to the emergency room. A quick medical history from Bill indicates that Emilio was working in the yard when the chest pain began several hours ago, but it didn't go away and it now seems worse. The physician on call tells Bill that she has ordered some blood tests from which she will be able to detect if a heart attack has occurred. Why would the levels of the following enzymes help indicate if a person suffered a myocardial infarction?

   Total serum CK: 70 U/mL

   Isoenzymes:
      CK-MM: 40%
      CK-MB: 60%
      CK-BB: 0%

   Troponin T: 0.13 ng/mL

3. Teresa is taken to the emergency room after a car accident. She sustained abdominal trauma during the accident and is very sore in the upper left abdominal quadrant. She shows no other signs or symptoms, and her vital signs are within normal limits. Thirty minutes after arrival, she is pale and anxious, and the pain has intensified. She is breathing rapidly, and her hands are clammy and cyanotic. Her pulse is 90 and weak; her blood pressure falls to 80/55, then to 75/50 within minutes. Teresa states that the pain is getting more severe and that it now extends to her left shoulder. While blood tests are being performed, she is put on IV fluids and supplemental oxygen. Inspection of the abdomen reveals abdominal distension and rigidity. A focused abdominal ultrasound to assess the trauma reveals fluid in the perisplenic area. Results from the initial blood work and the follow-up work are as follows:

	Test upon Arrival	Retest after 30 Minutes
RBC count	3.8 million/mm^3	3.2 million/mm^3
Hct	33	28
Hb	11 g/dL	9 g/dL
MCV	78 mm^3	78 mm^3
MCH	27 pg	27 pg
WBC count	8000/mm^3	8000/mm^3
Platelets	200,000/mm^3	200,000/mm^3

On the basis of this information, Teresa is rushed to surgery. Why?

NOTES:

Answers to these problems can be found on page 241.

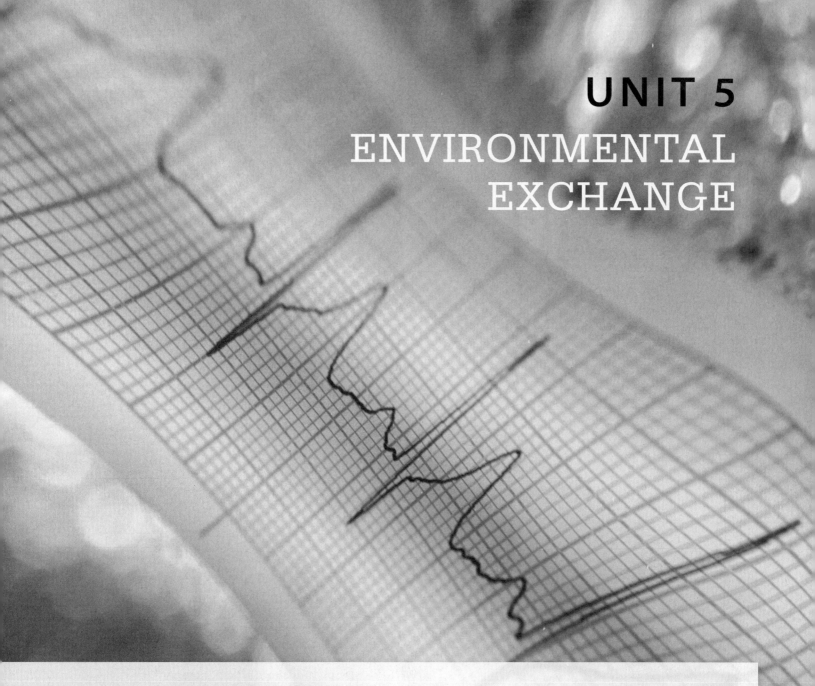

Homeostatic regulation ensures a stable environment for living tissues. The cardiovascular system transports metabolic fuels, waste products, dissolved gases, and heat from one place to another, preventing localized alterations in the extracellular environment. Yet these activities would serve no useful purpose without the cooperation of other systems. Cells are constantly consuming oxygen and nutrients, and generating carbon dioxide and waste products. Stabilizing the concentrations of these materials involves a continual exchange of materials with the outside world. Oxygen and nutrients must be absorbed, carbon dioxide and wastes excreted. Electrolytes and water must be ingested to keep pace with the rates of excretion in urine, fecal materials, and various glandular secretions. The homeostatic adjustments that balance these gains and losses are regulated by the nervous and endocrine systems, which work together to achieve a coordinated result.

This unit will examine representative disorders that impact environmental exchanges and disrupt the activities of the organ systems involved.

MAJOR SECTIONS INCLUDED WITHIN THIS UNIT:

The Respiratory System

The Digestive System and Metabolism

The Urinary System and Fluid, Electrolyte, and Acid–Base Balance

END-OF-UNIT CLINICAL PROBLEMS

The Respiratory System

▶ An Introduction to the Respiratory System and Its Disorders

The anatomical components of the respiratory system can be divided into two parts: an *upper respiratory system*, which includes the nose, nasal cavity, paranasal sinuses, and pharynx; and a *lower respiratory system*, composed of the larynx, trachea, bronchi, and lungs. The *respiratory tract* consists of the airways that carry air to and from the exchange surfaces of the lungs. The respiratory tract can be divided into a *conducting portion* and a *respiratory portion*. The conducting portion begins at the entrance to the nasal cavity and extends through the pharynx and larynx and along the trachea and bronchi to the terminal bronchioles. The respiratory portion of the tract includes the respiratory bronchioles and the alveoli, which are part of the respiratory membrane, where gas exchange occurs.

Symptoms of Lower Respiratory Disorders

Lower respiratory disorders generally cause one or two major symptoms—in particular, *chest pain* and *dyspnea*:

1. Chest pain associated with a respiratory disorder frequently worsens when the person takes a deep breath or coughs. This pain with breathing is somewhat distinct from the chest pain experienced by individuals with the heart problems of angina (pain appears during exertion) or myocardial infarction (pain is continuous, even at rest). Disorders affecting the pleural membranes may cause localized, inspiratory (pleuritic) chest pain in the specific region of the thorax where they occur. A person with such pleuritic pain may press against the sensitive area and avoid coughing or deep breathing in an attempt to reduce the local movement that brings on pain.

2. *Dyspnea*, or difficulty in breathing, can be a symptom of pulmonary disorders, cardiovascular disorders, metabolic disorders, or environmental factors such as hypoxia at high altitudes. Dyspnea may be a chronic problem, or it may develop only during exertion or when the person is lying down.

   Dyspnea due to respiratory problems generally indicates one of the following classes of disorders:

   • **Obstructive disorders** result from increased resistance to airflow along the respiratory passageways. The individual usually struggles to breathe, even at rest, and exhalation is more difficult than inhalation. Examples of obstructive disorders include *asthma* and *emphysema*.

   • **Restrictive disorders** include (1) arthritis; (2) paralysis or weakness of respiratory muscles as a result of trauma, muscular dystrophy, myasthenia gravis, multiple sclerosis, polio, or other factors; (3) physical trauma or congenital structural disorders, such as scoliosis, that limit lung expansion; and (4) pulmonary fibrosis, in which abnormal fibrous tissue in the alveolar walls slows oxygen diffusion into the bloodstream. Individuals with restrictive disorders initially experience dyspnea only during exertion, because pulmonary ventilation cannot increase enough to meet the respiratory demand.

   Cardiovascular disorders that produce dyspnea include *coronary artery disease*, *congestive heart failure*, and *pulmonary embolism*.

In *paroxysmal nocturnal dyspnea*, a person awakens at night, gasping for air. In most cases, the underlying cause is a reduced cardiac output due to advanced heart disease or heart failure. Periodic, or *Cheyne–Stokes respiration*, consists of alternating cycles of rapid, deep breathing and periods of respiratory arrest (*apnea*). This breathing pattern is most commonly seen in persons with CNS disorders and occasionally congestive heart failure.

Dyspnea may also be related to metabolic problems, such as the acute acidosis associated with *diabetes mellitus* and the *uremia* of kidney failure. A sudden, severe decline in blood pH can trigger *Kussmaul breathing*, which consists of rapid, deep breaths.

The Physical Examination and the Respiratory System

Several components of the physical examination will detect signs of respiratory disorders:

1. **Inspection** can reveal abnormal dimensions, such as the "barrel chest" that develops in emphysema or other chronic obstructive disorders, or *clubbing* of the fingers (p. 42). Clubbing is typically a late sign of disorders such as chronic lung infections or congestive heart failure. *Cyanosis*, a blue color of the skin and mucous membranes, generally indicates hypoxia (low tissue oxygen content). Laboratory testing of arterial blood gases will assist in determining the cause and severity of hypoxia. Postural changes, such as sitting up, leaning forward, and exhaling through pursed lips or gasping also indicate respiratory difficulties.

2. **Palpation** of the bones, muscles, and surface of the thoracic cage can detect structural problems or asymmetry. For example, a one-sided pleural effusion blunts the vibration of speech passed through the bronchi to the chest wall on that side.

3. **Percussion** on the surface of the thoracic cage over the lungs should yield sharp, resonant sounds. Dull or flat sounds can indicate structural changes that decrease air in the lungs, such as those accompanying pneumonia, or the collapse of part of a lung (*atelectasis*). Increased resonance can result from obstructive disorders, such as emphysema, due to hyperinflation of the lungs as the individual attempts to improve alveolar ventilation, or from air in the pleural space collapsing the lung (*pneumothorax*).

4. **Auscultation** of the lungs with a stethoscope yields the distinctive sounds of inhalation and exhalation. These sounds vary in intensity, pitch, and duration. Abnormal breath sounds accompany several pulmonary disorders:

   • **Rales** (rahlz) are hissing, whistling, scraping, or rattling sounds associated with increased airway resistance. The sounds are created by turbulent airflow past accumulated fluid or mucus or through airways narrowed by inflammation. Descriptions and interpretations of these sounds are highly subjective, but in general, *moist rales* are gurgling sounds produced as air flows over thin fluid layers within the respiratory tract. They are heard in conditions such as *bronchitis*, *tuberculosis*, and *pneumonia*. *Dry rales* are produced as air flows over thicker secretions, through inflamed airways, or into fluid-filled alveoli. Dry rales are characteristic of *asthma*, congestive heart failure, and *pulmonary edema*. *Rhonchi* are loud dry rales produced by mucus buildup in the air passages, which frequently clear after coughing.

   • **Stridor** is a very loud, high-pitched sound that can be heard without a stethoscope. Stridor generally indicates acute laryngeal airway obstruction, such as the partial blockage of the glottis by a foreign object.

   • **Wheezing** is a whistling, musical sound that can occur with inhalation or exhalation. It generally indicates airway obstruction due to epithelial inflammation, bronchospasms (bronchial smooth muscle contraction), or mucus buildup.

   • **Coughing** is a familiar sign of several respiratory disorders. Although primarily a reflex mechanism that clears the airway, coughing may also indicate irritation of the lining of the respiratory passageways. The duration, pitch, causative factors, and productivity associated with coughing may be important clues in the diagnosis of a respiratory disorder. (A *productive cough* ejects *sputum*, a mixture of mucus, cell debris, and pus; a *nonproductive cough*, or *dry cough*, does not.) If the cough is productive, the sputum can be collected and examined microscopically for cells, microorganisms, and debris. If squamous epithelial cells are rare, the sputum can be cultured to identify specific microorganisms. (Squamous epithelial cells are a marker for saliva, which contains oral microorganisms that make sputum culture results unreliable.)

   • A **friction rub** is a distinctive scratching sound produced by abrasion (which is frequently painful) between abnormal serous membranes. A *pleural rub* accompanies chest wall and lung respiratory movements. It indicates problems with the pleural membranes, such as *pleurisy*. A *pericardial rub* accompanies the heartbeat and indicates inflammation of the pericardium, as in *pericarditis*.

5. During the assessment of a person's vital signs, the respiratory rate (number of breaths per minute) is recorded, along with notations about the general rhythm and depth of respiration. *Tachypnea* is a respiratory rate faster than 20 breaths per minute in an adult; *bradypnea* is an adult respiratory rate below 12 breaths per minute.

Table 34 introduces important procedures and laboratory tests that are useful in diagnosing respiratory disorders.

▶ Disorders of the Respiratory System

The respiratory system provides a route for the movement of air into and out of the lungs and supplies a large, warm, moist surface area for the exchange of oxygen and carbon dioxide between the air and circulating blood. Disorders affecting the respiratory system may therefore involve any or all of the following three mechanisms:

1. **Interfering with the movement of air along the respiratory passageways.** Internal or external factors may be involved. Within the respiratory tract, the constriction of small airways, or bronchospasm, as in *asthma*, can reduce airflow to the lungs. The blockage of major airways (for example, by a swollen epiglottis or as a result of choking on a toy or a piece of food) can completely shut off the air supply. External factors that interfere with air movement include (1) the introduction of air (*pneumothorax*) or blood (*hemothorax*) into the

Table 34 Representative Diagnostic and Laboratory Tests for Respiratory Disorders

Diagnostic Procedure	Method and Result	Representative Uses
Pulmonary function studies	A spirometer is used to determine lung volumes and capacities, including V_T, IC, ERV, IRV, FRC, vital capacity, and total lung capacity on exertion	Differentiate obstructive from restrictive lung diseases; determine extent of pulmonary disease. Increased functional residual capacity (FRC) occurs in obstructive diseases, such as emphysema and chronic bronchitis
	Forced vital capacity (FVC) is the amount of air forcibly exhaled after maximal inhalation	FRC is normal or decreased in restrictive diseases, such as pulmonary fibrosis
Peak expiratory flow (PEF)	Measured by forceful exhalation into a PEF meter	Useful in evaluating asthma; permits self-monitoring; guides appropriate medication of airway obstruction
Pulse oximeter	External probe over fingernail bed or earlobe detects percentage of hemoglobin oxygen saturation (SaO_2) of cutaneous vessels.	Gives rapid, continuous noninvasive monitoring of arterial oxygen saturation. Relatively imprecise above $PaO_2 = 60$ mm Hg where $SaO_2 = 90\%$.
Bronchoscopy	Fiber-optic tubing is inserted through the mouth into the trachea, larynx, and bronchus for inspection	Detects abnormalities such as inflammation and tumors. Used to remove aspirated foreign objects or mucus from bronchi and to obtain samples of secretions or tissue for examination
Lung biopsy	Lung tissue is removed for pathological analysis via bronchoscopy, needle aspiration, or during exploratory surgery of the thoracic area	Differentiates pulmonary pathologies and determines the presence of malignancy
Chest x-ray study	Standard x-ray produces film sheet with radiodense tissues shown in white on a negative image	Detects abnormalities of lungs, such as tumors, inflammation of the lungs, rib or sternal fractures, or pneumothorax; detects fluid accumulation (pulmonary edema), pneumonia, and atelectasis ; determines heart size
Thoracentesis	A needle is inserted into the intrapleural space for removal of fluid for analysis or for relieving pressure due to fluid accumulation	See Pleural fluid analysis on next page
Pulmonary angiography	A catheter is inserted in the femoral vein and threaded through the right ventricle and into the pulmonary arteries. Contrast dye is intermittently injected as x-ray films are taken	Detects pulmonary embolism and other pulmonary vascular abnormalities; measures right atrial, ventricular, and pulmonary artery pressures
Lung scan	Radionuclides are inhaled and injected intravenously; radiation that is emitted is captured to create an image of the lungs	Determines areas with decreased blood flow or ventilation due to pulmonary embolism or other pulmonary disease
Computerized tomography (CT) scan	Standard or helical/spiral CT with or without intravenous contrast media may be used	Detects tumors, cysts, or other structural abnormalities

Laboratory Test	Normal Values in Blood Serum or Plasma	Significance of Abnormal Values
Arterial blood gases and pH		
pH	7.38–7.42	Lower than 7.35 indicates acidosis; higher than 7.45 indicates alkalosis
$PaCO_2$	38–42 mm Hg	Higher than 45 mm Hg with pH lower than 7.35 indicates respiratory acidosis present in disorders such as chronic bronchitis, in chronic obstructive pulmonary disease, and in CNS depression leading to irregular breathing; lower than 35 mm Hg with a pH higher than 7.45 indicates respiratory alkalosis that occurs during prolonged hyperventilation; also seen in pulmonary embolism
PaO_2	75–100 mm Hg	Lower than 75 mm Hg may occur in pneumonia, asthma, and COPD
HCO_3^-	22–28 mEq/L	Lower than 28 mEq/L (with elevated $PaCO_2$ and decreased pH) indicates renal compensation for respiratory acidosis; lower than 22 mEq/L (with decreased $PaCO_2$ and elevated pH) is characteristic of renal compensation for respiratory alkalosis
Blood tests (other)		
Alpha$_1$-antitrypsin determination	Adults: 100–300 mg/dL of serum	Decreased value ($\leqslant$80 mg/dL) indicates a possible genetic predisposition for emphysema
Sputum studies		
Cytology	No malignant cells are present	Sloughed malignant cells indicate cancerous process in lungs

(continued)

Table 34	Representative Diagnostic and Laboratory Tests for Respiratory Disorders (*Continued*)	
Laboratory Test	**Normal Values in Blood Serum or Plasma**	**Significance of Abnormal Values**
Culture and sensitivity (C&S)	Sputum sample is placed on growth medium	Identifies pathogenic organism and the organism's susceptibility to antibiotics
Acid-fast stain for bacilli	Staining technique reveals no acid-fast bacilli	Presence of stained rod-shaped microbes may indicate active infection with tuberculosis bacteria
Tuberculin skin test	Skin wheal, produced by intradermal application of tuberculin, at 48–72 hours should be ≤10 mm (area that is hardened, as opposed to the reddened area)	A skin papule at injection site that measures ≥10 mm after 48–72 hours is a positive test for immunity for tuberculosis. Does not prove active infection
Pleural fluid analysis		
Fluid color and clarity	No pus; fluid is clear, with WBC ≤1000 mm^3	WBC ≥ 1000 per mm^3 indicates probable infectious or inflammatory process; detects lung or pleural malignancy
Culture	No bacterial growth (fluid sample is placed on growth medium)	Presence of bacteria indicates infection; fungi can be present in immunocompromised person

pleural cavity, with subsequent lung collapse; (2) the buildup of fluid within the pleural cavities (a *pleural effusion*), which compresses and collapses the lungs; and (3) arthritis, muscular paralysis, or other conditions that prevent the normal skeletal or muscular activities responsible for moving air into and out of the respiratory tract.

2. **Damaging or otherwise impeding the diffusion of gases at the respiratory membrane.** The walls of the alveoli are part of the respiratory membrane, where gas exchange occurs. Any disease that affects the alveolar walls will reduce the efficiency of gas exchange. In *emphysema*, alveoli are destroyed. With the inflammation and obstruction of *lung cancer*, or infection of the lungs, as in the various types of pneumonia, respiratory exchange is disrupted by the buildup of tumor, fluid, or mucus within the alveoli and bronchi.

3. **Blocking or reducing the normal circulation of blood through the alveolar capillaries.** Blood flow to portions of the lungs may be prevented by the circulatory blockage of a *pulmonary embolism*. Not only does a pulmonary embolism prevent normal gas exchange in the affected regions of a lung, but also it results in tissue damage and, if the blockage persists for several hours, sustained alveolar collapse. Pulmonary blood pressure may rise (a condition called *pulmonary hypertension*), leading to *cor pulmonale*, or right heart failure, which may result from heart, lung, or arterial disease.

These types of problems can result from trauma, congenital or degenerative problems, the formation of tumors, inflammation, or infection of the lungs (▶**Figure 72**). Illnesses caused by infections of the upper respiratory tract include some of the most common diseases. Many respiratory infections are transmitted by droplets in the air, typically emitted in a sneeze or cough. Infections of the lower respiratory tract include two of the most deadly diseases in human history: pneumonia and tuberculosis. **Table 35**, p. 167, summarizes information about some of the most important infectious diseases of the respiratory system, along with their causative organisms.

Respiratory system disorders also occur secondarily, as a consequence of dysfunctions of other body systems. For instance, asthma may result from problems with immune function, and most pulmonary emboli originate as intravascular clots that embolize, pass through the right heart to reach the pulmonary arteries, and impair lung perfusion.

Overloading the Respiratory Defenses

Large quantities of airborne particles can overload the respiratory defenses and produce a variety of illnesses. Chemical or physical irritants that reach the lamina propria or underlying tissues promote the formation of scar tissue (*fibrosis*), reducing the elasticity of the lung, and can restrict airflow along the passageways. Irritants or foreign particles may also enter the lymphatic vessels of the lung, producing inflammation of the regional lymph nodes. Chronic irritation and the stimulation of the epithelium and its defenses cause changes in the epithelium that increase the likelihood of lung cancer.

Severe symptoms of such disorders develop slowly; they may take 20 years or more to appear. **Silicosis** (sil-i-KŌ-sis, produced by the inhalation of silica dust), **asbestosis** (as-bes-TŌ-sis, from the inhalation of asbestos fibers), and **anthracosis** (an-thra-KŌ-sis, the "black lung disease" of coal miners, caused by the inhalation of coal dust) are conditions caused by the overloading of the respiratory defenses (▶**Figure 73**). Rescue workers at the World Trade Center Towers site were exposed to enormous amounts of dust and smoke of undetermined composition. Although the primary component was undoubtedly concrete dust, samples contain asbestos, small shards of glass, heavy metals, and other products of combustion from the fires that burned for weeks. Many of the rescue workers have developed a persistent cough, hyperreactive airways with bronchospasm, and restrictive interstitial lung disease.

Tuberculosis

Tuberculosis (TB) is a major health problem throughout the world. With roughly 1.3 million deaths in 2012, it is the leading cause of death from infectious diseases. An estimated 30 percent of the world's human population, or 2.1 *billion* people are

Respiratory System

▶ **Figure 72** **Disorders of the Respiratory System.**

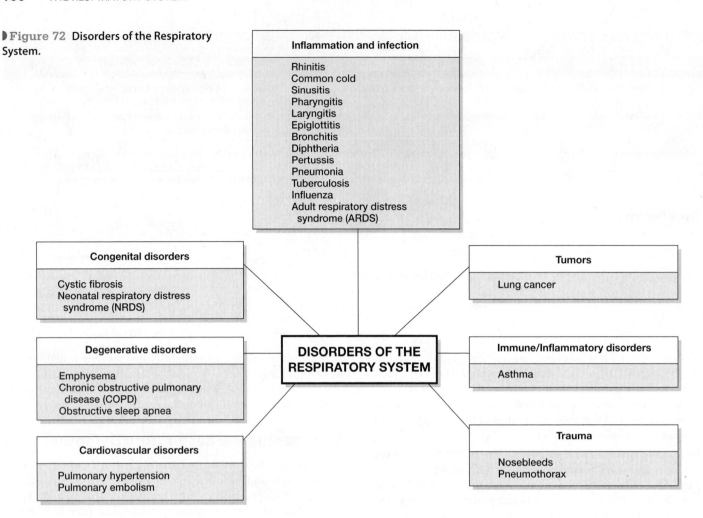

Inflammation and infection

Rhinitis
Common cold
Sinusitis
Pharyngitis
Laryngitis
Epiglottitis
Bronchitis
Diphtheria
Pertussis
Pneumonia
Tuberculosis
Influenza
Adult respiratory distress
  syndrome (ARDS)

Congenital disorders

Cystic fibrosis
Neonatal respiratory distress
  syndrome (NRDS)

Tumors

Lung cancer

DISORDERS OF THE RESPIRATORY SYSTEM

Degenerative disorders

Emphysema
Chronic obstructive pulmonary
  disease (COPD)
Obstructive sleep apnea

Immune/Inflammatory disorders

Asthma

Cardiovascular disorders

Pulmonary hypertension
Pulmonary embolism

Trauma

Nosebleeds
Pneumothorax

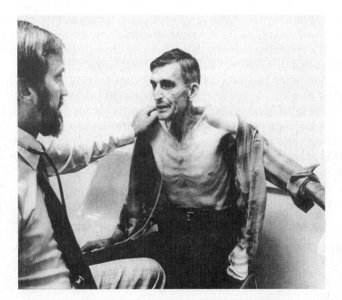

▶ **Figure 73** **A Person with Anthracosis (Black Lung Disease).**

infected, with over 8 million cases of active, potentially contagious disease each year. Unlike other deadly diseases, such as AIDS, TB can be transmitted through casual contact and through the air, and has been spread from one airplane passenger to other passengers. Anyone who breathes is at risk of contracting this disease; all it takes is exposure to the causative bacterium. Coughing, sneezing, or speaking by an actively infected individual spreads the pathogen through the air in the form of tiny droplets that can be inhaled by other people. Crowded living conditions and prolonged contact (usually days to weeks) increase the risk of infection.

Untreated, the disease progresses in stages. The primary site of infection is usually in the lungs. Other organs may become infected, and without antibiotic treatment the mortality rate is approximately 33 percent.

At the site of infection, macrophages and fibroblasts wall off the area, forming an abscess. For most persons in otherwise good health this containment leaves them with latent, noncontagious TB diagnosed by a few scarred nodules visible on chest x-ray, a positive TB skin test (see **Table 34**), and no chest x-ray, lab, or clinical signs of active disease. In 10 percent of latent TB cases, at some point after primary infection the scar-tissue barricade fails, and the bacteria multiply and move into the surrounding lung tissues, causing active TB with cough and infectious sputum production. Reactivation is most common in the first year after infection, but may appear decades later. Conditions that weaken the immune system such as aging, malnutrition, and infections such as HIV/AIDS lower resistance to reactivation. Worldwide, TB is the leading cause of death for HIV-infected persons.

Respiratory System

Table 35 Examples of Infectious Diseases of the Respiratory System

Disease	Organism(s)	Description
Bacterial diseases		
Sinusitis	*Streptococcus pneumoniae* or *Haemophilus influenzae*	Inflammation of the paranasal cavities; headaches, pain, purulent nasal discharge, pressure in facial bones
Pharyngitis	*Streptococcus pyogenes*	Inflammation of the throat; strep throat; sore throat, fever, less cough or sputum
Laryngitis	*Streptococcus pneumoniae* or *Haemophilus influenzae*	Inflammation of the larynx; dry, sore throat; hoarse voice or loss of voice
Epiglottitis	*Haemophilus influenzae*	Inflammation of the epiglottis; most common in children; fever, noisy breathing, drooling; swelling can block trachea and lead to suffocation
Bronchitis	Respiratory viruses	Initial viral infection with dry cough may lead to secondary bacterial infections and pneumonia
Diphtheria	*Corynebacterium diphtheriae*	Inflammation of the pharynx; pseudomembrane in pharynx; bacterial toxin affects heart and other tissues
Pertussis (whooping cough)	*Bordetella pertussis*	Highly contagious disease; nasal mucus production; severe coughing ends in a "whoop" sound during inhalation
Pneumonia	*Streptococcus pneumoniae, Mycoplasma pneumoniae*	Inflammation of the lungs; alveoli fill with fluids; chills, high fever; cough with sputum of mucus, pus, and blood. Mycoplasma, or atypical pneumonia, has a dry cough and is more prolonged
Tuberculosis	*Mycobacterium tuberculosis*	Highly contagious infection of the lungs; can spread to other tissues; abscesses, or tubercles, form in lungs; bacteria multiply in WBCs and alveolar macrophages and break down alveoli
Viral diseases		
Common cold	Rhinoviruses or coronaviruses	Mucosal edema, nasal obstruction, nasal discharge (coryza), sneezing, and headache
Influenza	Orthomyxoviruses (influenza virus A, B, C)	Frequent variations in Types A and B viruses can produce new epidemics; fever, headache, sore throat, nasal discharge, muscle pain (*myalgia*), fatigue, chest pain, and cough

Treatment for TB is complex, because the bacteria are slow growing, and can spread through the bloodstream to many different tissues. Prolonged treatment is required and resistance can develop to standard antibiotics, particularly if infected persons fail to take their antibiotics at appropriate intervals over the entire recommended treatment period. To combat antibiotic resistance, several drugs are used in combination over a period of six to nine months, often with "directly observed therapy" (DOT) where health personnel watch the patient swallow the medications. The most effective drugs now available include *isoniazid*, which interferes with bacterial replication, and *rifampin*, which blocks bacterial protein synthesis.

Tuberculosis was extremely common in the United States early in the 20th century. An estimated 80 percent of Americans born around 1900 became infected with tuberculosis during their lives. Although most were able to recover from the primary bacterial infection, it was the leading cause of death in 1906. These statistics have been dramatically changed with the advent of effective antibiotics in the 1940s and techniques for early detection of infection. Between 1906 and 2004 the death rate fell from 200 deaths per 100,000 population to 0.2 deaths per 100,000 population. Strong public health measures, including TB surveillance by TB skin testing and chest x-rays, lab identification through sputum culture and drug susceptibility testing, directly observed treatment (DOT), isolation of infectious patients, and investigation and treatment of close contacts has significantly reduced the number of new U.S. cases of TB, down to 10,528 in 2011. The TB problem is much more severe in developing nations. As of 2000 an estimated 10–15 million

people in the United States are infected with the bacterium, 50 percent of them immigrants who had latent infection when they arrived. These individuals with TB are not infectious unless the disease reactivates, causing symptoms of coughing, fever, and weight loss. This weight loss led to the common name for advanced TB, "consumption."

Cystic Fibrosis

Cystic fibrosis (CF) is the most common lethal inherited disease that affects Caucasians of Northern European descent; it occurs at a frequency of 1 in 2500 to 3500 births. The disease occurs with less frequency in individuals of Southern European ancestry, in the Ashkenazi Jewish population, and in African Americans and Asians. The condition results from defects in the *CFTR* gene located on chromosome 7. Within the U.S. Caucasian population, 1 person in 30 carries one abnormal copy of the gene for this disorder, and an infant receiving a copy from both parents will develop CF. In the United States, adults can be tested to see if they carry an abnormal *CFTR* gene, and prenatal testing can detect the disease. All states screen newborns for the disease. In the United States, 1000 babies are born with CF each year, and roughly 30,000 persons are living with this condition. In the 1950s, most patients died in infancy; with current medical care, the average life expectancy is almost 40 years. Death is generally the result of chronic, recurrent bacterial infection of the lungs and associated heart failure.

The gene involved carries instructions for a transmembrane protein responsible for the active transport of chloride ions. This

Respiratory System

membrane protein is abundant in exocrine cells that produce watery secretions. In persons with CF, the protein does not function normally. The secretory cells cannot transport salts and water effectively, and the secretions produced are thick and gooey. Mucous glands of the respiratory tract and secretory cells of the pancreas, salivary glands, and the digestive and reproductive tracts are affected.

The most serious symptoms appear because the respiratory defense system cannot transport such dense mucus. The mucus escalator stops working, and mucus plugs block the smaller respiratory passageways. This blockage reduces the diameter of the airways, and the inactivation of the normal respiratory defenses leads to frequent bacterial infections. The most dangerous infections involve the bacterium *Pseudomonas aeruginosa*, which colonizes the stagnant mucus and further stimulates mucus production by epithelial cells.

Treatment has been limited primarily to aggressive pulmonary therapy and antibiotic therapy to control infections. In a few instances, lung transplants have provided relief, but the technical and logistical problems involved with that approach are formidable. The structure of the normal gene has been determined, along with more than 1000 different mutations of this gene that cause cystic fibrosis. Variation in disease severity in different families probably results from this multitude of abnormal mutations. The current goal is to correct the defect by inserting the normal gene into the cells in critical areas of the body. In the meantime, one of the factors contributing to the thickness of the mucus has been discovered to be the presence of DNA released from degenerating cells in areas of inflammation. Inhaling an aerosol spray containing an enzyme that breaks down DNA has proven to be effective in improving respiratory performance.

Tracheal Blockage

We sometimes inadvertently breathe in foreign objects, a process called *aspiration*. Foreign objects that become lodged in the larynx or trachea are generally expelled by coughing. If the individual can speak or make a sound, the airway is still open, so emergency measures may not be necessary. If the person is silent, and can neither breathe nor speak, an immediate threat to life exists. Unfortunately, many victims become acutely embarrassed by this situation; instead of seeking assistance, they run to the nearest restroom and quietly die there.

In the **Heimlich** (HĬM-lik) **maneuver**, or *abdominal thrust* (which was first tried and refined in experiments on anesthetized dogs) a rescuer applies brief, strong compressions to the victim's abdomen just inferior to the diaphragm. This action elevates the diaphragm forcefully and may generate enough pressure to expel objects blocking the trachea and upper airway. The maneuver must be performed properly to avoid damage to internal organs. Organizations such as the American Red Cross and local fire departments periodically hold brief training sessions in the proper performance of the Heimlich maneuver.

If blockage results from a swelling of the epiglottis or tissues surrounding the glottis, a trained rescuer may insert a curved tube through the pharynx and glottis to permit airflow. This procedure is called *intubation*. If the blockage is immovable or the larynx has been crushed, a **tracheostomy** (trā-kē-OS-to-mē) may be performed. In a tracheostomy, an incision is made through the anterior tracheal wall, inferior to the larynx, and a tube is inserted to permit air to flow directly into the trachea.

Bronchoscopy

One method of investigating the status of the respiratory passageways is the use of a **bronchoscope**, a tube with a fiber-optic bundle small enough to be inserted through the mouth and larynx into the trachea and steered along the conducting passageways to the level of the smaller bronchioles. This procedure is called **bronchoscopy** (brong-KOS-ko-pē). In addition to permitting direct visualization of bronchial structures, the bronchoscope can collect tissue or mucus samples from the respiratory tract. In **bronchography** (brong-KOG-ra-fē), a bronchoscope or catheter introduces a radiopaque material into the bronchi. This technique can permit detailed x-ray analysis of bronchial masses, such as tumors or other obstructions, but has been largely replaced by lung CT scans.

Asthma

Asthma (AZ-muh) is defined as chronic inflammation of the lung airways with hyper-reactive air-conducting passageways that respond to irritants by muscular bronchial constriction, called *bronchospasm. Asthmatic bronchitis*, is the term that is commonly applied when symptoms of bronchospasm are acute, reversible, and intermittent.

Wheezing is the sound made as air whistles through constricted airways. Asthma affects an estimated 3–6 percent of the U.S. population. There are various inciting factors that lead to the coughs, wheezing, and dyspnea (labored breathing) that help diagnose asthma. In many cases, the trigger appears to be an immediate hypersensitivity reaction to an allergen in the inhaled air. Drug reactions, air pollution, chronic respiratory infections, exercise, and emotional stress can also induce an asthma attack in sensitive individuals.

The most obvious and potentially dangerous symptoms include the constriction of smooth muscles all along the bronchial tree, edema and swelling of the lining mucosa of the respiratory passageways, and the accelerated production of mucus. The combination makes breathing very difficult. Exhalation is affected more than inhalation; the narrowed passageways may collapse before exhalation is completed leaving air trapped and hyperexpanding the lungs. Mucus transport slows, so fluids accumulate. Dyspnea, coughing and wheezing then develop. The bronchoconstriction and mucus production may occur in a few minutes, in response to the release of histamine and prostaglandins by mast cells. The activated mast cells also release interleukins, leukotrienes, and platelet-activating factors. As a result, over a period of hours, neutrophils and eosinophils migrate into the area, which then becomes inflamed, further reducing airflow and damaging respiratory tissues. Because the inflammation compounds the problem, antihistamines, bronchodilators, and anti-inflammatory steroids may be needed to control a severe asthma attack.

When a severe attack occurs, it reduces the functional capabilities of the respiratory system. Peripheral tissues become starved for oxygen, a condition that can prove fatal. Asthma fatalities have been increasing in recent years. Both the number of persons with asthma (25 million) and the number who die from asthma (3400 in 2010) are increasing. Asthma attacks can be triggered by tobacco and wood smoke, air pollution, various allergies, and respiratory infections.

Respiratory System

The treatment of asthma involves the dilation of the respiratory passageways by administering **bronchodilators** (brong-kō-DĪ-lā-torz) (drugs that relax bronchial smooth muscle) and by reducing inflammation and swelling of the respiratory mucosa through anti-inflammatory medication. Important bronchodilators include *albuterol* (usually inhaled), *theophylline, epinephrine*, and other beta-adrenergic drugs. Although the strongest beta-adrenergic drugs are quite useful in a crisis, they are effective only for relatively brief periods, and overuse of them can lead to reduced efficiency. Asthmatic individuals must be closely monitored, due to the drugs' potential effects on cardiovascular function. Anti-inflammatory medications, such as inhaled or ingested steroids, are becoming increasingly important.

COPD, Chronic Bronchitis, and Emphysema

Chronic obstructive pulmonary disease (COPD) is a progressive disorder of the airways that restricts airflow and reduces alveolar ventilation with episodic exacerbations, often from infection. COPD shares many of the symptoms, anatomic and physiologic changes, and treatments of asthma discussed earlier, with the significant difference that progressive, permanent lung damage occurs. Two patterns of symptoms may appear in persons with COPD, but most people with COPD suffer from both disorders. In chronic *bronchitis*, the lining of bronchial airways is inflamed and thickened with excess mucous production. In *emphysema*, alveolar walls and terminal bronchioles are weakened and destroyed, resulting in large air spaces, called *bullae*, that provide less surface area for oxygen exchange. In both conditions, there is obstructed airflow. Symptoms are cough, often with excess mucus; shortness of breath with exercise; and *wheezing*, the whistling sound of air passing through partially obstructed airways. Over 12 million Americans have COPD; most are over 40 years of age, and it is the fourth leading cause of death.

On physical exam, individuals with COPD commonly expand their chests permanently in an effort to enlarge their lung capacities and make the best use of the remaining functional alveoli. This adaptation may give them a distinctive "barrel-chested" appearance. The respiratory rate with advanced disease may increase dramatically. The lungs are fully inflated at each breath, and expirations are forced. Two clinical patterns of advanced disease occur.

Bronchitis (brong-KĪ-tis) is an inflammation and swelling of the bronchial lining, leading to overproduction of mucous secretions. Usually an acute, infrequent self-limited lung infection, it may require antibiotic treatment if severe or prolonged. The most characteristic symptom is frequent coughing with copious sputum production. Arbitrarily, three or more episodes a year are called *chronic bronchitis*. This condition is most commonly related to cigarette smoking, but also results from other environmental irritants, such as chemical vapors. Over time, the increased mucus production can block smaller airways and reduce air exchange and respiratory efficiency. Chronic bacterial infections leading to more lung damage are common. Treatment involves stopping smoking and the administration of bronchodilators, inhaled corticosteroids, antibiotics, and supplemental oxygen as necessary.

Persons with chronic bronchitis may have symptoms of heart failure, including widespread edema. Their blood oxygenation is low, and their skin may have a bluish color. The combination of widespread edema and bluish coloration has led to the descriptive term *blue bloaters* for individuals with this condition.

Emphysema (em-fi-SĒ-muh) is a chronic, progressive condition characterized by shortness of breath and an inability to tolerate physical exertion. The underlying problem is the destruction of respiratory exchange surfaces. In essence, respiratory bronchioles and alveoli are functionally eliminated. The alveoli gradually expand and capillaries deteriorate, leaving large nonfunctional air cavities in the lungs where gas exchange is severely decreased or eliminated.

Emphysema has been linked to the inhalation of air that contains fine particulate matter or toxic vapors, such as those in cigarette smoke. Early in the disease, local regulation shunts blood away from the damaged areas, and the individual may not notice problems, even with strenuous activity. As the condition progresses, the reduction in exchange surface limits the ability to provide adequate oxygen. However, obvious clinical symptoms typically fail to appear until the damage is extensive.

Alpha$_1$-antitrypsin, an enzyme that is normally present in the lungs, helps prevent degenerative changes in lung tissue. Most people requiring treatment for emphysema are adult smokers, a group that includes both individuals with alpha$_1$-antitrypsin deficiency and those with normal tissue enzymes. In the United States, 1 person in 1000 carries two copies of a gene that codes for an abnormal, inactive form of this enzyme. A single change in the amino acid sequence of the enzyme appears responsible for this defect. At least 80 percent of nonsmokers with abnormal alpha$_1$-antitrypsin will develop emphysema, generally at ages 45–50 years. *All* smokers will develop at least some emphysema, typically by ages 35–40 years.

Unfortunately, the loss of alveoli and bronchioles in emphysema is permanent and irreversible. Further progression can be limited by stopping smoking. Other aspects of pulmonary structure and function may be relatively normal. The only effective treatment for severe cases is the administration of supplemental oxygen, but lung transplants have helped some patients, as has the surgical removal of nonfunctional lung tissue. For persons with alpha$_1$-antitrypsin deficiency who are diagnosed early, attempts are under way to provide enzyme supplements by daily or periodic infusion.

Persons with emphysema tend to maintain near-normal arterial P$_{O_2}$. Their respiratory muscles work hard, and they use a lot of energy just breathing. As a result, they tend to be thin. Because their blood oxygenation is near normal, skin color in Caucasians in this group is pink. The combination of heavy breathing and pink coloration has led to the descriptive term *pink puffers* for individuals with this condition.

Respiratory Distress Syndrome (RDS)

Septal cells among the lining cells of the alveoli begin producing surfactant at the end of the sixth fetal month. By the eighth month, surfactant production has risen to the level required for normal respiratory function. **Neonatal respiratory distress syndrome (NRDS)**, also known as *hyaline membrane disease* (*HMD*), develops when surfactant production fails to reach normal levels. Although some forms of HMD are inherited, the condition most commonly accompanies premature births (37 weeks gestation or earlier), especially before 28 weeks gestation. If a preterm birth cannot be prevented, giving the mother corticosteroids before delivery reduces the risk and severity of NRDS after birth.

In the absence of surfactant, the alveoli tend to collapse during exhalation. Although the conducting passageways remain open, the newborn infant must then inhale with extra force to reopen the alveoli on the next breath. In effect, every breath must approach the power of the first, so the infant rapidly becomes exhausted. Respiratory movements become progressively weaker; eventually, the alveoli fail to expand and gas exchange ceases.

One method of treatment involves assisting the infant by administering air under continuous pressure via the nostrils so that the alveoli are held open. This procedure, known as **continuous positive airway pressure (CPAP)**, can keep the newborn alive until surfactant production increases to normal levels. CPAP treatment, often combined with surfactant from other sources administered via the trachea on the first day of life, has reduced the death rate from NRDS from nearly 100 percent to less than 10 percent. In the United States, 25,000 babies died of NRDS in 1960, whereas only 825 died in 2006.

Surfactant abnormalities can also develop in adults as the result of severe respiratory infections or other sources of pulmonary injury. Alveolar collapse follows, producing a condition known as **adult respiratory distress syndrome (ARDS)**. A more intense form of CPAP called positive end expiratory pressure (PEEP) is typically used to maintain life until the underlying problem can be corrected, but at least 30 percent of ARDS cases end in death.

Obstructive Sleep Apnea

Obstructive sleep apnea occurs when there is intermittent blockage of respiratory airflow past the mouth and nose during sleep. Muscles in the pharynx that keep the upper airway open normally relax during sleep and this passage may briefly close, especially if it is narrow to begin with. Loud snoring and even labored breathing may occur. With deep sleep, air blockage may persist for 10 to 20 seconds causing a brief period of apnea (the absence of breathing). As oxygen levels fall, breathing is resumed, often requiring a loud "snort and gasp" to overcome the obstruction. These episodes may occur 20 to 30 times an hour.

The result is poor-quality sleep that leads to excessive daytime drowsiness. Fatigue, headaches, irritability, and poor concentration may result, as well as increased risk of driving and work accidents. Severe sleep apnea stresses the heart and can worsen cardiovascular disease. Family members who observe the afflicted person sleeping may help in diagnosis.

Any condition that narrows the upper airway, including obesity, a large neck, or large tonsils, adenoids, and/or tongue increase the incidence of obstructive sleep apnea. Sleeping on one's back and taking alcohol or sedatives before sleep aggravate the condition. In most cases, nasal CPAP, similar to that used to treat infant respiratory distress syndrome, keeps the upper airway open and gives almost immediate relief.

Boyle's Law and Air Overexpansion Syndrome

Swimmers descending 1 m (~3 feet) or more beneath the surface experience significant increases in pressure, due to the weight of the overlying water. The air in spaces throughout the body decreases in volume. The increase in pressure normally produces mild discomfort in the middle ear, but some people experience acute pain and disorientation. The pressure first collapses the auditory tubes. As the volume of air in the middle ears decreases, the tympanic membranes are forced inward. This uncomfortable situation can be remedied by closing the mouth, pinching the nose, and exhaling gently; elevating the pressure in the nasopharynx forces air through the auditory tubes and into the middle ear spaces. As the volume of air in each middle ear increases, the tympanic membrane returns to its normal position. When the swimmer returns to the surface, the pressure drops and the air in the middle ear expands. This expansion usually goes unnoticed, because the air simply forces its way out along the auditory tube and into the nasopharynx.

Scuba divers breathe air under pressure, and that air is delivered at the same pressure as that of their surroundings. A descent to a depth of 10 m (~33 feet) doubles the pressure on a diver, due to the weight of the overlying water. Consider what happens if that diver then takes a full breath of air and heads for the surface. As the pressure of the air declines, its volume increases (Boyle's law). Thus, at the surface, the volume of air in the lungs will have doubled. Such a drastic increase cannot be tolerated. The usual result is a tear in the wall of the lung. The symptoms and severity of this *air overexpansion syndrome* depend on where the air ends up. If the air flows into the pleural cavity, the lung may collapse; if it enters the mediastinum, it may compress the pericardium and produce symptoms of cardiac tamponade. Worst of all, the air may rupture blood vessels and enter the bloodstream. The air bubbles then form emboli that can block blood vessels in the heart or brain, producing a heart attack or stroke. These are all serious conditions, so divers are trained to avoid holding their breath and to exhale when swimming toward the surface.

Artificial Respiration

Artificial respiration, or *rescue breathing*, is a technique used to provide air to an individual whose respiratory muscles are no longer functioning. In **mouth-to-mouth resuscitation**, a rescuer provides ventilation by exhaling into the mouth, or mouth and nose, of the victim. After each breath, contact is broken to permit passive exhalation by the victim. Air provided in this way contains adequate oxygen to meet the needs of the victim. If the victim's cardiovascular system is nonfunctional (cardiopulmonary arrest), a technique called *cardiopulmonary resuscitation* (*CPR*) may temporarily maintain adequate blood flow and tissue oxygenation. **Cardiopulmonary resuscitation**, or **CPR**, restores some blood flow and ventilation to an individual whose heart has stopped beating. Compression applied to the rib cage over the sternum reduces the volume of the thoracic cavity, squeezing the heart and propelling blood into the aorta and pulmonary trunk. When the pressure is removed, the thorax expands and blood moves into the great veins. Cycles of compression are interspersed with cycles of mouth-to-mouth breathing that maintain pulmonary ventilation.

CPR, supplemented by electric shock(s) to normalize cardiac rhythm, has been credited with saving many lives. Basic techniques can be mastered in about eight hours of intensive training, using special equipment. Yearly recertification courses must be taken, because the skills fade with time and because CPR techniques cannot be practiced on a living person without causing severe injuries, such as broken ribs. Training is available at minimal cost through charitable organizations such as the Red Cross and the American Heart Association.

In adults, heart disease is the main cause of cardiopulmonary arrest, and in terms of restoring cardiac function, chest compression alone is probably as effective as full CPR, and easier to perform. However, in children respiratory problems cause most arrests, and rescue breathing is essential.

Pulmonary Function Tests

Pulmonary function tests monitor several aspects of respiratory function. A **spirometer** (spī-ROM-e-ter) measures parameters such as vital capacity, expiratory reserve, and inspiratory reserve. A **peak flow meter** records the maximum rate of air movement during forced exhalation. Tests using these instruments are relatively simple to perform, and they have considerable diagnostic significance.

For example, in people with asthma, the constricted airways tend to close before an exhalation is completed. As a result, pulmonary function tests show a reduction in vital capacity, expiratory reserve volume, and peak flow rate. Asthma patients are taught to measure their peak flow rates to detect incipient problems.

Conditions that restrict the maximum distensibility of the lungs also reduce vital capacity, because they lower the inspiratory reserve. However, because the airways are not affected, the expiratory reserve and expiratory flow rates are relatively normal.

Decompression Sickness

Decompression sickness can develop when an individual experiences a sudden change in pressure. Nitrogen is the gas responsible for this condition. Nitrogen, which accounts for 78.6 percent of the atmospheric gas mixture, has a relatively low solubility in body fluids. Under normal atmospheric pressures, blood contains few nitrogen molecules, but at higher-than-normal pressures, additional nitrogen molecules diffuse across the alveolar surfaces and into the bloodstream.

As more nitrogen enters the blood, the gas is distributed throughout the body. Over time, nitrogen diffuses into peripheral tissues and into body fluids such as the cerebrospinal fluid, aqueous humor, and synovial fluids. If the atmospheric pressure decreases, the change must occur slowly enough that the excess nitrogen can diffuse out of the tissues, into the blood, and across the alveolar surfaces. If the pressure falls suddenly, this gradual movement of nitrogen from the periphery to the lungs cannot occur. Instead, the nitrogen leaves solution and forms bubbles of nitrogen gas in the blood, tissues, and body fluids.

A few bubbles in peripheral connective tissues may not be particularly dangerous, at least initially. However, these bubbles can fuse together to form larger bubbles that distort tissues, causing pain. Bubbles typically develop in joint capsules first, producing severe pain, and the afflicted individual tends to bend over or curl up. This symptom accounts for the popular name of the condition: "the bends." Bubbles in the systemic or pulmonary circulation can cause infarcts, and those in the cerebrospinal circulation can cause strokes or spinal cord ischemia, leading to sensory losses, paralysis, or respiratory arrest.

Treatment consists of recompression in a hyperbaric (*hyper*, high + *baro*-, pressure) chamber—exposing the individual to elevated pressures that force the nitrogen back into solution, thereby alleviating the symptoms. Pressures are then reduced gradually over a period of hours to a day or more. Breathing air with more oxygen and less nitrogen than are in atmospheric air accelerates the removal of excess nitrogen from the blood. This *hyperbaric oxygen treatment* raises blood and tissue oxygenation and is used for some serious wounds and infections involving bacteria that thrive in low-oxygen environments.

Today, most cases of the bends involve scuba divers who have gone too deep or stayed at depth too long. The condition is not restricted to divers, however: The first reported cases involved construction crews who worked in pressurized surroundings. Although such accidents are exceedingly rare, the sudden loss of cabin pressure in a commercial airplane can also produce symptoms of decompression sickness.

Mountain Sickness

Acute mountain sickness (AMS), also called *high altitude illness*, may develop after a rapid ascent to altitudes of 2500 to 3500 m (8000 to 11,500 feet). Symptoms, which appear within a day after arrival, include severe headache, insomnia, fatigue, shortness of breath, light-headedness, and nausea. The underlying cause is the combination of hypoxia and lower atmospheric pressure, which affects the nervous system and lungs to bring on a combination of mild cerebral edema and pulmonary insufficiency. The symptoms, which may persist for a week before subsiding, can be prevented or reduced by acclimatization, which involves slow ascent (no more than 300 m or 1000 ft per day) and the administration of acetazolamide (*Diamox*) or a corticosteroid, dexamethasone. Diamox increases the respiratory rate and inhibits carbonic anhydrase activity to promote bicarbonate loss at the kidneys, while dexamethasone reduces edema. Airplanes are pressurized to 2000 m (7000 ft), so mountain sickness is not a problem for air travelers.

If severe, High Altitude Cerebral Edema (HACE) and/or High Altitude Pulmonary Edema (HAPE) may occur. These are life-threatening conditions that can affect over 10 percent of persons doing strenuous physical activities shortly after arriving at high altitudes of 3000 m or more. Fatalities result from the combination of cerebral edema and pulmonary edema. Treatment includes rest, breathing pure oxygen, dexamethasone, and immediate transport to lower altitudes. A portable chamber pressurized by a foot-pump may be used to buy time until descent is possible.

Shallow-Water Blackout

In preparing to dive under water, misguided swimmers may attempt to outwit their chemoreceptor reflexes. The usual method involves taking some extra deep breaths before submerging. These individuals are intentionally hyperventilating, usually with the stated goal of "taking up extra oxygen."

From the hemoglobin saturation curve, it should be obvious that this explanation is incorrect. At a normal alveolar P_{O_2} hemoglobin is 97.5 percent saturated; no matter how many breaths the swimmer takes, the alveolar P_{O_2} cannot rise significantly. But the P_{CO_2} will be affected, because the increased ventilation rate lowers the carbon dioxide concentrations of the alveoli and blood. This lowering produces the desired effect by temporarily shutting off the chemoreceptors that monitor P_{CO_2}. As long as CO_2 levels remain depressed, the swimmer does not feel the need to breathe, despite the fact that his or her P_{O_2} continues to fall.

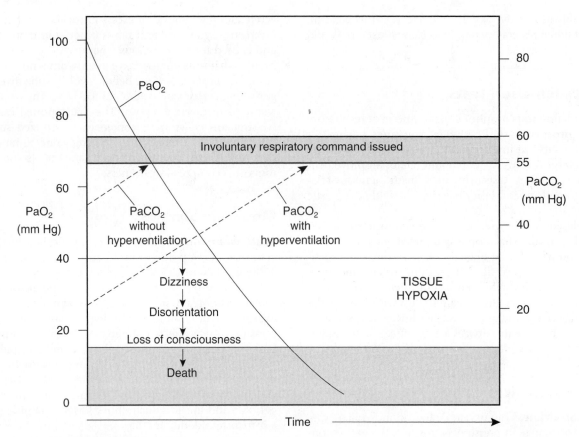

▶ **Figure 74 Carbon Dioxide and Respiratory Demand.** The stimulation of respiration by rising $PaCO_2$ normally prevents a dangerous decline in oxygen levels. After hyperventilation, the $PaCO_2$ may be so low that respiratory stimulation will not occur before low PaO_2 impairs neural function.

Under normal circumstances, holding one's breath causes a decline in PaO_2 and a rise in $PaCO_2$. As indicated in ▶ **Figure 74**, by the time the PaO_2 has fallen to 60 percent of normal levels, carbon dioxide levels will have risen enough to make breathing an unavoidable necessity. But after hyperventilation, oxygen levels can fall so low that the swimmer becomes unconscious (generally at a PaO_2 of 15–20 mm Hg) before the $PaCO_2$ rises enough to stimulate breathing. This *shallow-water blackout* has fatal consequences in many cases. Hence, swimmers and divers should not hyperventilate.

Chemoreceptor Accommodation and Opposition

Carbon dioxide receptors accommodate sustained P_{CO_2} levels above or below normal. Although these receptors register the initial change quite strongly, over a period of days they adapt to the new levels, as "normal." As a result, after several days of elevated P_{CO_2} levels, the effects on the respiratory centers begin to decline. Fortunately, the response to a low arterial P_{O_2} remains intact, and the response to arterial oxygen concentrations becomes increasingly important.

Some individuals with severe chronic lung disease, however, cannot maintain normal P_{O_2} and P_{CO_2} levels in the blood. The arterial CO_2 level rises to 50–55 mm Hg, and the P_{O_2} falls to 45–50 mm Hg. At these levels, the carbon dioxide receptors accommodate, and most of the respiratory drive comes from the arterial oxygen receptors. If individuals with this condition are given too much oxygen, they may simply stop breathing. Vigorous stimulation to encourage breathing or a mechanical respirator may be required.

Lung Cancer

Lung cancers are estimated to cause 14 percent of new cancer cases in both men and women. This condition is the primary cause of cancer death in the U.S. population, accounting for 26 percent of all female cancer deaths, and 30 percent of all male cancer deaths. Lung cancer kills more people than colon, breast, and prostate cancer combined. Despite advances in the treatment of other forms of cancer, the survival statistics for lung cancer have not changed significantly. Even with early detection, the five-year survival rates are only 30 percent for men and 50 percent for women, and over 50 percent of lung cancer patients die within a year of diagnosis.

Detailed statistical and experimental evidence has shown that *85–90 percent of all lung cancers are the direct result of cigarette smoking.* Claims to the contrary are simply unjustified and insupportable. The data are far too extensive to detail here, but the incidence of lung cancer for nonsmokers is 3.4 per 100,000 population, whereas the incidence for smokers ranges from 59.3 per 100,000 for those who smoke between a half-pack and a pack per day to 217.3 per 100,000 for those who smoke one to two packs per day. Before about 1970, this disease affected primarily

Respiratory System

middle-aged men, but as the number of women smokers has increased (a trend that started in the 1940s), so has the number of women who die from lung cancer.

Smoking changes the quality of the inspired air, making it drier and contaminated with several carcinogenic compounds and particulate matter. The combination overloads the respiratory defenses and damages the epithelial cells throughout the respiratory system. Whether lung cancer develops appears to be related to the total cumulative exposure to the carcinogens. The more cigarettes smoked, the greater the risk, whether those cigarettes are smoked over a period of weeks or years. Up to the point at which tumors form, the histological changes induced by smoking are reversible; a normal epithelium will return if the carcinogens are removed. At the same time, the statistical risks decline to significantly lower levels. Ten years after quitting, a former smoker stands only a 10 percent greater chance of developing lung cancer than a nonsmoker does. Fewer people take up smoking now, and more are quitting. The lung cancer death rates started to decline in 1991 for men and in 2003 for women.

The fact that cigarette smoking typically causes cancer is not surprising in view of the toxic chemicals contained in the smoke. What is surprising is that *more* smokers do not develop lung cancer. Evidence suggests that some smokers have a genetic predisposition to developing one form of lung cancer. Dietary factors may also play a role in preventing lung cancer, although the details are controversial. In terms of their influence on the risk of lung cancer, there is a general agreement that (1) vitamin A has no effect; (2) vegetables containing beta-carotene reduce the risk, but pills of beta-carotene increase the risk; and (3) a high-cholesterol, high-fat diet increases the risk.

Respiratory System

The Digestive System and Metabolism

▶ An Introduction to the Digestive System and Its Disorders

The digestive system consists of the *digestive tract* and *accessory digestive organs*. The digestive tract is divided into the oral cavity, pharynx, esophagus, stomach, small intestine, and large intestine. Most of the absorptive functions of the digestive system occur in the small intestine, with lesser amounts in the stomach and large intestine. The accessory digestive organs provide the chewing, acids, enzymes, and buffers that assist in the mechanical and chemical breakdown of food. The accessory organs include the teeth, salivary glands, the liver, the gallbladder, and the pancreas. The salivary glands produce saliva, a liquid lubricant containing enzymes that aid in the digestion of carbohydrates. The liver produces bile, which is concentrated in the gallbladder and released into the small intestine for fat emulsification. The pancreas secretes enzymes and buffers that are important to the digestion of proteins, carbohydrates, and lipids.

The activities of the digestive system are controlled through a combination of local reflexes, autonomic innervation, and the release of gastrointestinal hormones such as *gastrin*, *secretin*, and *cholecystokinin*.

Symptoms and Signs of Digestive System Disorders

The functions of the digestive organs are varied, and the symptoms and signs of digestive system disorders are equally diverse. Common symptoms of digestive disorders include the following:

- **Pain** may occur in a number of locations. Widespread pain in the oral cavity can result from (1) trauma; (2) infection of the oral mucosa by bacteria, fungi, or viruses; or (3) a deficiency in vitamin C (see *scurvy*, p. 5) or in one or more of the B vitamins. Focal pain in the oral cavity accompanies (1) the infection or blockage of salivary gland ducts; (2) tooth disorders, such as fractures, *dental caries*, *pulpitis*, abscesses, and *gingivitis*; and (3) oral lesions, such as those produced by the *herpes simplex* virus.

  Abdominal pain is characteristic of a variety of digestive disorders. If the pain is acute, a surgical emergency may exist. Abdominal pain can cause rigidity in the abdominal muscles in the painful area. The rigidity is easily felt on palpation. The muscle contractions (*guarding*) may be voluntary, in an attempt to protect a painful area, or an involuntary spasm resulting from irritation of the peritoneal lining, as in *peritonitis*. Persons with peritoneal inflammation also experience *rebound tenderness*, in which pain appears when fingertip pressure on the abdominal wall is suddenly removed.

  Abdominal pain can result from disorders of the digestive, circulatory, urinary, or reproductive system. Digestive tract disorders producing abdominal pain include *appendicitis*, *peptic ulcers*, *pancreatitis*, *cholecystitis*, *hepatitis*, intestinal *diverticulitis*, *peritonitis*, and certain cancers.

- **Indigestion** is pain or discomfort in the epigastric region. Associated digestive tract disorders include *esophagitis, gastritis, peptic ulcers, gastroesophageal reflux* (p. 181), and *cholecystitis* (p. 185).

- **Nausea** is a sensation that usually precedes or accompanies vomiting. Nausea results from digestive disorders or from disturbances of the ear or central nervous system function.

- **Anorexia** is a decrease in appetite that, if prolonged, is accompanied by weight loss. Digestive disorders that cause anorexia include *stomach cancer*, pancreatitis, hepatitis, and *diarrhea*. Anorexia may also accompany disorders that involve other systems. *Anorexia nervosa*, an eating disorder with a strong psychological basis, is discussed on pages 194–195.

- **Dysphagia** is difficulty in swallowing. The difficulty may result from trauma, infection, inflammation, or a blockage of the posterior oral cavity, pharynx, or esophagus. For example, the infections of *tonsillitis* or *pharyngitis*, and *esophageal tumors* may cause dysphagia.

The Physical Examination and the Digestive System

Physical assessment can provide information that is useful in the diagnosis of digestive system disorders. The abdominal region is particularly important, because most of the digestive system is located within the abdominopelvic cavity. Four methods of physical assessment of the digestive system are inspection, palpation, percussion, and auscultation:

1. **Inspection** can provide a variety of diagnostic clues:

   - Bleeding of the gums, as in gingivitis, and characteristic oral lesions can be seen on inspection of the oral cavity. Examples of distinctive lesions include those of oral herpes simplex infections and *thrush*—lesions produced by infection of the mouth by *Candida albicans*. This fungus (also called a yeast) is a normal resident of the digestive tract. However, the fungus may cause widespread oral and esophageal infections in immunodeficient persons, such as individuals who have AIDS or who are undergoing immunosuppressive therapies. Healthy infants can also get a mild thrush infection.

   - Peristalsis in the stomach and intestines may be seen as waves passing across the abdominal wall in persons who do not have a thick layer of abdominal fat. The waves become very prominent during the initial stages of intestinal obstruction.

   - A general yellow discoloration of the skin and sclera, a sign called *jaundice*, results from excessive levels of bilirubin in blood and other body fluids. Jaundice is commonly seen in individuals with cholecystitis (p. 185) or liver diseases such as hepatitis (pp. 183–184) and *cirrhosis* (p. 184).

   - Abdominal distention is caused by (1) accumulation of fluid in the peritoneal cavity, as in *ascites*; (2) air or gas (flatus) within the digestive tract; (3) obesity; (4) abdominal masses, such as tumors, or enlargement of visceral organs; (5) pregnancy; (6) the presence of an abdominal *hernia*; or (7) fecal impaction, as in severe and prolonged constipation.

   - **Striae** are multiple separated dermal scars, 1–6 cm in length, that are visible through the epidermis. Striae develop in damaged dermal tissues after stretching; they are typically seen in the abdominal region after a pregnancy or some other rapid weight gain. Abnormal striae may develop after ascites or in cases of subcutaneous edema. Purple striae are a sign of *Cushing's disease* (p. 112).

2. **Palpation** of the abdomen may reveal specific details about the status of the digestive system, including:

   - The presence of abnormal masses, such as tumors, within the abdominal cavity

   - Abdominal distention from (1) excess fluid within the digestive tract or peritoneal cavity or (2) gas within the digestive tract

   - Herniation of digestive organs through the inguinal canal or weak spots in the abdominal wall (p. 69)

   - Changes in the size, shape, or texture of visceral organs. For example, in several liver diseases, the liver becomes enlarged and firm, and these changes can be detected on palpation of the right upper quadrant of the abdomen.

   - Voluntary or involuntary abdominal muscle contractions (called guarding)

   - Rebound tenderness

   - Specific areas of tenderness and pain. Dividing the abdominal area into Left and Right and Upper and Lower quadrants helps focus the diagnostic possibilities. For example, someone with acute hepatitis, a liver disease, generally experiences pain on palpation of the right upper quadrant. In contrast, a person with appendicitis generally experiences pain when the right *lower* quadrant is palpated.

3. **Percussion** of the abdomen is less instructive than percussion of the chest, because the visceral organs do not contain extensive air spaces that reflect the sounds conducted through surrounding tissues. However, the stomach usually contains a small air bubble, and percussion over this area produces a sharp, resonant sound. The sound becomes dull or disappears when the stomach fills, the spleen enlarges, or the peritoneal cavity contains abnormal quantities of peritoneal fluid, as in ascites.

4. **Auscultation** can detect gurgling abdominal sounds or bowel sounds, produced by peristaltic activity along the digestive tract. Increased bowel sounds occur in persons with acute diarrhea, and bowel sounds may disappear in persons with (1) advanced intestinal obstruction; (2) peritonitis, an infection of the peritoneum; or (3) spinal-cord injuries that damage normal innervation.

Diagnostic procedures, such as endoscopy, are commonly used to provide additional information. Information on representative diagnostic procedures and laboratory tests is given in Tables 36 and 37 (pp. 176–178).

Disorders of the Digestive System

Because of the relatively large number and diversity of digestive organs, there are many types of digestive system disorders. One common method of categorizing these disorders uses a combination of four anatomical and functional characteristics

Digestive System

Table 36	Examples of Tests Used in the Diagnosis of Gastrointestinal Disorders	
Diagnostic Procedure	**Method and Result**	**Representative Uses**
Oral Cavity		
Sialography	X-ray film is taken after contrast medium is injected into salivary ducts while patient's salivary glands are stimulated	Identifies calculi (stones), inflammation, and tumors in salivary duct or gland
Periapical (PA) x-rays	Periapical film is taken of crown and root area of several teeth	Detects tooth decay, tooth impactions, fractures, progression of bone loss with periodontal disease, inflammation of periodontal ligament, and periapical abscesses (at end of root)
Bitewing x-rays of teeth	Bitewing film is taken. X-ray cone is pointed perpendicularly toward the spaces between the crowns of adjacent posterior teeth	Reveal tooth decay between teeth and early bone loss in periodontal (gingival) disease
Upper GI (esophagus, stomach, and duodenum)		
Esophagogastroduodenoscopy, esophagoscopy, gastroscopy	Fiber-optic endoscope is inserted into oral cavity and further into esophagus, stomach, and duodenum to permit visualization of lining and lumen	Detects tumors, ulcerations, polyps, esophageal varices, inflammation, and obstructions; provides biopsy of tissues in upper GI tract
X-ray of abdomen	May be taken with person supine, on side, or standing upright; radiodense tissues appear in white on negative film image	Detects abdominal masses, some kidney stones, bowel obstructions, and presence of free air in peritoneal cavity (usually from perforated ulcer)
Barium swallow	Series of x-rays of esophagus is taken after barium sulfate is ingested to increase contrast of structures; normally done with upper GI series	Determines abnormalities of pharynx and esophagus, especially to determine cause of dysphagia; identifies tumors, hiatal hernia, and diverticuli
Upper GI series	Series of x-rays of stomach and duodenum is taken after barium sulfate is ingested to increase contrast	Investigates epigastric pain; detects ulcers, polyps, gastritis, tumors, and inflammation in upper GI tract
Arteriography (celiac and mesenteric)	Catheter is inserted into femoral artery and threaded to celiac trunk or a mesenteric artery, where contrast medium is injected; x-ray films are then taken	Determines site of bleeding in gastrointestinal tract or blockage of blood supply
Lower GI (colon)		
Barium enema	Barium sulfate enema is administered to provide contrast in intestinal lumen, and x-ray film series is taken	Detects causes of abdominal pain and bloody stools; detects tumors and obstructions of bowel; identifies polyps and diverticula
Sigmoidoscopy, colonoscopy	Fiber-optic tube is inserted into rectum for viewing of anus, rectum, sigmoid colon; colonoscopy allows viewing of the entire large intestine to cecum; biopsy can be performed	Detects tumors, polyps, and ulcerations of intestinal lining. Polyps and biopsies may be removed with scope. Recommended screening test for colon cancer in persons over 50
Liver, Pancreas, Gallbladder		
Cholangiography Intravenous or percutaneous	Intravenously administered dye is concentrated in the liver and released into bile duct; x-ray films are taken. Percutaneous method involves insertion of needle with catheter into bile duct, using ultrasonography for guidance; dye is then administered through catheter	Detects biliary calculi and other obstructions of biliary tract
Liver biopsy	Needle is inserted through small skin incision and into liver; small plug of liver tissue is removed	Liver tissue is examined for evidence of cirrhosis, hepatitis, tumors, and granuloma. Often guided by CT scan during procedure
Endoscopic retrograde cholangiopancreatography (ERCP)	Duodenoscope (fiber-optic tube) is inserted into oral cavity and threaded through stomach to the duodenum; through scope, catheter is inserted into duodenal ampulla for dye injection; x-rays are then taken to visualize the bile duct system	Detects calculi, tumors, or cysts of pancreatic or bile ducts; determines presence of pancreatic tumor

(continued)

Table 36	Examples of Tests Used in the Diagnosis of Gastrointestinal Disorders *(Continued)*	
Diagnostic Procedure	**Method and Result**	**Representative Uses**
Abdomen		
Computerized tomography	Cross-sectional radiation is applied to provide three-dimensional images of liver, gallbladder, pancreas, and bowel	Identifies tumors, pancreatitis (acute, chronic), hepatic cysts and abscesses, and biliary calculi (stones); diagnosis of appendicitis
Ultrasonography	Liver, pancreas, gallbladder, and bowel are examined by means of sound waves emitted by transducer placed on abdomen. Sound waves reflect off dense structures and produce echoes, which are amplified and graphically recorded	Detects polyps, tumors, abscesses, hepatic cysts, and gallstones

Table 37	Examples of Laboratory Tests Used in the Diagnosis of Gastrointestinal Disorders	
Laboratory Test	**Normal Values in Blood Plasma or Serum**	**Significance of Abnormal Values**
Upper and Lower GI		
Serum electrolytes		
Potassium Sodium Magnesium	Adults: 3.5–5.0 mEq/L 135–145 mEq/L 1.5–2.5 mEq/L	Vomiting, diarrhea, and nasogastric intubation and drainage can cause electrolyte losses and create imbalances
Serum gastrin level	Adults: <100 Pg/mL	Elevated levels occur with pernicious anemia and some gastric ulcers; very high levels in Zollinger–Ellison syndrome (usually an intestinal tumor that produces excessive gastrin)
Lactose tolerance test	Fasting patient receives 50 mg of lactose. The hydrogen gas content of exhaled air is measured; a rise of more than 20 ppm within 90 minutes is a positive result	Unabsorbed lactose is metabolized by bacteria, producing hydrogen gas that is exhaled at the lungs. Blood glucose levels do not rise to normal levels after lactose ingestion if patient lacks lactase
Carcinoembryonic antigen (CEA) (in plasma)	Adults: Nonsmokers <3.0 ng/mL Smokers <5.0 ng/mL Not used to detect or screen for cancer	Colorectal cancer is a possibility with elevated levels. This antigen is not specific for colorectal cancer; elevated levels may occur in other types of carcinomas and in ulcerative colitis. Used for monitoring response to treatment, and to detect recurrent disease
Fecal Analysis		
Stool culture		
Culture and sensitivity (C&S). Stool sample is cultured for bacterial content	Only normal intestinal microbial inhabitants should be isolated	Bacteria such as *Shigella, Campylobacter*, enterotoxic *Escherichia coli, Vibrio* cholera, and *Salmonella* can cause acute diarrhea
Ova and parasites (O&P). A stool sample is microscopically examined for parasite eggs or adults	No pathogenic organisms detected	Typical parasites found in feces that cause intestinal disturbances are tapeworms, entamoeba, and some protozoans, such as *Giardia*
Occult blood FOBT: detects blood chemically, has diet and drug restrictions FIT: immunochemical test, detects human hemoglobin from lower GI, fewer false positive results	None found	Hidden (occult) blood in feces can occur when bleeding from GI tract is minimal due to inflammation, ulceration, or a tumor that is causing small amounts of bleeding
Fat content	Adults: ≤6 g/24 hours	Fat content >6 g/24 hours, a condition called steatorrhea, often indicates malabsorption syndrome; occurs in cystic fibrosis and pancreatic diseases
Pancreas		
Amylase		
Urine	Adults: 4–400 U/L	Elevated levels occur for 7–10 days after pancreatic disease begins
Serum	Adults: 60–180 U/L	Elevated levels occur with pancreatic disease and with obstruction of pancreatic duct
Lipase (serum)	Adults: 0–160 U/L	Elevated levels most commonly occur in acute pancreatitis

Digestive System

(continued)

Table 37	Examples of Laboratory Tests Used in the Diagnosis of Gastrointestinal Disorders *(Continued)*	
Laboratory Test	**Normal Values in Blood Plasma or Serum**	**Significance of Abnormal Values**
Liver and Gallbladder		
Serum bilirubin (total)	Total serum bilirubin: 0.1–1.2 mg/dL	Can be elevated with hepatitis (infectious or toxic) and biliary obstruction
	Indirect bilirubin: 0.2–1.0 mg/dL (unconjugated)	Excessive hemolysis as in hemolytic anemias or erythroblastosis fetalis; elevation also occurs with hepatitis
	Direct bilirubin: 0.1–0.3 mg/dL (conjugated)	Elevated levels occur with obstructions of bile duct system, hepatitis, cirrhosis of liver, and liver cancer
	Infant (newborn) total serum bilirubin: 1–12 mg/dL	Bilirubin higher than 15 mg/dL in newborns may result in serious neurological problems
Urine bilirubin	None	Detectable bilirubin is present when direct bilirubin is elevated, as in obstruction of bile duct system, hepatitis, cirrhosis of liver, and liver cancer
Liver enzyme tests		
Aspartate aminotransferase (AST or SGOT)	Adults: 0–35 U/L	Elevated levels occur with hepatitis, acute pancreatitis, and cirrhosis
Alanine aminotransferase (ALT or SGPT)	Adults: 0–35 U/L	Elevated levels occur with hepatitis and other liver diseases
Alkaline phosphatase (isoenzyme ALP_1)	Adults: 30–120 mU/mL	Elevated levels occur in biliary obstructions, hepatitis, liver cancer, and bone disease
5′-nucleotidase	Adults: 0–11 U/L	Elevated levels are useful for early detection of liver disease
Gamma-glutamyl transferase/ transpeptidase (GGT or GGTP)	Adults: 1–94 U/L	Elevated levels indicate liver damage; levels also become elevated after ingestion of alcohol
Serum protein	Adults: Albumin 3.5–5.5 g/dL (or 52–68% of the total protein) Adults: Globulin 2.0–3.0 g/dL (or 32–48% of the total protein)	Decreased levels of albumin and increased levels of globulin occur in chronic liver disease
Ammonia (plasma)	Adults: 10–80 mg/dL	Elevated level occurs with liver failure and hepatic encephalopathy
Hepatitis virus studies	No viral specific antibodies; viral counts negative	Antibodies can be detected against viruses that cause hepatitis A (HAV); hepatitis B (HBV) can be detected by testing for both viral antigen and antibodies; hepatitis C (HCV) can be detected by testing for antibodies against the virus. Quantitative viral counts are done for hepatitis B and C
Prothrombin time (PT) test	Reported as INR (International normalized ratio) see **Table 27c** in Unit 4 (p. 124)	Prolonged time occurs with liver disease, vitamin K deficiency, and warfarin (coumadin) administration
Sweat chloride test		
	Adult: ≤70 mEq/L Children: ≤40 mEq/L	Elevated levels of chloride above 60 mEq/L in children and 90 mEq/L in adults are diagnostic of cystic fibrosis; levels do not reflect severity of disorder and intermediate levels require different confirmatory tests

(▶ **Figure 75a**). The largest category includes inflammation and infection of the digestive organs. This group is usually broken down regionally (by oral cavity, esophagus, stomach, and so forth). **Table 38** summarizes information about infectious diseases of the digestive system. Digestive system cancers are also relatively common and diverse. Malabsorption disorders are characterized by problems with the absorption of one or more nutrients. Congenital disorders result from developmental problems affecting the structure of the digestive tract or accessory organs. Nutritional and metabolic disorders (▶ **Figure 75b**), which affect the entire body, are often considered together with disorders of the digestive system, because they typically accompany digestive system disorders. However, nutritional and metabolic disorders may also reflect (1) disorders involving the endocrine or nervous system, (2) congenital metabolic problems, or (3) dietary abnormalities.

Dental Problems and Solutions

The mass of a plaque deposit on a tooth protects the bacteria that normally reside in the mouth from salivary secretions. As the pathogenic bacteria digest nutrients, they generate acids that erode the enamel and dentin of the teeth. The result is **dental caries**, or "cavities." Vaccines are now being developed to prevent dental caries by promoting specific resistance to *Streptococcus mutans*, the most abundant bacterium at these sites.

In 1901, a dentist in Colorado Springs, CO, observed that many people there had mottled brown teeth that were strong and

Malabsorption disorders

Biliary obstruction
Pancreatic obstruction/insufficiency
Lactose intolerance
Celiac disease

Tumors

Leukoplakia (white raised patches
  on oral mucosa)
Pharyngeal cancer
Esophageal cancer
Stomach cancer
Colorectal cancer
Pancreatic cancer
Liver cancer
  Hepatoma
Gallbladder cancer

Congenital disorders

Cleft palate
Pyloric stenosis
Intestinal atresia, stenosis
  volvulus, or meconium obstruction

DIGESTIVE SYSTEM DISORDERS

Inflammation and infection

Oral cavity: Dental caries
  Pulpitis
  Gingivitis
  Vincent's disease
  Periodontitis
  Mumps
  Stomatitis
  Thrush

Esophagus: Esophagitis
  GERD (gastro-
  esophageal
  reflux disease)

Intestines: Enteritis
  Dysentery
  Gastroenteritis
  Traveler's diarrhea
  Typhoid
  Cholera
  Viral enteritis
  Giardiasis
  Amebiasis
    Amoebic dysentery
  Ascariasis
  Colitis
    Irritable bowel syndrome
    Inflammatory bowel disease
  Diverticulitis
  Hemorrhoids

Stomach: Gastritis
  Peptic ulcer
    Gastric ulcer
    Duodenal ulcer
  Dyspepsia

Accessory organs: Pancreatitis
  Hepatitis
    Non alcoholic
      fatty liver disease
  Cirrhosis
  Viral hepatitis
  Parasitic infection
  Cholecystitis

(a)

Eating disorders

Inadequate food intake
  Anorexia
  Anorexia nervosa
Excessive food intake
  Bulimia
  Obesity
  Binge eating

NUTRITIONAL AND METABOLIC DISORDERS

Thermoregulatory disorders

Elevated body temperature
  Fever
  Heat exhaustion
  Heat stroke
  Hyperthermia
Lowered body temperature
  Accidental hypothermia
  Induced hypothermia

Metabolic disorders

Catabolic problems
  Ketosis
Congenital disorders
  Phenylketonuria (PKU)
Protein deficiency diseases
  Marasmus
  Kwashiorkor
Mineral disorders
  Deficiencies
  Excesses
Vitamin disorders
  Hypervitaminosis
  Avitaminosis
Water balance disorders
  Dehydration
  Overhydration
    Water intoxication

(b)

▶ **Figure 75 Disorders of the Digestive System. (a)** Major categories of digestive system disorders. **(b)** Nutritional and metabolic disorders.

Table 38 Examples of Infectious Diseases of the Digestive System

Disease	Organism(s)	Description
Bacterial Diseases		
Dental caries	*Streptococcus mutans* and other oral bacteria	Tooth decay; bacteria in dental plaque on teeth produce acids that dissolve tooth enamel, leading to cavities
Pulpitis	As above	Infection of the pulp of the tooth
Gingivitis	As above	Infection of the gums
Vincent's disease	As above	Acute necrotizing ulcerative gingivitis, or trench mouth; bacterial infection and ulcer formation
Periodontitis	As above	Infection of gums, gastric and duodenal lining, and bone; results in loosening and loss of teeth
Peptic ulcers	*Helicobacter pylori*	Ulcers in gastric and duodenal lining
Traveler's diarrhea	Enterotoxic form of *Escherichia coli*	Mild to severe watery diarrhea, nausea, vomiting, abdominal pain, and general lack of energy
Typhoid	*Salmonella typhi*	Infection of the intestines and gallbladder; abdominal pain, abdominal distention, low WBC count, and enlarged spleen
Cholera	*Vibrio cholerae*	Intestinal infection; symptoms include nausea, vomiting, abdominal pain, and diarrhea; causes severe dehydration
Viral Diseases		
Mumps	Mumps virus (paramyxovirus)	Infection of the salivary glands; can spread to the pancreas, meninges, and gonads
Viral enteritis	Rotaviruses	Intestinal infection; causes watery diarrhea, especially in young children
Viral hepatitis	Hepatitis A virus (HAV)	Infectious hepatitis; transmitted by fecal-contaminated water, milk, shellfish, or other food, no chronic disease
	Hepatitis B virus (HBV)	Serum hepatitis; transmitted by exchange of body fluid through transfusions, wounds, shared needles (intravenous drug use), or sexual contact; pregnant carrier of the virus can pass it on to her baby; may become chronic
	Hepatitis C virus (HCV)	Formerly non-A, non-B hepatitis; transmitted through blood, injecting drug use, and rarely sexual contact; may be chronic
	Hepatitis D virus	Occurs only in persons infected with HBV; transmission as for HBV
	Hepatitis E virus	Transmission as for HAV; no chronic disease
Fungal Diseases		
Candidiasis (thrush)	*Candida albicans*	Yeast infection of the oral mucosa and/or esophagus; forms white, milky patches in the mouth
Parasitic Diseases		
Protozoa		
Giardiasis	*Giardia lamblia*	Intestinal infection; symptoms include diarrhea, bloating, and weight loss; more common in children than adults
Amebiasis, or amoebic dysentery	*Entamoeba histolytica*	Infection of the large intestine; can produce ulcers and peritonitis; diarrhea contains blood
Helminths		
Ascariasis	*Ascaris lumbricoides*	Roundworm infestation of the intestines; larval movement to pharynx causes cough and damage to intestinal wall; adult worms eat contents of the intestine
Enterobiasis, or pinworms	*Enterobius vermicularis*	Small (1 cm) roundworm, common in small children, lives in colon, deposit eggs on perianal skin, causes itching. Eggs survive in the environment for up to two weeks. Ingested eggs continue the life cycle

resistant to cavities. Over the next 30 years, research found that only children who had drunk the water while under 8 years old, especially those under 4 (when most permanent teeth are developing under the gums) developed brown tooth enamel. Too high natural levels of fluoride in the water caused this fluorosis of the teeth. Fluoride inhibits mouth bacteria and enhances remineralization of enamel erosion, resulting in healthier teeth, even in adults. Further research showed that if water had over 2 mg/L, or 2 ppm (parts per million), of fluoride, fluorosis was more likely, and that 0.7 ppm (equal to 0.7 mg/L) of fluoride still provided the benefit of fewer cavities. In 2010, 89 percent of Americans had community water systems, and 74 percent of those had fluoridation. These numbers have steadily increased as the public health benefits are recognized. For those without fluoridated water, there are fluoride supplements available, and fluoride is in toothpaste and various gels and tooth sealants, but fluoridated water is the most cost effective way to prevent dental caries. If children under age 4 swallow too much toothpaste or other supplemental fluoride, they are at higher risk of fluorosis.

If *S. mutans* (or another bacterium) reaches the pulp and infects it, *pulpitis* (pul-PĪ-tis) results. Treatment generally involves the complete removal of the pulp tissue, especially the sensory innervation and all areas of decay; the pulp cavity is then packed with appropriate materials. This potentially painful procedure is called a *root canal*.

Digestive System

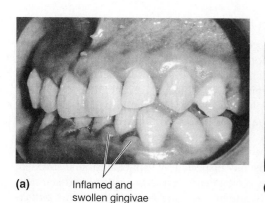

(a) Inflamed and swollen gingivae

Enlarged interproximal spaces

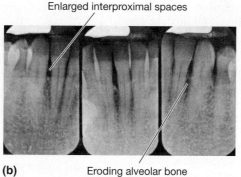

(b) Eroding alveolar bone

▶**Figure 76 Periodontal Disease.**
(a) Notice the inflammation and swelling of the gingivae. This is an indication of serious periodontal disease. **(b)** An x-ray of three teeth in a person with severe periodontal disease. The gums have receded from the necks of the teeth, enlarging the interproximal spaces, and the alveolar bone is being eroded.

Brushing the exposed surfaces of your teeth after you eat helps prevent the settling of bacteria and the entrapment of food particles, but bacteria between your teeth (in the region known as the *interproximal space*) and within the gingival sulcus may elude the brush. Dentists recommend the daily use of dental floss to clean these spaces and stimulate the gingival epithelium. If bacteria and plaque remain within the gingival sulcus for extended periods, the acids generated begin eroding the connections between the neck of the tooth and the gingiva. The gums appear to recede from the teeth, and **periodontal disease** develops. As this disease progresses, the bacteria attack the cementum, eventually destroying the periodontal ligament and eroding the bone of the alveolus (▶**Figure 76**). This deterioration loosens the tooth, and it falls out or must be pulled. Periodontal disease is the most common cause of tooth loss. The chronic inflammation and smoldering infection have also been linked to increased complications with heart disease and possibly diabetes.

Lost or broken teeth have commonly been replaced by "false teeth" attached to a plate or frame inserted into the mouth (dentures). An alternative that uses *dental implants* was developed in the 1980s. A ridged titanium cylinder is inserted into the alveolus, and osteoblasts lock the ridges into the surrounding bone. After four to six months, an artificial tooth is screwed into the cylinder.

Dental implants are not suitable for everyone. Enough alveolar bone must be present to provide a firm attachment, for example, and complications may occur during or after surgery. Nevertheless, as the technique evolves, dental implants will become increasingly important. Roughly 27.3 percent of individuals over age 65 have lost all their teeth; the rest have lost an average of 10 teeth.

Achalasia and Esophagitis

In the condition known as **achalasia** (ak-a-LĀ-zē-uh), a swallowed bolus descends along the esophagus relatively slowly, as a result of abnormally weak peristaltic waves, and its arrival does not trigger the opening of the lower esophageal sphincter. Materials then accumulate at the base of the esophagus like cars at a stoplight. Secondary peristaltic waves may occur repeatedly, causing the individual discomfort. The most successful treatment involves weakening the lower esophageal sphincter muscle by either cutting the circular muscle layer at the base of the esophagus or expanding a balloon in the lower esophagus until the muscle layer tears. Injections of Botox (botulinus toxin) to paralyze the sphincter may be helpful.

Brief, limited reflux of stomach contents into the lower part of the esophagus often occurs after meals, and the incidence is increased with abdominal obesity, pregnancy, and reclining after large meals. Up to 15 percent of adults report weekly symptoms of heartburn or regurgitation. A weakened or permanently relaxed sphincter can cause frequent, prolonged reflux, leading to **gastroesophageal reflux disease (GERD)**, or **esophagitis** (ē-sof-a-JĪ-tis) (inflammation of the esophagus), from contact with powerful stomach acids. The esophageal epithelium has few defenses against attack by acids and enzymes; inflammation, epithelial erosion, and intense discomfort result. This condition supports a multimillion-dollar industry devoted to producing and promoting antacids and medications that suppress acid production. Simply elevating the head of one's bed reduces GERD symptoms as much as medication in many people and weight reduction helps as well. Some persons with GERD may cough and suffer throat and lung problems presumably from flow of stomach fluids up the esophagus and aspiration into the trachea. However, coughing may promote esophageal reflux, so the cause-and-effect link is uncertain.

Esophageal Varices

The veins draining the inferior portion of the esophagus empty into tributaries of the hepatic portal vein. If the venous pressure in the hepatic portal vein (portal hypertension) becomes abnormally high due to liver damage, blood will pool in the submucosal veins of the esophagus. The veins may become grossly distended and create bulges in the esophageal wall. These distorted **esophageal varices** (VAR-i-sēz; *varices*, dilated veins) may rupture and cause life-threatening massive bleeding into the submucosal tissues or into the lumen of the esophagus and stomach. Esophageal varices commonly develop in individuals with advanced *cirrhosis*, a chronic liver disorder that restricts hepatic blood flow (p. 184).

Peptic Ulcers

A **peptic ulcer** is a sore that erodes the lining of the stomach or adjacent proximal small intestine (duodenum). The specific location of the ulcer is indicated by the terms **gastric ulcer** (stomach) and **duodenal ulcer** (duodenum). Peptic ulcers can result from the excessive production of digestive acid and enzymes or the inadequate production of the alkaline mucus that protects the epithelium against that acid. Ulcers may also be caused by regular or prolonged use of irritants to the gastro-intestinal mucosa, including alcohol, tobacco, aspirin, and other nonsteroidal anti-inflammatories (NSAIDS), and some medicines used for osteoporosis.

The treatment of peptic ulcers has changed radically since the identification of *Helicobacter pylori* as a likely causative agent in roughly 80 percent of peptic ulcers. These bacteria are able to resist gastric acids long enough to penetrate the mucous coating of the epithelium. Once within the protective layer of mucus, they bind to the epithelial surfaces, where they are safe from

Digestive System

the action of gastric acids and enzymes. Over time, the bacteria release toxins that damage the epithelial lining. Ultimately, the epithelium is eroded, and the lamina propria is destroyed by gastric juices. Individuals whose stomachs harbor *H. pylori* are also at higher-than-normal risk of gastric cancer, although the reason is not known.

Regardless of the primary cause of a gastric or duodenal ulcer, once gastric juices have destroyed the epithelial layers, the virtually defenseless lamina propria will be exposed to digestive attack. Sharp abdominal pain results. In severe cases, the damage to the mucosa of the stomach can cause significant bleeding, and the acids can even perforate the wall of the stomach into the peritoneal cavity. This condition, called a **perforated ulcer**, requires immediate surgical correction.

The administration of antacids can help control gastric or duodenal ulcers by neutralizing the acids and allowing time for the mucosa to regenerate. *Cimetidine* (sī-MET-i-dēn) and related drugs inhibit the secretion of acid by the parietal cells, while *omeprazole*, or *Prilosec*, and related proton pump inhibitor drugs (PPIs) block the pump that secretes gastric acid from the parietal cells. Both classes of drugs dramatically improve symptoms and healing rates. Dietary and medicine restrictions that limit the intake of foods that promote acid production (caffeine and pepper) or that damage unprotected mucosal cells (alcohol, aspirin, and nonsteroidal anti-inflammatories [NSAIDs] such as ibuprofen) also aid healing.

Current treatment for *H. pylori*–related ulcers consists of combinations of up to three antibiotics (*tetracycline, amoxicillin,* and *metronidazole,* among others), possibly Pepto-Bismol, and treatment with drugs that suppress acid production for 14 days. Because strains of *H. pylori* that are resistant to at least one of these antibiotics have already appeared, research is under way to find alternative methods of controlling the infections.

When people have upper abdominal symptoms but diagnostic testing does not find an ulcer or other disease, non-ulcer **dyspepsia** (Greek for "difficult digestion") may be diagnosed. Variably associated with eating habits, stress, medications, and other irritants linked to other gastrointestinal diseases, treatment results are also variable. Some persons respond to acid suppression, some to *H. pylori* eradication, and others to changes in diet and stress reduction. In many people, the problems resolve without any specific therapy. For some, persistent symptoms and test results may progress to a more specific diagnosis.

Stomach Cancer

Stomach, or *gastric*, **cancer** is one of the most common lethal cancers, responsible for roughly 10,000 deaths in the United States in 2013. The incidence is higher in Japan, Korea, and other countries where the typical diet includes large quantities of pickled, fermented, or smoked foods. Because the signs and symptoms can resemble those of gastric ulcers, the condition may not be recognized in its early stages. Diagnosis may involve examining x-rays of the stomach at various degrees of distension. The mucosa can also be visually inspected by using a flexible instrument called a *gastroscope*. Attachments permit the biopsy of tissue samples for analysis. Treatment of stomach cancer starts with *gastrectomy* (gas-TREK-to-mē), the surgical removal of part or all of the stomach. People can survive even a total gastrectomy, because the loss of such functions as food storage and acid production is not life-threatening.

Protein breakdown can still be performed by the small intestine, although at reduced efficiency, and the diet can be supplemented with vitamin B_{12} which usually requires intrinsic factor from the stomach to be absorbed.

Drastic Weight-Loss Techniques

At any moment, an estimated 20 percent of the U.S. population is dieting to promote weight loss. In addition to the appearance of "fat farms" and exercise clubs, the use of surgery to promote weight loss—called **bariatric surgery** (▶ Figure 77)—has been on the rise. Some surgeries are restrictive, reducing the stomach volume in an attempt to reduce intake and bring on early satiety. Others bypass absorptive surfaces of the proximal small intestine to reduce dietary absorption, and others combine both approaches. Recent studies suggest that both the gastric mucosa and proximal small bowel secrete peptides that affect absorption, insulin levels, and satiety (the sensation of being full).

Surgical Remodeling of the Gastrointestinal Tract

A *gastric bypass* (▶ Figure 77a) is a surgical procedure that connects the proximal small intestine to a small pouch formed by a superior portion of the stomach, with the same goal of reducing gastric capacity, plus bypassing the duodenum and proximal jejunum. This procedure seems more effective than gastric stapling, but involves more complicated surgery. *Gastric banding* (▶ Figure 77b)

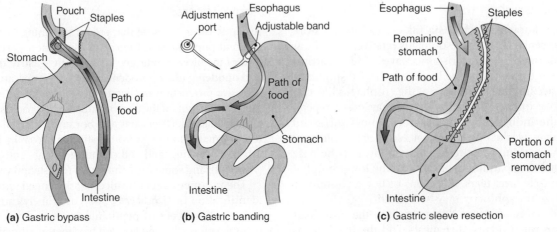

(a) Gastric bypass   **(b)** Gastric banding   **(c)** Gastric sleeve resection

▶ **Figure 77 Bariatric Surgery. (a)** Gastric bypass. **(b)** Gastric banding. **(c)** Gastric sleeve resection.

Digestive System

allows control over the passage of food from the esophagus into the stomach, and regulates the amount of food that can be swallowed. *Gastric sleeve resection* (▶ **Figure 77c**) involves the removal of much of the body (fundus) of the stomach. The effects are similar to those of an earlier procedure, called *gastric stapling*, wherein a large portion of the gastric lumen is stapled shut. These procedures leave a small gastric pouch or tube in contact with the esophagus and duodenum. After this surgery, the individual can eat only a small amount before the stretch receptors in the gastric wall become stimulated and a feeling of fullness results. Removing the fundus may also reduce hunger by lowering levels of ghrelin, a hormone that promotes appetite.

An even more drastic approach, seldom performed today, involves the surgical bypass of a large portion of the jejunum. This procedure reduces the effective absorptive area, producing marked diarrhea and weight loss. After the operation, the individual must take dietary supplements to ensure that all the essential nutrients and vitamins can be absorbed before the chyme enters the large intestine. Because chronic diarrhea and serious liver disease are potential complications, ileal-jejunal bypass surgery has largely been replaced by gastric bypass or gastric sleeve resection. Abdominal surgery and anesthesia in obese persons has higher risks of complications and death than in persons with normal BMIs, but obesity has its own increased risks as well. Certainly, in obese diabetics, significant, sustained weight loss can improve control and even "cure" diabetes mellitus. Some case studies of bariatric surgery in morbidly obese diabetic patients claim a 60 percent to 80 percent reversal of diabetes.

Liposuction

One much-publicized method of getting rid of fat is **liposuction**, a surgical procedure for the removal of unwanted adipose tissue. But it is not an effective treatment for generalized obesity. Cosmetic body shaping is perhaps a better description. Adipose tissue is flexible, but not as elastic as areolar tissue, and it tears relatively easily. In liposuction, a small incision is made through the skin, and a tube is inserted into the underlying adipose tissue. Suction is then applied and chunks of tissue containing adipocytes, other cells, fibers, and ground substance can be vacuumed away. Approximately 202,000 liposuction surgeries (lipoplasties) were performed in 2013, a 43 percent decrease from 2000, making this the fourth most common cosmetic surgery performed today after breast augmentation, nose, and eyelid surgery.

Liposuction has received a lot of news coverage, and many advertisements praise the technique as easy, safe, and effective. In fact, it is not always easy, and it can be dangerous and have limited effectiveness. The density of adipose tissue varies from place to place in the body and from individual to individual, and it is not always easy to suck through a tube. An anesthetic must be used to control pain, and anesthesia always poses risks; also, blood vessels are stretched and torn, and extensive bleeding can occur. Heart attacks, pulmonary embolism, and fluid balance problems can develop, with fatal results. Finally, adipose tissue can repair itself, and adipocyte populations recover over time. The only way to ensure that fat lost through liposuction will not return is to adopt a lifestyle that includes a proper diet and adequate exercise. Over time, such a lifestyle can produce the same weight loss, *without liposuction*, avoiding the surgical expense and risk.

Vomiting

The responses of the digestive tract to chemical or mechanical irritation are predictable: Fluid secretion accelerates all along the digestive tract, and the intestinal contents are eliminated as quickly as possible. The *vomiting reflex* occurs in response to irritation of the fauces, pharynx, esophagus, stomach, or proximal portions of the small intestine. These sensations are relayed to the *vomiting center* of the medulla oblongata, which coordinates the motor responses. In preparation for vomiting, the pylorus relaxes and the contents of the duodenum and proximal jejunum are discharged into the stomach by strong peristaltic waves that travel toward the stomach rather than toward the ileum.

Vomiting, or *emesis* (EM-ē-sis), then occurs as the stomach regurgitates its contents through the esophagus and pharynx. During regurgitation, the uvula and soft palate block the entrance to the nasopharynx. Increased salivary secretion assists in buffering the stomach acids, thereby slowing erosion of the teeth. In conditions marked by repeated vomiting, severe tooth damage can occur; we will discuss one example—the eating disorder *bulimia*—in a later section. Most of the force of vomiting comes from expiratory movements that elevate intra-abdominal pressures and force the stomach against the tensed diaphragm.

Liver Disease

Assessing Liver Structure and Function

The liver is the largest and most important visceral organ, and liver disorders affect almost every other vital system in the body. A variety of clinical tests are used to check the functional and physical state of the liver (see **Tables 36** and **37**):

- **Liver function tests** can assess specific functional capabilities.
- A **serum bilirubin level** indicates how efficiently the liver has been able to modify and excrete bilirubin.
- **Serum** or **plasma protein assays** can detect changes in the liver's rate of protein synthesis, and **serum enzyme tests** can reveal liver damage by detecting intracellular enzymes from liver cells in the circulating blood.
- **Liver scans** involve the injection of radioisotope-labeled compounds into the bloodstream. Compounds are chosen that will be selectively absorbed by Kupffer cells, liver cells, or abnormal tissues.
- **CT** and **ultrasound scans** of the abdominal region are commonly used to provide information about cysts, abscesses, tumors, or hemorrhages in the liver.
- A **liver biopsy** can be taken by a long needle, commonly guided by CT scans to avoid large blood vessels. The needle is inserted through the abdominal wall. Laparoscopic examination can also reveal gross structural changes in the liver or gallbladder, and collect a biopsy as well.

Hepatitis

Hepatitis (hep-a-TĪ-tis) is inflammation of the liver. Viruses that target the liver are responsible for most cases of hepatitis, although some environmental toxins can cause similar symptoms. Five forms of viral hepatitis have been identified, and some cases cannot be attributed to any of them. These mystery cases are currently called non-A-E hepatitis. Effective vaccines are

Digestive System

Digestive System

available against hepatitis A and B. The primary forms of hepatitis are the following:

1. **Hepatitis A**, or *infectious hepatitis*, results from the ingestion of water, milk, shellfish, or other food contaminated by virally infected fecal wastes. The disease has a relatively short incubation period of two to six weeks and generally runs its course in a matter of months. Fatalities are rare among individuals under age 40. There is no ongoing chronic infection. A vaccine to prevent hepatitis A is available and recommended for infants and travelers to areas with high risk of exposure. Immunity may last 14 to 25 years.

2. **Hepatitis B**, or *serum hepatitis*, is primarily transmitted by the exchange of body fluids during intimate contact. For example, infection can occur through the transfusion of infected blood products, needle sharing by injecting drug users, and body fluid contact through a break in the skin or mucosa, or by sexual contact. (Blood products have been screened for hepatitis B since the 1970s.) The incubation period for hepatitis B ranges from one to six months. If a pregnant woman is infected, the newborn baby may become infected at birth. Presumably because of the newborn's weaker immune system, up to 90 percent of infants infected at birth become chronically infected (chronic carriers), whereas only 1 percent of newly infected adults become chronic carriers. Hepatitis B vaccinations recommended for all newborns have essentially broken the cycle of maternal–infant transmission in the United States. Chronic carriers are infectious and may experience cumulative liver damage, including increased risk of liver cancer. Treatment with interferon and antiviral medication may slow the progression of chronic hepatitis B. New cases in the United States primarily involve adults. Most healthcare workers should be vaccinated.

3. **Hepatitis C**, originally designated *non-A, non-B hepatitis*, is most commonly transmitted by contact with infected blood and blood products, especially by needle sharing by injecting drug users, and in the past, the transfusion of unscreened infected blood. Since 1992, blood-screening procedures have been used to lower the incidence of transfusion-related hepatitis C to 1 case per 100,000 units transfused. The disease can also be transmitted (but rarely is) by sexual contact. Up to 75 percent of hepatitis C infections become chronic, and the individuals remain infectious. The chronic infectious carrier state produces significant liver damage in roughly half the individuals infected with the virus. Interferon treatment, combined with antiviral medication early in the course of the disease, can reduce the viral load to undetectable levels and slow the progression of the disease in many patients.

4. **Hepatitis D** is caused by a virus that produces symptoms only in persons already infected with hepatitis B. The transmission of hepatitis D resembles that of hepatitis B. In the United States, the disease is most common among intravenous drug users. The combination of hepatitis B and hepatitis D causes progressive and severe liver disease.

5. **Hepatitis E** resembles hepatitis A in that it is transmitted by the ingestion of contaminated food or water and is not a chronic disease. Hepatitis E is the most common form of hepatitis worldwide, but cases seldom occur in the United States. Hepatitis E infections are most acute and are potentially lethal for pregnant women. A vaccine is currently being developed but is not available commercially. An early version was successful in preventing infection in members of the Nepalese army.

The hepatitis viruses disrupt liver function by attacking and destroying liver cells. An infected individual may develop a high fever, and the liver may become inflamed and tender. As the disease progresses, several hematological parameters change markedly. For example, intracellular liver enzymes, normally present in blood in low levels, leak into the blood in large amounts. Normal metabolic regulatory activities become less effective, and blood glucose levels decline. Plasma protein synthesis slows, and the clotting time becomes unusually long. The injured hepatocytes stop removing bilirubin from the circulating blood, and signs of jaundice appear.

Hepatitis is either acute or chronic. *Acute hepatitis* is characteristic of hepatitis A and E. Almost everyone who contracts hepatitis A or hepatitis E (except in pregnancy) eventually recovers, although full recovery can take several months. Once recovered, infected individuals cannot transmit the disease. Symptoms of acute hepatitis include severe fatigue and jaundice. *Chronic hepatitis* is a progressive disorder that can lead to severe medical problems as liver function deteriorates and cirrhosis develops. Common complications of cirrhosis include the following:

- The formation of *esophageal varices*, due to portal hypertension
- Ascites, caused by portal hypertension compounded by abnormally low levels of plasma proteins, especially albumins (hypoalbuminemia), which reduces the blood colloid osmotic pressure (BCOP)
- Bacterial peritonitis, which may recur for unknown reasons
- Hepatic encephalopathy, which is characterized by disorientation and confusion, is probably caused by CNS-toxic metabolites from protein digestion (such as ammonia) that either bypass the liver or are not detoxified by a failing liver.

Hepatitis B, C, and D infections can produce both acute and chronic hepatitis. Individuals with chronic forms are potentially infectious and may eventually experience fatal liver failure or liver cancer. Roughly 10 percent of hepatitis B patients develop potentially dangerous complications; the percentage is higher for hepatitis C.

Passive immunization with pooled immunoglobulins is available for persons exposed to the hepatitis A and B viruses. (Active immunization for hepatitis A and B is also available and is preferred.) No vaccines are available for hepatitis C, D, or E.

Cirrhosis

The underlying cause of liver **cirrhosis** (sir-Ō-sis) appears to be the widespread destruction of hepatocytes and scarring of the liver most commonly by years of chronic injury from exposure to drugs (especially alcohol), viral infection, ischemia, or a blockage of the hepatic ducts. In *Non Alcoholic Fatty Liver Disease* (*NAFLD*) fat builds up in the liver and results in cirrhosis. Obesity and diabetes are risk factors for NAFLD. Initially, the damage to hepatocytes leads to the formation of extensive areas of scar tissue that branch throughout the liver. The surviving hepatocytes then undergo repeated cell divisions, but the fibrous tissue prevents the new hepatocytes from achieving a normal arrangement of lobules. As a result, the liver gradually converts from an organized assemblage of lobules to a fibrous aggregation of poorly functioning cell clusters. Jaundice, ascites, and other symptoms appear as the condition progresses.

Cholecystitis

An estimated 16–20 million people in the United States have gallstones that go unnoticed; small stones are commonly flushed down the bile duct and eliminated. If gallstones enter and jam the cystic duct or bile duct, the painful symptoms of **cholecystitis** (kō-lē-sis-TĪ-tis) appear. Approximately 1 million people develop acute symptoms each year. The gallbladder becomes swollen and inflamed, serious infections can ensue, and symptoms of *obstructive jaundice* arise.

In most cases, surgery is required to remove large gallstones because the gallbladder can become infected, inflamed, or perforated. Under these conditions, the gallbladder is surgically removed in a procedure known as a *cholecystectomy*; this surgery is performed roughly 500,000 times each year in the United States. Most surgeries involve using a laparoscope, inserted through three or four small abdominal incisions. Without a gallbladder, bile flows directly to the duodenum, and digestion proceeds normally.

Colon and Rectal Cancers

Colon and **rectal cancers** are relatively common. Approximately 140,000 cases occurred in the United States in 2013 with 51,000 deaths. The mortality rates for these cancers remain high, and the best defense appears to be early detection and prompt treatment. It is believed that most colorectal cancers begin as small, localized mucosal tumors, or **polyps** (POL-ips), that grow from the intestinal wall. The prognosis improves dramatically if cancerous polyps are removed before metastasis has occurred. An early sign of polyp formation may be the appearance of blood in the feces. Unfortunately, many people ignore small amounts of blood in fecal materials, because they attribute the bleeding to "harmless" hemorrhoids. This offhand diagnosis should always be professionally verified.

One screening test involves checking the feces for blood, a simple procedure that can easily be performed on a stool (fecal) sample as part of a routine physical. For those at increased risk because of family history, associated disease, or older age (over 50), visual inspection of the colon lumen by fiber-optic colonoscopy is recommended. A flexible **colonoscope** (ko-LON-o-skōp) permits direct visual inspection of the lining of the large intestine. The colonoscope can take a biopsy of the mucosal lining and remove small polyps.

Primary risk factors for colorectal cancer include (1) a diet rich in animal fats and low in fiber and vegetables, (2) *inflammatory bowel disease*, and (3) a number of inherited disorders that promote epithelial tumor formation along the intestines.

Diverticulosis and Irritable Bowel Syndrome

In **diverticulosis** (dī-ver-tik-ū-LŌ-sis), pockets (*diverticula*) form in the mucosa, generally in the sigmoid colon. The pockets get forced outward, probably by the pressures generated during defecation. If they push through weak points in the muscularis externa, the pockets form semi-isolated chambers that are subject to recurrent infection and inflammation. The infections cause pain and occasional bleeding, a condition known as *diverticulitis* (dī-ver-tik-ū-LĪ-tis).

Irritable bowel syndrome (IBS) is characterized by diarrhea, constipation, or an alternation between the two. When constipation is the primary problem, the condition is sometimes called a *spastic colon*, or *spastic colitis*. Persons with IBS have frequent symptoms but not a disease, because the GI tract is not damaged.

Irritable bowel syndrome may have a partly neurological basis; the mucosa is normal in appearance, but the motility can be abnormal. Hypersensitivity and reactivity of the "enteric nervous system" of the bowel has been implicated. Bulking agents such as fiber or psyllium (*Metamucil*) and drugs that affect the enteric nervous system can help control symptoms.

Inflammatory Bowel Disease

Inflammatory bowel disease, or **colitis** (kō-LĪ-tis), is much more serious than irritable bowel syndrome. Of unknown cause, it seems to have genetic, environmental, and autoimmune components. It involves chronic inflammation of the digestive tract. Two main forms are *Crohn's disease* and *ulcerative colitis*. In **Crohn's disease**, all layers of the intestinal wall may be involved, particularly in the ileum, but the disease is focal, with areas of healthy bowel between areas of disease. Intestinal blockage is frequent in Crohn's disease and may require surgery. In **ulcerative colitis**, the colonic mucosa becomes diffusely inflamed, and ulcerated, with areas of tissue death and deterioration of colonic function. Acute bloody diarrhea, cramps, fever, and weight loss may occur with either disease.

The treatment of inflammatory bowel disease involves anti-inflammatory drugs and corticosteroids that reduce inflammation. In cases that do not respond to other therapies, immunosuppressive drugs, such as *azathioprine*, may be effective. Patients have high levels of an immune system protein called Tumor Necrosis Factor (TNF), but whether this is a cause or a result of the disease is unknown. Monoclonal antibody medicines (p. 157) have been made that bind TNF and reduce the inflammation. Inflammatory bowel disease also increases risk of intestinal cancer.

The treatment of prolonged severe inflammatory bowel disease involving the colon may involve a **colectomy** (ko-LEK-to-mē)—the removal of all or a portion of the colon. If a large part or even all of the colon must be removed, normal connection with the anus cannot be maintained. Instead, the end of the intact digestive tube is sutured to the abdominal wall, and wastes then accumulate in a plastic pouch or sac attached to the opening, or a surgically created internal pouch. If the attachment involves the colon, the procedure is a **colostomy** (ko-LOS-to-mē); if the ileum is involved, it is an **ileostomy** (il-ē-OS-to-mē). Surgeons can often construct an internal pouch of healthy intestine attached to the rectum, preserving bowel function at the cost of diarrhea.

Diarrhea

In **diarrhea** (dī-a-RĒ-uh), an individual has frequent, watery bowel movements. The condition can result from small bowel hypersecretion or decreased levels of colonic mucosal absorption. Bacterial, viral, or protozoan infection of the colon or small intestine can cause acute bouts of diarrhea lasting several days. Severe diarrhea is life-threatening due to cumulative fluid and ion losses.

Many conditions result in diarrhea, and we will consider only a representative sampling here. Most infectious diarrhea involves organisms that are shed in the stool and then spread from person to person (or from person or animal to food to person). Proper food preparation and storage, hand washing after defecation, clean drinking water, and good sewage disposal are the best preventive measures.

Gastroenteritis

An irritation of the small intestine can lead to a series of powerful peristaltic contractions that eject the contents of the small intestine into the large intestine. An extremely powerful irritating

Digestive System

stimulus produces a "clean sweep" of the absorptive areas of the digestive tract. Vomiting clears the stomach, duodenum, and proximal jejunum, and peristaltic contractions evacuate the distal jejunum, ileum, and colon. Bacterial toxins, viral infections, and various poisons may produce these extensive gastrointestinal responses. Conditions affecting primarily the small intestine are usually referred to as a form of **enteritis** (en-ter-Ī-tis). If both vomiting and diarrhea are present, the term **gastroenteritis** (gas-trō-en-ter-Ī-tis) may be used instead.

Traveler's Diarrhea

Traveler's diarrhea, a form of infectious diarrhea generally caused by a bacterial or viral infection, develops because the irritated or damaged mucosal cells hypersecrete or are unable to maintain normal absorption levels. The irritation stimulates the production of mucus, and the damaged cells and mucous secretions add to the volume of feces produced. Despite the inconvenience, this type of diarrhea is usually temporary, and limited diarrhea is probably a reasonably effective method of rapidly removing an intestinal irritant. Drugs, such as *Lomotil*, that prevent peristaltic contractions in the colon slow the diarrhea, but leave the irritant intact, and the symptoms may return with a vengeance when the effects of the drug fade. A three- to five-day course of antibiotics may be effective, but not always necessary, in controlling diarrhea due to bacterial infection.

Giardiasis

Giardiasis is an infection caused by the protozoans *Giardia intestinalis* and *G. lamblia*. These pathogens can colonize the duodenum and jejunum and interfere with the normal absorption of lipids and carbohydrates. Many people do not develop acute symptoms but act as carriers who can spread the disease. Acute symptoms usually appear within three weeks of initial exposure. Diarrhea, abdominal cramps, bloating, nausea, and vomiting are the primary complaints. These symptoms persist for five to seven days or longer, and some patients are subject to relapses, with chronic bloating, diarrhea, and weight loss. Treatment typically consists of the oral administration of antibiotics, such as *metronidazole*, that can kill the protozoan.

The transmission of giardiasis requires that food or water be contaminated with feces that contain *cysts*—resting stages of the protozoan that are produced during its passage through the large intestine. Rates of infection are highest (1) in developing countries with poor sanitation, (2) among campers drinking surface water, (3) among individuals with impaired immune systems (as in AIDS), and (4) among toddlers and young children. The cysts can survive in cold, fresh water for months, and they are not killed by the chlorine treatment used to kill bacteria in drinking water. Travelers are advised to boil or ultrafilter water and to heat food properly before eating it, as these preventive measures will destroy the cysts.

Cholera Epidemics

Cholera (KOL-e-ruh) epidemics are most common in areas where sanitation is poor, drinking water is contaminated by fecal wastes, and eating raw fish or shellfish is popular. After an incubation period of one to two days, the symptoms of nausea, vomiting, and diarrhea persist for two to seven days. The cholera bacteria bind to the intestinal lining and release toxins that stimulate a massive secretion of fluids across the intestinal epithelium. Fluid loss during the worst stage of the disease can approach 1 liter per hour. This dramatic loss causes a rapid drop in blood volume, leading to acute hypovolemic shock and damage to the kidneys and other organs. Without treatment, a person with cholera can die of acute dehydration in a matter of hours. Treatment consists of *prompt* oral or intravenous fluid replacement while the disease runs its course. Antibiotic therapy may also prove beneficial. A vaccine is available, but its low success rate (40–60 percent) and short-lived effectiveness (four to six months of protection) make it relatively ineffective in preventing or controlling cholera outbreaks. Many hundreds of thousands of cases of cholera were reported during an epidemic that spread through Central and South America in 1991. Relatively small outbreaks continue to occur in Sub-Saharan Africa, South and Southeast Asia, and in 2010 in Haiti. Recent epidemic death rates are around 1–2 percent, compared with death rates of up to 60 percent in outbreaks in the early 20th century.

Lactose Intolerance

Lactose intolerance is a malabsorption syndrome, revealed by eating lactose-containing dairy foods. It results from the lack of the enzyme *lactase* at the brush border of the intestinal epithelium. It is not a milk allergy, which is an immune response to milk protein. The condition poses more of a problem than would be expected if the only outcome were the inability to use the disaccharide lactose found in milk. Undigested lactose provides a particularly stimulating energy source for the bacterial inhabitants of the colon. The result is increased intestinal gas, cramps, and diarrhea, problems that can develop within 30 minutes to 2 hours after the lactose-intolerant person drinks more than 8 oz of milk or eats similar amounts of other dairy products. Fermented milk products such as yogurt or hard cheeses have less lactose.

Lactose intolerance appears to have a genetic basis. Infants produce lactase to digest milk, but older children and adults may stop producing this enzyme. In some populations, lactase production continues throughout adulthood. Only about 15 percent of Caucasians develop lactose intolerance, whereas estimates ranging from 80 to 90 percent have been suggested for the adult African and Asian populations. These differences reflect and affect dietary preferences in those groups, and food relief efforts must take these facts into account. For example, shipping powdered milk to famine-stricken areas in Africa can make matters worse if supplies are distributed to adults rather than to children. Temporary lactose intolerance lasting several days may occur if gastroenteritis affects the brush border epithelium of the small intestine. Normal cell turnover usually resolves the problem in several days.

Constipation

Constipation is infrequent defecation of small, dry, hard feces, usually less than three times a week. It is not the frequency of defecation, and there is not a "right" number of bowel movements a week. Constipation occurs when fecal materials move through the colon so slowly that excessive reabsorption of water occurs. The feces become extremely compact, difficult to move, and highly abrasive. Inadequate dietary fiber and water, coupled with a lack of exercise, is a common cause of constipation.

Constipation can generally be treated by more fiber and water in the diet and the oral administration of stool softeners such as Colace. This promotes movement of water into the feces, increasing fecal mass and softness. Indigestible fiber adds bulk to the feces, retaining moisture and stimulating stretch receptors that promote peristalsis. Thus, the promotion of peristalsis is a benefit of high-fiber cereals. Active movement during exercise also assists in the movement of fecal materials through the colon. Laxatives, or **cathartics** (ka-THAR-tiks), that promote defecation by irritating the lining of the colon to stimulate peristalsis should be used sparingly to avoid nerve damage in the colon. Such damage, which may occur with prolonged use, ultimately interferes with normal peristalsis. Osmotic laxatives, such as magnesium hydroxide (Milk of Magnesia), sodium phosphate (Phosphate Soda), or PEG 3350 (Miralax) may be used chronically if necessary.

Hemorrhoids

Varicose veins (p. 139, blood vessel disorders) are not limited to the extremities. Another common site involves a network of veins in the walls of the anus. Pressures within the abdominopelvic cavity rise dramatically when the abdominal muscles are tensed. Constipation and straining to force defecation can force blood into these veins, and repeated incidents leave them permanently distended. These distended veins, known as **hemorrhoids** (HEM-o-roydz), can be uncomfortable and, in severe cases, extremely painful. Both hemorrhoids and leg varicose veins are often associated with pregnancy, as a result of changes in blood flow and abdominal pressures on the inferior vena cava. Minor hemorrhoids can be treated by the topical application of drugs that promote the contraction of smooth muscles within the venous walls. More severe cases may benefit from the surgical removal or destruction of the distended veins.

Malabsorption

Malabsorption is a disorder characterized by impaired nutrient absorption. Difficulties in the absorption of all classes of compounds will result from damage to the main accessory digestive organs (liver and pancreas) or the intestinal mucosa. If the accessory organs are functioning normally but their secretions cannot reach the duodenum, the condition is called *biliary obstruction* (bile duct blockage) or *pancreatic obstruction* (pancreatic duct blockage). Alternatively, the ducts may remain open, but the glandular cells may be damaged and unable to continue normal secretory activities. Two examples, cholecystitis and cirrhosis, were mentioned earlier in this chapter.

Even with the normal enzymes present in the lumen, absorption will not occur if the mucosa cannot function properly. The inability to manufacture specific enzymes (frequently a genetic disorder) will result in discrete patterns of malabsorption; lactose intolerance is a good example. Mucosal damage due to ischemia, radiation exposure, toxic compounds, the immune disorder of *celiac disease*, or infection will adversely affect absorption and, as a result, will deplete nutrient and fluid reserves.

Celiac Disease

Celiac disease, or *gluten-sensitive enteropathy*, is a chronic inflammatory disorder. If an affected person eats foods containing the protein gluten, antibodies are made that attack and damage the villi of the small intestinal mucosa. This intestinal damage affects nutritional absorption, particularly of vitamins, iron, and calcium. Abdominal symptoms of pain and diarrhea are common, along with poor growth (in children), and nonspecific symptoms of fatigue, irritability, and musculoskeletal pain. In adults, anemia and osteoporosis may occur. Diagnosis can be made by blood tests for anti-tissue antibodies and confirmed by small bowel biopsy.

Genetic inheritance is involved, with up to 10 percent of family members also affected, and gluten sensitivity is more frequent in European populations (up to 1 percent incidence) than in east Asian or African populations, possibly related to less use of gluten-containing cereal grains in these populations. (Rice has no gluten.) Other immune disorders affecting the skin, thyroid, and joints may occur.

Gluten is a protein found in the cereal grains wheat, barley, and rye, and less so in oats. Treatment is a gluten-free diet with symptomatic improvement starting in days to weeks, while intestinal mucosal healing may take months. Other starches including rice, corn (maize), potatoes, cassava, all vegetables, fruits, meat, fish, and chicken do not contain gluten. The difficulty in maintaining a gluten-free diet comes from the use of wheat flour in many prepared food items, spice mixtures, and additives (including some medicines!). With increased awareness of this condition, access to gluten-free prepared foods has improved, the listing of ingredients has become more comprehensive, and updated lists of allowed foods are available.

▶ Disorders Affecting Metabolism and Energetics

Aerobic Metabolism: A Closer Look

Mitochondria break down organic substrates, such as the pyruvic acid produced by glycolysis (▶ Figure 78a). Aerobic metabolism includes the citric acid cycle (▶ Figure 78b) and the oxidation–reduction reactions of the electron transport system, or ETS.

Phenylketonuria

Phenylketonuria (fen-il-kē-to-NOO-rē-uh), or **PKU**, is one of about 6000 disorders that have been traced to abnormality in a single gene. Individuals with PKU have an abnormal PAH gene and are deficient in a key enzyme, *phenylalanine hydroxylase*, responsible for the conversion of the amino acid phenylalanine to tyrosine. This reaction is a necessary step in the synthesis of tyrosine, an important component of many proteins and the structural basis for a pigment (melanin), two hormones (epinephrine and norepinephrine), and two neurotransmitters (dopamine and norepinephrine). In addition, the conversion must occur before the carbon chain of a phenylalanine molecule can be recycled or broken down in the citric acid cycle.

If PKU is undetected and untreated at birth, plasma concentrations of phenylalanine gradually escalate from normal (about 3 mg/dL) to levels above 20 mg/dL. High plasma concentrations of phenylalanine affect overall metabolism, and a number of unusual by-products are excreted in the urine. The synthesis and degradation of proteins and other amino acid derivatives are affected as well. Developing neural tissue is most strongly influenced by these metabolic changes, and severe brain damage and mental retardation result.

Fortunately, the condition is detectable shortly after birth, when the infant has digested milk or formula, because this digestion produces elevated levels of phenylalanine in the blood and

Digestive System

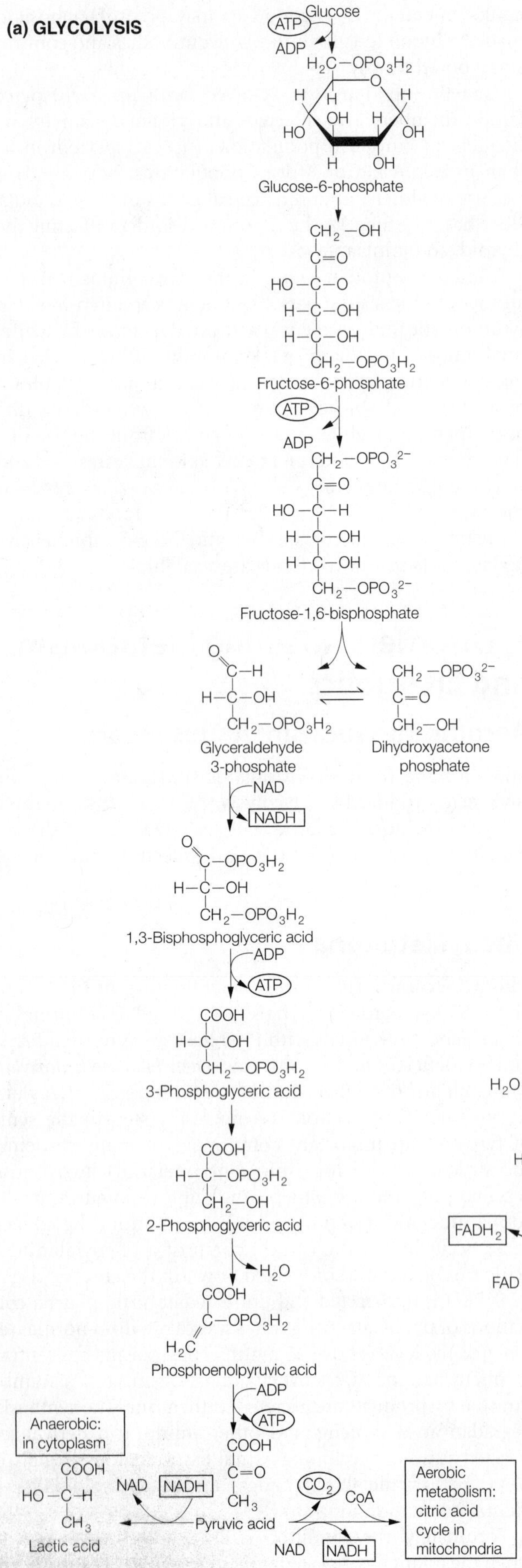

(a) GLYCOLYSIS

phenylketone, a metabolic by-product, in the blood and urine. (During pregnancy, the normal mother metabolizes phenylalanine for the PKU fetus, so fetal levels are normal prior to delivery.) Treatment consists of controlling the amount of phenylalanine in the diet while plasma concentrations are monitored. This treatment is most important in infancy and childhood, when the nervous system is developing. Because tyrosine cannot be synthesized from dietary phenylalanine, the diet of these children must contain adequate amounts of tyrosine. It is difficult to adhere to the PKU diet because food choices are limited, and some of the baby foods have an unpleasant taste. While normal phenylalanine levels are desirable even in adults, dietary restriction of phenylalanine in adults with PKU is usually relaxed, except during pregnancy. A pregnant woman with PKU must protect the fetus from high levels of phenylalanine by following a strict phenylalanine-restricted diet that must begin before the pregnancy occurs.

One popular artificial sweetener, *Nutrasweet*, consists of phenylalanine and aspartic acid. The consumption of food or beverages that contain this sweetener can therefore cause problems for persons with PKU.

In its most severe form, where there are two abnormal copies of the gene, PKU affects approximately 1 infant in 15,000. Individuals who carry only a single gene copy for PKU will produce the affected enzyme, but in lesser amounts. These individuals are asymptomatic, but have slightly elevated phenylalanine levels in

▶ **Figure 78** **Aerobic Metabolism.**

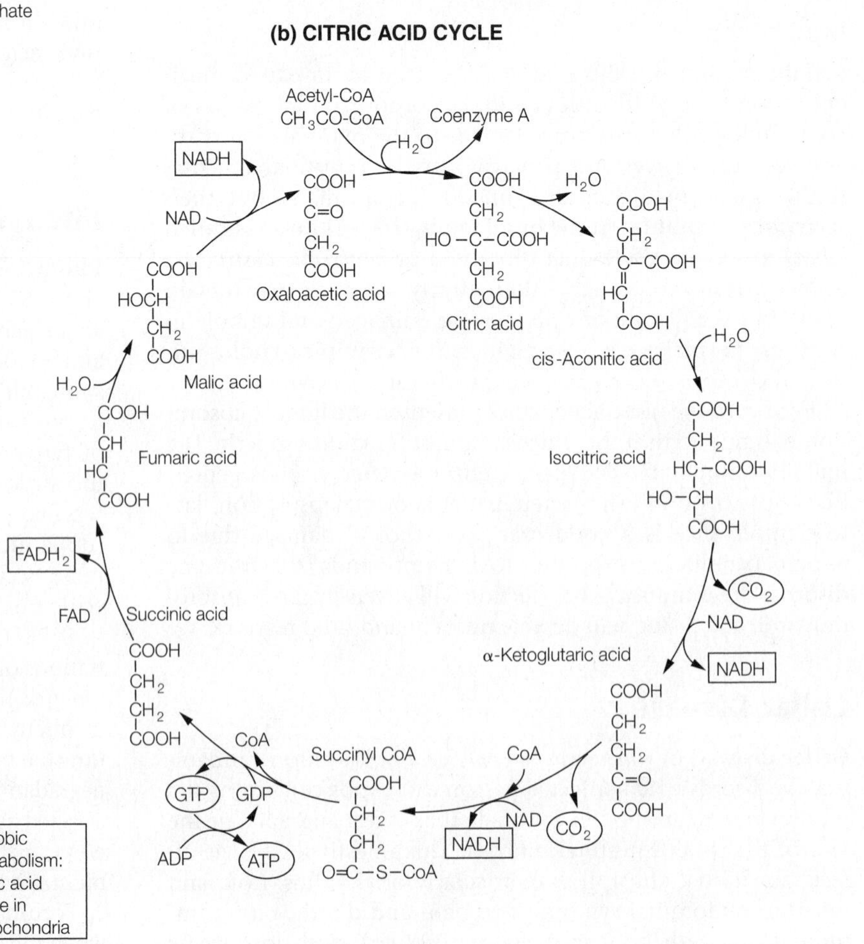

(b) CITRIC ACID CYCLE

their blood. Statistical analysis of the incidence of fully developed PKU indicates that as many as 1 person in 70 may carry an abnormal gene for the condition.

Protein Deficiency Diseases

Regardless of the energy content of the diet, if it is deficient in essential amino acids, the individual will be malnourished to some degree. In a **protein deficiency disease**, protein synthesis decreases throughout the body. As protein synthesis in the liver fails to keep pace with the breakdown of plasma proteins, plasma osmolarity falls. The reduced osmolarity results in a fluid shift as more water moves out of the capillaries and into interstitial spaces, the peritoneal cavity, or both. The longer the individual remains in this state, the more severe are the ascites and edema that result.

This clinical scenario occurs more often in developing countries, where poverty, drought, and wars often make dietary protein scarce or prohibitively expensive. Growing children suffer from **marasmus** (ma-RAZ-mus) when they are deprived of adequate proteins and calories. **Kwashiorkor** (kwash-ī-OR-kor) occurs in children whose protein intake is inadequate, even if the caloric intake is acceptable (⟩ **Figure 79**). In either case, additional complications include damage to the developing brain. The World Health Organization estimates that in 2011, 35 percent of all deaths in children under 5 years old were partially caused by these conditions, known collectively as malnutrition. War and civil unrest that disrupt local food production and distribution have been more instrumental than a shortage of food in producing recent famines. However, the World Health Organization reports that the incidence of underweight children (below age 5) is declining in developing countries, falling from 28 percent in 1990 to 17 percent in 2011. The highest number (56 million in 2011) are in South-Central Asia.

Metabolic Interactions and Adaptations to Starvation

To understand how the body responds to the presence or absence of nutrients, we need to take a closer look at how energy reserves are stored and mobilized. The nutrient requirements of each

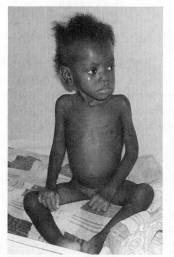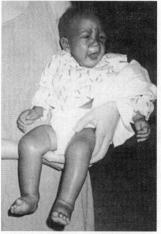

⟩ **Figure 79 Children with Marasmus (left), Kwashiorkor (right).**

tissue vary with the types and quantities of enzymes present in the cytoplasm of cells. From a metabolic standpoint, we can consider the body to have five distinctive components: the liver, adipose tissue, skeletal muscle, neural tissue, and other peripheral tissues.

1. **The Liver.** The liver is the focal point of metabolic regulation and control, and it contains significant energy reserves in the form of glycogen deposits.
2. **Adipose Tissue.** Adipose tissue stores lipids, primarily as triglycerides.
3. **Skeletal Muscle.** Skeletal muscles maintain substantial glycogen reserves, and their contractile proteins can be broken down and the amino acids used as an energy source.
4. **Neural Tissue.** Neurons must be provided with a reliable supply of glucose, because they are generally unable to metabolize other molecules. If blood glucose levels become too low, neural tissue in the central nervous system cannot continue to function.
5. **Other Peripheral Tissues.** Other peripheral tissues do not maintain large metabolic reserves, but they may be able to metabolize glucose, fatty acids, or other substrates, depending on what signals they receive from the endocrine system.

The interrelationships among these five components can best be understood by considering events during the *absorptive state* and the *postabsorptive state*.

The Absorptive (Anabolic) State

The **absorptive state** is the period following a meal, when nutrient absorption is under way along the digestive tract. After a typical meal, the absorptive state continues for about 4 hours. If you eat three meals a day, you spend 12 out of every 24 hours in the absorptive state.

A typical meal contains proteins, lipids, and carbohydrates in various proportions. While you are in the absorptive state, the intestinal mucosa is absorbing these nutrients. Glucose and amino acids enter the bloodstream, and the hepatic portal vein carries them to the liver. Most of the absorbed fatty acids are packaged in chylomicrons, which enter the lacteals of the lymphatic system.

Some of the carbohydrates, lipids, and amino acids are broken down (catabolized) immediately to provide energy for cellular operations. The remainder is stored, lessening the impact of future shortages. Insulin is the primary hormone of the absorptive state, although various other hormones stimulate amino acid uptake (growth hormone, or GH) and protein synthesis (GH, androgens, and estrogens). Next we consider the activities under way at specific sites throughout the body.

The Liver

The liver regulates the levels of glucose and amino acids in the blood arriving in the hepatic portal vein before that blood reaches the inferior vena cava. Despite the continuous absorption of glucose at the intestinal mucosa, blood glucose levels do not skyrocket, because liver cells (hepatocytes), under insulin stimulation, remove glucose from blood delivered by the hepatic portal circulation. Blood glucose levels rise, but only from about 90 mg/dL to perhaps 150 mg/dL, even after a meal

Digestive System

rich in carbohydrates. The liver uses some of the absorbed glucose to generate the ATP required to perform synthetic operations. Glycogenesis (glycogen formation) continues until glycogen accounts for about 5 percent of total liver weight. If excess glucose remains in the bloodstream, hepatocytes use glucose to synthesize triglycerides. Although small quantities of lipids are normally stored in the liver, most of the synthesized triglycerides are bound to transport proteins and released into the bloodstream as VLDLs. Peripheral tissues—primarily adipose tissues—then absorb these lipids for storage.

The liver does not control circulating levels of amino acids as precisely as it does glucose concentrations. Plasma amino acid levels normally range between 35 and 65 mg/dL, but they may become elevated after a protein-rich meal. The absorbed amino acids are used to support the synthesis of proteins, including plasma proteins and the proenzymes of the clotting system. Liver cells can also synthesize many amino acids, and an amino acid present in abundance may be converted to another, less-abundant amino acid and released into the bloodstream.

Most of the lipids absorbed by the digestive tract bypass the hepatic portal vein and liver. Triglycerides, cholesterol, and large fatty acids reach the general venous circulation in chylomicrons that are picked up by lacteals, transported in the thoracic duct, and released into the left subclavian vein. Most of these lipids are absorbed by other tissues.

Adipose Tissue

During the absorptive state, adipocytes remove fatty acids and glycerol from the bloodstream. Lipids continue to be removed from the blood for 4–6 hours after you have eaten a fatty meal. Over this period, the presence of chylomicrons in the plasma, a characteristic called **lipemia** (lip-Ē-mē-uh), gives it a milky appearance.

Adipocytes are particularly active in absorbing these lipids and in synthesizing new triglycerides for storage. At normal blood glucose concentrations, any glucose entering adipocytes is catabolized to provide the energy needed for lipogenesis (lipid synthesis). Adipocytes also absorb amino acids as needed for protein synthesis. Although these cells can use glucose or amino acids to manufacture triglycerides, they do so only if circulating concentrations are unusually high.

If, on a daily basis, you take in more nutrients during the absorptive state than you catabolize during the postabsorptive state, the fat deposits in your adipose tissue will enlarge. Most of the enlargement represents an increase in the size of individual adipocytes. An increase in the total number of adipocytes does not ordinarily occur, except in children before puberty and in extremely obese adults.

Skeletal Muscle, Neural Tissue, and Other Peripheral Tissues

When blood glucose and amino acid concentrations are elevated, insulin is released from the pancreatic islets, and all tissues increase their rates of absorption and utilization. Glucose is catabolized for energy, and the amino acids are used to build proteins.

Glucose is normally retained in the body, because the kidneys prevent the loss of glucose molecules in urine. The kidneys' ability to conserve glucose breaks down only when blood glucose concentrations are extraordinarily high—in excess of 180 mg/dL. The amino acid content of urine is not as carefully regulated; amino acids commonly appear in the urine after a protein-rich meal.

When blood glucose levels are elevated, most cells ignore the circulating lipids, so the adipocytes absorb most of these lipids. In resting skeletal muscles, a significant portion of the metabolic demand is met through the catabolism of fatty acids. Glucose molecules are used to build glycogen reserves, which may account for 0.5–1 percent of the weight of each muscle fiber.

Major relationships in the absorptive state are summarized in ▶ **Figure 80**.

The Postabsorptive (Catabolic) State

The **postabsorptive state** is the period when your body must rely on internal energy reserves to meet its energy demands. You spend roughly 12 hours each day in the postabsorptive state. Over this period, energy reserves are mobilized to support cellular activities while maintaining normal blood glucose levels. These activities are coordinated by several hormones, including glucagon, epinephrine, glucocorticoids, and growth hormone.

Metabolic reserves are organic substrates in the body that can be catabolized to obtain the ATP needed to sustain life. The metabolic reserves of a typical 70-kg (154-lb) individual include carbohydrates, lipids, and proteins (▶ **Figure 81**). Due to its high energy content, adipose tissue represents a disproportionate percentage of the total reserve in the form of triglycerides. Most of the available protein reserve is located in the contractile proteins of skeletal muscle. Carbohydrate reserves are relatively small and sufficient for only a few hours or, at most, overnight.

Next we examine the events occurring in specific tissues and organs during the postabsorptive state.

The Liver

As the absorptive state ends, intestinal cells stop providing glucose to the portal circulation. At first, the peripheral tissues continue to remove glucose from the blood, so blood glucose levels begin to decline. The liver responds by reducing its synthetic activities. When plasma concentrations fall below 80 mg/dL, hepatocytes begin breaking down glycogen reserves and releasing glucose into the bloodstream. This glycogenolysis occurs in response to a rise in circulating levels of glucagon and epinephrine. The liver contains 75–100 g of readily available glycogen, a reserve that is adequate to maintain blood glucose levels for about four hours.

As glycogen reserves decline and plasma glucose levels fall to about 70 mg/dL, hepatocytes begin to make glucose in an attempt to stabilize blood glucose levels. The shift from glycogenolysis to gluconeogenesis occurs under stimulation by *glucocorticoids*, steroid hormones from the adrenal cortex.

Gluconeogenesis

By means of gluconeogenesis, hepatocytes synthesize glucose molecules from smaller carbon fragments. In effect, any carbon fragment that can be converted to pyruvic acid or one of the three-carbon compounds involved in glycolysis in the cytoplasm can be used to synthesize glucose. With glucose already in short supply, lipids and amino acids must be catabolized to provide the ATP molecules needed for these syntheses.

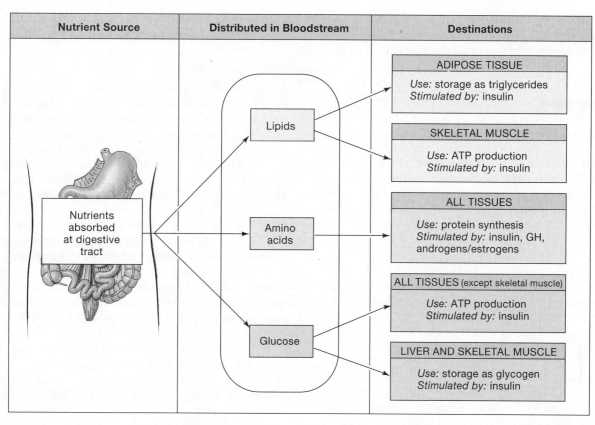

Nutrient Source	Distributed in Bloodstream	Destinations

Nutrients absorbed at digestive tract

Lipids

ADIPOSE TISSUE
Use: storage as triglycerides
Stimulated by: insulin

SKELETAL MUSCLE
Use: ATP production
Stimulated by: insulin

Amino acids

ALL TISSUES
Use: protein synthesis
Stimulated by: insulin, GH, androgens/estrogens

Glucose

ALL TISSUES (except skeletal muscle)
Use: ATP production
Stimulated by: insulin

LIVER AND SKELETAL MUSCLE
Use: storage as glycogen
Stimulated by: insulin

▶ **Figure 80 The Absorptive State.** During the absorptive state, the primary metabolic goal is anabolic activity, especially growth and the storage of energy reserves.

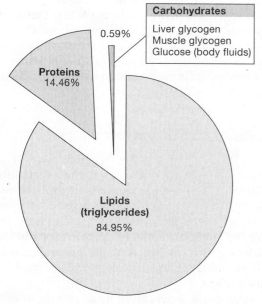

Carbohydrates
Liver glycogen
Muscle glycogen
Glucose (body fluids)

0.59%

Proteins
14.46%

**Lipids
(triglycerides)**
84.95%

▶ **Figure 81 Metabolic Reserves.** The distribution of the estimated metabolic reserves of a 70-kg individual. For the caloric value of these reserves, see Figure 83.

Utilization of Lipids

In the postabsorptive state, the liver absorbs fatty acids and glycerol from the blood. The glycerol molecules are converted to glucose. Fatty acids are broken down through beta-oxidation to produce large quantities of acetyl-CoA. However, because the enzymatic reaction that converts pyruvic acid to acetyl-CoA is not reversible, acetyl-CoA cannot be used to synthesize glucose. Instead,

- Some of the acetyl-CoA molecules deliver their two-carbon acetyl fragments to the citric acid cycle, during which they are broken down. The ATP generated can then be used to support gluconeogenesis.

- In addition, some of the molecules of acetyl-CoA are converted to special compounds that can be utilized by peripheral tissues. These compounds, called **ketone bodies**, are organic acids that are also produced during the catabolism of amino acids.

Utilization of Amino Acids

Before an amino acid can be used for either gluconeogenesis or energy production through its breakdown in the citric acid cycle, the amino group ($-NH_2$) must be removed through deamination. The structure of the remaining carbon chain determines its subsequent fate. After deamination, some amino acids can be converted to molecules of pyruvic acid or to one of the intermediary molecules of the citric acid cycle. Other amino acids—including most of the essential amino acids—can be converted only to acetyl-CoA and must be either broken down further or converted to ketone bodies.

The liver is the most active site of amino acid breakdown. There, the ammonia generated by deamination is converted to urea,

Digestive System

which circulates in the bloodstream until it is excreted in urine. The urea concentration of blood rises during the postabsorptive period, because the rate of amino acid catabolism increases.

Ketone Bodies

During the postabsorptive state, hepatocytes conserve glucose and break down lipids and amino acids. Both lipid catabolism and amino acid catabolism generate acetyl-CoA. As the concentration of acetyl-CoA rises, ketone bodies begin to form. There are three such compounds: (1) **acetoacetate** (as-ē-tō-AS-e-tāt), (2) **acetone** (AS-e-tōn), and (3) **betahydroxybutyrate** (bā-ta-hī-droks-ē-BŪ-te-rāt). Hepatocytes do not metabolize ketone bodies, and these compounds diffuse through the cytoplasm and into the circulation. Cells in peripheral tissues then absorb the ketone bodies and reconvert them to acetyl-CoA for introduction into the citric acid cycle.

The normal concentration of ketone bodies in the blood is about 30 mg/dL, and at this concentration very few of these compounds appear in urine. During even a brief period of fasting, the increased production of ketone bodies results in **ketosis** (kē-TŌ-sis), a high concentration of ketone bodies in body fluids. In ketosis, the concentration of ketone bodies is elevated in blood, a condition called **ketonemia** (kē-tō-NĒ-mē-uh), and in urine, a condition called **ketonuria** (kē-tō-NOO-rē-uh). Ketonemia and ketonuria are clear indications that the catabolism of proteins and lipids is under way. Acetone, which diffuses out of the pulmonary capillaries and into the alveoli very readily, can be smelled on the individual's breath.

In summary, during the postabsorptive state, the liver attempts to stabilize blood glucose concentrations, first by breaking down glycogen reserves and later by gluconeogenesis. Over the remainder of the postabsorptive state, the combination of lipid and amino acid catabolism provides the necessary ATP and generates large quantities of ketone bodies that diffuse into the bloodstream.

In *diabetes mellitus*, most peripheral tissues cannot transport and use glucose normally. Under these circumstances, cells survive by catabolizing lipids and proteins. The result is the production of large quantities of ketone bodies. This condition leads to *diabetic ketoacidosis*, the most common and life-threatening form of ketoacidosis.

Adipose Tissue

Adipose tissue contains a tremendous storehouse of energy in the form of triglycerides. Fat accounts for approximately 15 percent of the body weight of the average, normal-weight man and 21 percent of the average, normal-weight woman. Although adipose tissue is rare in the eyelids, the nose, and the backs of the hands and feet, other regions are preferential sites of deposition. Typically, 50 percent of a normal-weight individual's adipose tissue is located in the hypodermis, 15–20 percent in the greater omentum, 5–8 percent between muscles, and the rest distributed around the kidneys and reproductive organs.

As blood glucose levels decline, the rate of triglyceride synthesis falls. Under the stimulation of epinephrine, glucocorticoids, and growth hormone, adipocytes soon begin breaking down their lipid reserves, releasing fatty acids and glycerol into the bloodstream. This process, called *fat mobilization*, continues for the duration of the postabsorptive state. A normal-weight individual retains about a 2-month supply of energy in the triglycerides of adipose tissue. The evolutionary advantages are obvious: The retention of an energy reserve provides a buffer against daily, monthly, and even seasonal changes in the available food supply.

Skeletal Muscle

At the start of the postabsorptive state, skeletal muscles obtain energy by breaking down their glycogen reserves and catabolizing the glucose that is released. As the concentrations of fatty acids and ketone bodies in the bloodstream increase, these substrates become increasingly important as energy sources.

Skeletal muscle as a whole contains twice as much glycogen as the liver, but it is distributed throughout the muscular system. These glucose reserves are not directly available to other tissues, because the lack of a key enzyme prevents skeletal muscle cells from releasing glucose into the bloodstream. Even as skeletal muscle fibers metabolize fatty acids as an energy source, they continue to break down their glycogen reserves and convert the resulting pyruvic acid molecules to lactic acid. Lactic acid then diffuses out of the muscle fibers and into the bloodstream. However, even if all of the available glycogen reserves in your body were mobilized as glucose or as lactic acid, the energy provided would be only enough to get you through a good night's sleep. If the postabsorptive state continues for an unusually long period—long enough that lipid reserves are being depleted—muscle proteins will be broken down by enzymes called *cathepsins*. The amino acids that are released diffuse into the blood for use by the liver in gluconeogenesis.

Other Peripheral Tissues

With rising plasma concentrations of lipids and ketone bodies and falling blood glucose levels, peripheral tissues gradually decrease their reliance on glucose. Circulating ketone bodies and fatty acids are absorbed and converted to acetyl-CoA for entry into the citric acid cycle.

Neural Tissue

Neurons are unusual in that they continue "business as usual" during the postabsorptive state. Neurons depend on a reliable supply of glucose, and changes in the activity of the liver, skeletal muscle, and other tissues ensure that glucose remains available throughout the postabsorptive state. The situation changes only when the postabsorptive state is unusually prolonged, as in starvation.

Major relationships in the postabsorptive state are diagrammed in ▶ **Figure 82**.

Adaptations to Starvation

▶ **Figure 83** shows changes in the metabolic stores of a 70-kg individual during prolonged starvation. Carbohydrate utilization declines almost immediately as the stores are depleted. As blood glucose levels decline, gluconeogenesis accelerates, using glycerol, amino acids, and lactic acid. The glycerol is provided by adipocytes; the amino acids and lactic acid are provided primarily by skeletal muscle. At this point, the kidneys begin to assist the liver by deaminating amino acids and generating additional glucose molecules.

Gluconeogenesis is accompanied by an increase in circulating ketone bodies, some derived from ketogenic amino acids, others from the catabolism of fatty acids. As the stress due to starvation continues, peripheral tissues further restrict their glucose utilization.

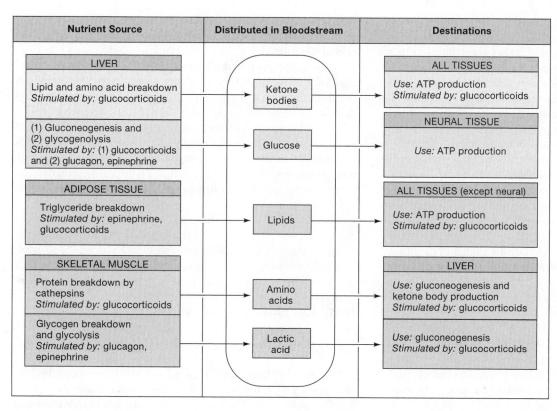

▶ **Figure 82** **The Postabsorptive State.** In the postabsorptive state, energy reserves are mobilized, and peripheral tissues (except neural tissues) shift from glucose catabolism to fatty acid or ketone body catabolism to obtain energy.

The ketone bodies generated by fatty acid catabolism become the primary energy source.

The fasting individual gradually becomes weak and lethargic as peripheral systems are weakened by protein catabolism and stressed by changes in pH. Buffer systems are challenged by the circulating amino acids, lactic acid, and ketone bodies, and ketoacidosis becomes a potential problem. Under these circumstances, most tissues begin catabolizing ketone bodies almost exclusively, and in extreme starvation more than 90 percent of the daily energy demands are met by the oxidation of ketone bodies. At this stage, even neural tissue relies on ketone bodies to supplement declining glucose supplies.

Structural proteins are the last to be mobilized, with other forms, such as the contractile proteins of skeletal muscle, more readily available. When peripheral tissues catabolize proteins, the amino acids are exported to the liver, where they can be safely deaminated. The carbon fragments are then catabolized to provide ATP, or they are used to manufacture

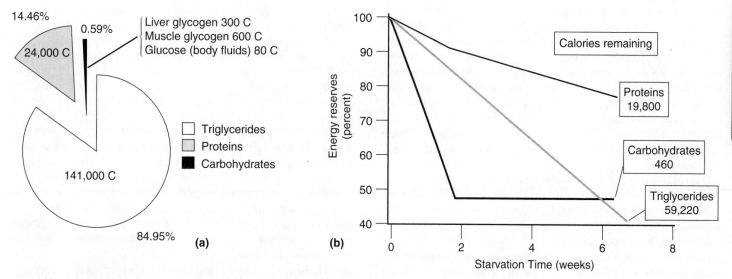

▶ **Figure 83** **Metabolic Reserves and the Effects of Starvation.** **(a)** Estimated metabolic reserves, in calories (C), of a 70-kg individual. **(b)** Projected effects of prolonged starvation on the metabolic reserves of the same individual.

glucose molecules or ketone bodies that can be broken down by peripheral tissues.

When lipid reserves are exhausted, crises soon follow. On a gram-for-gram basis, cells must catabolize almost twice as much protein as lipid to obtain the same energy benefits. Making matters worse, by this time most of the easily mobilized proteins have already been broken down. As structural proteins are disassembled, a variety of dangerous effects may appear. Accelerated protein catabolism causes problems with fluid balance, because the nitrogenous and acidic wastes must be eliminated in urine. These waste products are excreted in solution, and the more waste products that are eliminated, the greater is the associated water loss. An increase in urinary water losses can lead to dehydration, and the combination of dehydration and acidosis can cause kidney damage.

When glucose concentrations can no longer be sustained above 40–50 mg/dL, the individual becomes disoriented and confused. The eventual cause of death is kidney failure, ketoacidosis, protein deficiency, and/or hypoglycemia.

How long does it take to reach this critical state? That essentially depends on the size of the person's lipid reserves. Prolonged starvation for most people would last about eight weeks, but the truly obese can hold out far longer. With adequate water and vitamin supplements, an eight-month fast has been used as a weight-loss technique. (This technique was an emergency treatment rather than a diet plan, because prolonged fasting can result in kidney damage or severe ketoacidosis.)

Metabolic adjustments during starvation are coordinated primarily by the endocrine system. The glucocorticoids produced by the adrenal cortex are the most important hormones, aided by growth hormone from the pituitary gland.

RNA Catabolism and Gout

In the breakdown of RNA, the bonds between nucleotides are broken and the molecule is disassembled into individual nucleotides. The nucleotides are usually recycled into new nucleic acids. However, they can be catabolized to simple sugars and nitrogenous bases.

RNA catabolism makes a relatively insignificant contribution to the total energy budget of the cell. Proteins account for 30 percent of the cell's weight, and much more energy can be provided through the catabolism of proteins. Even when RNA is broken down, only the sugars and pyrimidines provide energy. The sugars can enter the glycolytic pathways. The pyrimidines (cytosine and uracil, in RNA) are converted to acetyl-CoA and metabolized through the citric acid cycle. The purines (adenine and guanine) cannot be catabolized. Instead, they are deaminated and excreted as **uric acid**. Like urea, uric acid is a relatively nontoxic waste product, but it is far less soluble than urea. Urea and uric acid are called **nitrogenous wastes**, because they are waste products that contain nitrogen atoms.

Normal uric acid concentrations in plasma average 2.7–7.4 mg/dL, depending on gender and age. When plasma concentrations exceed 7.4 mg/dL, *hyperuricemia* (hī-per-ū-ri-SĒ-mē-uh) exists. This condition may affect 18 percent of the U.S. population. At concentrations over 7.4 mg/dL, body fluids are saturated with uric acid. Although symptoms may not appear immediately, uric acid crystals may begin to form in body fluids, often after a minor injury. The condition that then develops is called *gout*. Most cases of hyperuricemia and gout are linked to problems

with the excretion of uric acid by the kidneys. The severity of the resulting symptoms depends on the amount and location of the crystal deposits.

Initially, the joints of the limbs, especially the metatarsophalangeal joint of the great toe, are likely to be affected. This intensely painful condition, called *gouty arthritis*, may persist for several days and then disappear for a period of days to years. Recurrences often involve other joints and may produce generalized fevers. Precipitates may also form within cartilages, in synovial fluids, in tendons or other connective tissues, or in the kidneys and urine. At serum concentrations of over 12–13 mg/dL, half the patients will develop kidney stones, and kidney function can be affected to the point of kidney failure.

The incidence of gout, ranging from 0.13 to 0.37 percent of the population, is much lower than that of hyperuricemia. Only about 5 percent of persons with gout are women, and most affected males are over 50. Foods high in purines, such as organ meat, oily fish, or fats, may aggravate or initiate the onset of gout. These foods tend to cost more than carbohydrates, so "rich foods" have often been associated with the condition.

Eating Disorders

Eating disorders have traditionally been considered mainly psychological problems that result in either inadequate or excessive food consumption. The most common conditions are **anorexia nervosa**, characterized by self-induced starvation, and **bulimia**, characterized by feeding binges followed by vomiting, the use of laxatives, or both. Most cases of anorexia nervosa and bulimia occur in adolescent females; only 5–15 percent occur in males. A common thread in the two conditions is an obsessive concern about food and body weight plus a distorted body image.

According to current estimates, the incidence of anorexia nervosa in the United States ranges from 0.4 to 1.5 per 100,000 population (up to 3.7 percent of females). A typical person with this condition is an adolescent Caucasian female whose weight is roughly 30 percent below normal. Although underweight, she is convinced that she is too fat, so she refuses to eat normal amounts of food. The incidence among Caucasian women age 12–18 is estimated to be 1 percent.

The psychological factors responsible for anorexia are complex. Young women with the condition tend to be high achievers who are attempting to reach an "ideal" weight that will be envied and admired and thereby provide a sense of security and accomplishment. The factors thus tend to be a combination of their view of society ("thin is desirable or demanded"), their view of themselves ("I am not yet thin enough"), and a desire to be able to control their fate ("I can decide when to eat"). Female models, dancers, figure skaters, gymnasts, and theater majors of any age may feel forced to lose weight to remain competitive. The few male anorexics diagnosed typically face comparable stresses. They tend to be athletes, such as jockeys, wrestlers, or actors who need to maintain a minimal weight to succeed in their careers.

Young anorexic women may starve themselves down to a weight of 30–35 kg (66–77 lb; ◗ **Figure 84**). Dry skin, peripheral edema, an abnormally low heart rate and blood pressure, a reduction in bone and muscle mass, and a cessation of menstrual cycles are relatively common signs. Some of these changes, especially in bone mass, can be permanent. Treatment is difficult, and only 50–60 percent of anorexics who regain normal weight retain it for

Digestive System

▶ **Figure 84 Anorexia Nervosa.**

five years or more. Death rates from severe, prolonged anorexia nervosa range from 10 to 15 percent.

Bulimia is more common than anorexia nervosa. In this condition, the individual goes on an "eating binge" that may last an hour or two and include 20,000 or more calories. Bingeing is followed by induced vomiting, commonly accompanied by the use of laxatives (drugs that promote the movement of the material through the digestive tract) and diuretics (drugs that promote fluid loss through urination). These often expensive binges may occur several times each week, separated by periods of either normal eating or fasting.

Bulimia generally involves females of the same age group as those who suffer from anorexia nervosa. The actual incidence is difficult to determine; published estimates for young college-age women range from 5 to 18 percent. However, many bulimics are not diagnosed until age 30–40. Because they ingest plenty of food, bulimics may have normal body weight, so the condition is harder to diagnose than anorexia nervosa. The health risks of bulimia result from (1) cumulative damage to the stomach, esophagus, oral cavity, and teeth from repeated exposure to stomach acids; (2) electrolyte imbalances resulting from the loss of sodium and potassium ions in the gastric juices, diarrhea, and urine; (3) edema; and (4) cardiac arrhythmias.

The underlying cause of bulimia remains unclear. Societal factors are undoubtedly involved, but bulimia has also been strongly correlated with depression and with elevated levels of antidiuretic hormone (ADH) in cerebrospinal fluid.

Binge eating disorder consists of recurrent episodes of out-of-control eating, but without self-induced vomiting/purging. Often the person may be overweight or obese. It is frequently done in secret and associated with lack of hunger, depression, anxiety, and sometimes substance abuse. The extra calories may lead to extreme obesity, and promotes a vicious cycle of binge eating → depression → binge eating. What part this disorder plays in the current epidemic of obesity has not been determined.

Body Mass Index and Obesity

Affluent societies and affluent members of most societies with easy access to food and a sedentary lifestyle are getting fatter every year. The most widely accepted measure of overweight and obesity is the body mass index (BMI), calculated by dividing a person's weight in kilograms by his or her height in meters squared. In the nonmetric United States, the BMI chart (▶ **Figure 85**) is helpful. For almost 100 years now, life insurance companies have used various charts of weight versus height to predict life expectancy. Generally, desirable weight–height ratios correspond to a BMI between 19 and 24. Somewhat arbitrarily, overweight is defined as a BMI of 25 or over, obesity as a BMI of 30 or over, and extreme obesity as a BMI of 40 or over. Using these criteria, over 66 percent of Americans are overweight; almost half of these people are obese. More than 5 percent are "super-obese" with BMIs over 40. A useful definition of **obesity** is "20 percent above ideal weight," because that is the point at which serious risks to health increase. Obesity can lead to diabetes, hypertension, hypercholesterolemia, and cancer.

This alarming problem has developed rapidly over the last 20 years, and has included the rest of the world. The World Health Organization estimates that in 2008, 1.5 billion people were overweight, and 800 million of these persons were obese. The incidence of obesity in the general population ranges from 5 percent (China, Japan, and much of Africa) to 75 percent (Samoa). Simply stated, obese individuals are taking in more energy in food than they use. Although genetic factors play some role, the societal changes associated with urbanization, mechanized transportation, and less manual labor, combined with economic growth and globalized food markets have increased obesity in all ethnic groups. Up to 20 percent of urban dwellers in China are obese. Our physiological systems are designed to store energy when it's available, so that it can be used when food is in short supply (the mechanisms involved were examined in earlier sections). These physiological adaptations, which are a benefit when you are living hand-to-mouth in the wild, become problematic when there *are* no lean times. Being overweight or obese, with the resulting increased risk for heart disease, has become fifth on the list of causes of death worldwide.

The two major categories of obesity are *regulatory obesity* and *metabolic obesity*. In **metabolic obesity**, the condition is secondary to some underlying organic malfunction that affects cell and tissue metabolism. For example, some cases of obesity have been linked to reduced insulin sensitivity due to a reduction in the number of insulin receptors in adipose tissue and in skeletal muscle. Metabolic obesity is commonly associated with chronic hypersecretion or hyposecretion of metabolically active hormones, such as insulin, glucocorticoids, or thyroxine.

Regulatory obesity results from a failure to regulate one's food intake so that appetite, diet, and activity remain in balance. Most instances of obesity fall within this category. In the majority of cases, there is no obvious organic cause, although in rare cases the problem arises because some disorder, such as a tumor, affects the hypothalamic centers that deal with appetite and satiation. Physiologically, individuals with regulatory obesity are overeating for the amount of physical activity they perform, thereby extending the duration and magnitude of the absorptive state.

Categorizing an obesity problem is less important in a clinical setting than is determining the degree of obesity and the number and severity of the related complications. The affected individuals are at a high risk of developing diabetes,

HEIGHT	5'0"	5'1"	5'2"	5'3"	5'4"	5'5"	5'6"	5'7"	5'8"	5'9"	5'10"	5'11"	6'0"	6'1"	6'2"	6'3"	6'4"
WEIGHT (lbs)																	
100	20	19	18	18	17	17	16	16	15	15	14	14	14	13	13	12	12
105	21	20	19	19	18	17	17	16	16	16	15	15	14	14	13	13	13
110	21	21	20	19	19	18	18	17	17	16	16	15	15	15	14	14	13
115	22	22	21	20	20	19	19	18	17	17	17	16	16	15	15	14	14
120	23	23	22	21	21	20	19	19	18	18	17	17	16	16	15	15	15
125	24	24	23	22	21	21	20	20	19	18	18	17	17	16	16	16	15
130	25	25	24	23	22	22	21	20	20	19	19	18	18	17	17	16	16
135	26	26	25	24	23	22	22	21	21	20	19	19	18	18	17	17	16
140	27	26	26	25	24	23	23	22	21	21	20	20	19	18	18	17	17
145	28	27	27	26	25	24	23	23	22	21	21	20	20	19	19	18	18
150	29	28	27	27	26	25	24	23	23	22	22	21	20	20	19	19	18
155	30	29	28	27	27	26	25	24	24	23	22	22	21	20	20	19	19
160	31	30	29	28	27	27	26	25	24	24	23	22	22	21	21	20	19
165	32	31	30	29	28	27	27	26	25	24	24	23	22	22	21	21	20
170	33	32	31	30	29	28	27	27	26	25	24	24	23	22	22	21	21
175	34	33	32	31	30	29	28	27	27	26	25	24	24	23	22	22	21
180	35	34	33	32	31	30	29	28	27	27	26	25	24	24	23	22	22
185	36	35	34	33	32	31	30	29	28	27	27	26	25	24	24	23	23
190	37	36	35	34	33	32	31	30	29	28	27	26	26	25	24	24	23
195	38	37	36	35	33	32	31	31	30	29	28	27	26	26	25	24	24
200	39	38	37	35	34	33	32	31	30	30	29	28	27	26	26	25	24
205	40	39	37	36	35	34	33	32	31	30	29	29	28	27	26	26	25
210	41	40	38	37	36	35	34	33	32	31	30	29	28	28	27	26	26
215	42	41	39	38	37	36	35	34	33	32	31	30	29	28	28	27	26
220	43	42	40	39	38	37	36	34	33	32	32	31	30	29	28	27	27

KEY

18 or less	Underweight
19 to 24	Normal
25 to 29	Overweight
30 or more	Obese

▶ **Figure 85 A Chart of Body Mass Index.**

hypertension, hypercholesterolemia, and coronary artery disease, as well as gallstones, thrombi or emboli, hernias, degenerative arthritis, varicose veins, and some forms of cancer. A variety of treatments may be considered, ranging from behavior modification, nutritional counseling, psychotherapy, and exercise programs to surgical gastric reduction, or a partial gastric bypass. The simplest and most natural treatment involves a reduced caloric intake (diet) and more caloric expenditure (exercise). Eating smaller portions, especially of foods high in caloric density, is helpful. (Fats have 9 calories per gram, proteins 4 calories per gram, and carbohydrates 4 calories per gram.) Eating more fruits and vegetables, which tend to have lower caloric densities, is beneficial but tends to be more expensive. Eating awareness (which may involve daily or twice-weekly food diaries of what one has eaten, in what quantities, and when) has proven effective for some people. Unfortunately, long-term compliance and sustained weight loss are difficult and rare. For one thing, losing weight is seldom fun—animal behaviorists have long noted that for both wild and domesticated animals, a fat animal is a happy animal. For another, there are physiological adjustments that occur in obesity that make weight loss more difficult than it would be for someone of normal body weight.

Fevers

When interleukin-1 increases the "thermostat setting" of the preoptic center of the hypothalamus, the heat-gain center is activated. The individual feels cold and may curl up in a blanket. Shivering may begin and may continue until the temperature at the preoptic area corresponds to the new setting. The fever passes when the thermostat is reset to normal. The **crisis phase** then ensues as the heat-loss center is stimulated. The individual feels unbearably warm and discards the blanket; the skin is flushed, and the sweat glands work furiously to bring the temperature down. Repeated cycles of this type constitute the "chills and fever" pattern of many febrile illnesses. Some fevers may help the body in overcoming an infection, but if a fever is high and uncomfortable, treatment with **antipyretic drugs,** such as aspirin, ibuprofen, or acetaminophen may be beneficial.

Chronic fevers may persist for weeks or months as the result of infections, cancers, or thermoregulatory disorders. In some cases, a discrete cause cannot be determined, leading to a classification as a *fever of unknown origin (FUO)*. Acute hyperthermia, such as that seen during heat stroke, is life-threatening. Immediate treatment may involve cooling the individual in an ice bath (to increase conduction) or giving alcohol rubs (to increase evaporation).

Digestive System

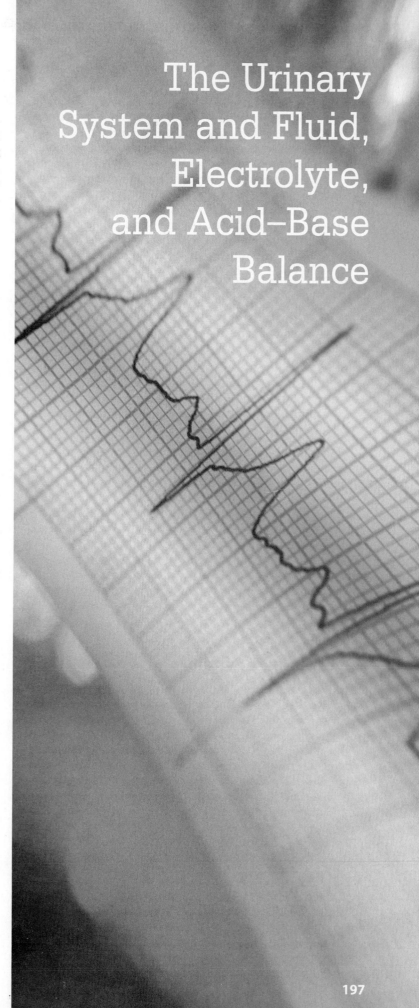

▶ An Introduction to the Urinary System and Its Disorders

The urinary system consists of the kidneys, where urine production occurs, and the conducting system, which transports and stores urine prior to its elimination from the body. The conducting system comprises the ureters, the urinary bladder, and the urethra. Although the kidneys perform all the vital homeostatic functions of the urinary system, problems with the conducting system can have direct and immediate effects on renal function.

The History and Physical Examination of the Urinary System

Pain: The primary symptoms of urinary system disorders are pain and changes in the volume and appearance of urine and the frequency of urination. The nature and location of the pain can provide clues to the source of the problem. For example,

- Pain in the superior pubic region may be associated with urinary bladder disorders.
- Pain in the superior lumbar region or in the flank that radiates to the right or left abdomen can be caused by kidney infections such as *pyelonephritis*, or by kidney stones.
- **Dysuria** (painful or difficult urination) can occur with *cystitis* or *urethritis* or with *urinary obstructions*. In males, enlargement of the prostate gland can lead to compression of the urethra and dysuria.

Individuals with urinary system disorders may urinate more or less frequently than usual and may produce normal or abnormal amounts of urine:

- **Urgency and/or Frequency.** An irritation of the lining of the ureters or urinary bladder can lead to the desire to urinate with increased frequency, although the total amount of urine produced each day remains normal. Unsuppressable detrusor muscle contractions may also lead to urinary urgency and frequency. When these problems exist, the individual feels the urge to urinate when the urinary bladder volume is very small. The irritation may result from trauma, urinary bladder infection (cystitis) or tumors, increased acidity of the urine, or hormonal and aging changes causing detrusor hyperreflexia.
- **Incontinence**, an inability to control urination voluntarily, may involve periodic involuntary leakage (stress incontinence), or inability to delay urination (urge incontinence)—or a continual, slow trickle of urine from a bladder that is always full (overflow incontinence). Incontinence results from urinary bladder or urethral problems, damage or weakening of the muscles of the pelvic floor, or interference with normal sensory or motor innervation in the region. Renal function and daily urinary volume are usually unaffected.
- In **urinary retention**, renal function is normal, at least initially, but urination does not occur. Urinary retention in males commonly results from enlargement of the prostate and compression of the prostatic urethra. In both genders, urinary retention can result from the obstruction of the outlet of the urinary bladder or from central nervous system damage involving control of

the detrusor muscle, such as might be caused by a stroke or damage to the spinal cord.

- Changes in the volume of urine produced by a person with average fluid intake indicates problems either at the kidneys or with the control of renal function. **Polyuria**, the production of excessive amounts of urine, results from hormonal or metabolic problems, such as those associated with *diabetes*, or from damage to the glomeruli, as in *glomerulonephritis*. **Oliguria** (a urine volume of 50–500 mL/day) and **anuria** (0–50 mL/day) are conditions that indicate serious kidney problems and potential renal failure. Renal failure can occur with *heart failure*, renal ischemia, *circulatory shock*, burns, pyelonephritis, hypovolemia, and a variety of other disorders.

Important clinical signs of urinary system disorders include the following:

- **Edema.** Renal disorders often lead to protein loss in the urine (proteinuria) and, if severe, result in generalized edema in peripheral tissues. Facial swelling, especially around the eyes, is common.

- **Fever.** A fever commonly develops when the urinary system is infected by pathogens. Urinary bladder infections (cystitis) may result in a low-grade fever; kidney infections, such as pyelonephritis, can produce very high fevers.

During the physical assessment, examiners may use percussion or palpation to check the status of the kidneys and urinary bladder. The kidneys lie in the costovertebral area, the region bounded by the lumbar spine and the 12th rib on either side. To detect tenderness due to kidney inflammation, the examiner gently thumps a fist over each flank posterior to the kidneys. This usually does not cause pain, unless the underlying kidney is inflamed.

The urinary bladder can be palpated just superior to the pubic symphysis. However, on the basis of palpation alone, urinary bladder enlargement due to urine retention can be difficult to distinguish from obesity or the presence of an abdominal mass.

Laboratory and Imaging Tests and the Urinary System

Many procedures and laboratory tests are used in the diagnosis of urinary system disorders. The functional anatomy of the urinary system can be examined with the use of a variety of sophisticated procedures. For example, an x-ray of the kidneys, ureter, and urinary bladder is taken after the administration of a radiopaque compound that will enter the urine. The resulting image is called an **intravenous pyelogram** (PĪ-el-ō-gram), or **IVP**. This procedure, sometimes called an *excretory urogram (EU)*, permits the detection of unusual kidney, ureter, or urinary bladder structures and masses. CT, MRI, and ultrasound scans also provide useful information about localized abnormalities.

Unusual laboratory findings may also provide clues as to the nature of a urinary system disorder.

- **Hematuria**, the presence of red blood cells in urine, indicates bleeding at either the kidneys or the conducting system. Hematuria producing dark red or tea-colored urine typically indicates bleeding in the kidney, and hematuria producing bright red urine indicates rapid bleeding in the kidney or bleeding in the inferior portion of the urinary tract (the urinary bladder or the urethra). Hematuria most commonly occurs with trauma to the kidneys, calculi (kidney stones), tumors, or urinary tract infections.

- **Hemoglobinuria** is the presence of hemoglobin in urine. Hemoglobinuria indicates increased hemolysis of red blood cells in the bloodstream due to cardiovascular or metabolic problems. Conditions that result in hemoglobinuria include the *thalassemias*, *sickle cell anemia*, *hypersplenism*, and some autoimmune disorders.

- Changes in the color of urine accompany some renal disorders. For example, urine becomes (1) cloudy due to the presence of bacteria, lipids, crystals, or epithelial cells; (2) red or brown from hemoglobin or myoglobin; (3) brown-black from hepatitis; or (4) red, orange, brown, blue, or green from various ingested medicines or chemicals. Not all color changes are abnormal, however; some foods can cause changes in urine color. For example, a serving of beets can give urine a reddish color, whereas eating rhubarb can give urine an orange tint, and B vitamins turn it a vivid yellow. Dehydration reduces urinary volume and increases the concentration of excreted waste products resulting in more intense dark amber color and stronger odor. Asparagus gives urine a distinctive odor that not everyone can detect.

Other diagnostic procedures and important laboratory tests are detailed in **Table 39.** ▶**Figure 86** outlines the major classes of disorders of the urinary system.

PAH and the Calculation of Renal Blood Flow

Although seldom used in clinical practice, *para-aminohippuric acid*, or *PAH*, can be administered to determine the rate of blood flow through the kidneys. PAH enters the filtrate through filtration at the glomerulus. As blood flows through the peritubular capillaries, any remaining PAH diffuses into the peritubular fluid, and the tubular cells actively secrete it into the filtrate. By the time blood leaves the kidney, virtually all the PAH has been removed from the bloodstream and filtered or secreted into the urine. You can therefore calculate renal blood flow if you know the PAH concentrations of the arterial plasma and urine. The calculation proceeds in a series of steps. The first step is to determine the plasma flow through the kidney by using the formula

$$P_f = \frac{PAH_u \times V_u}{PAH_p}$$

where P_f is the plasma flow, also known as the *effective renal plasma flow*, or *ERPF*; PAH_u is the concentration of PAH in the urine, usually expressed in milligrams per milliliter (mg/mL); V_u is the volume of urine produced, usually in terms of milliliters per minute; and PAH_p is the concentration of PAH in arterial plasma, in milligrams per milliliter.

Now consider the following example: A person producing urine at a rate of 1 mL per minute has a urinary PAH concentration of 15 mg/mL with an arterial PAH concentration of 0.02 mg/mL. The person's hematocrit is normal (Hct = 45%). The plasma flow is

$$P_f = \frac{15 \text{ mg/mL} \times 1 \text{ mL/min} = 750 \text{ mL/min}}{0.02 \text{ mg/mL}}$$

This value is an estimate of the plasma flow through the glomeruli and around the kidney tubules each minute. However, plasma accounts for only part of the volume of whole blood;

Urinary System

Table 39	Examples of Tests Used in the Diagnosis of Urinary System Disorders	
Diagnostic Procedure	**Method and Result**	**Representative Uses**
Cystoscopy	A small tube (cystoscope) is inserted through the urethra into the urinary bladder to view the lining of the urethra, urinary bladder, and ureteral openings within the bladder	Used to obtain a biopsy specimen or to remove stones (calculi) and small tumors; provides direct visualization of urethra, urinary bladder, and ureteral openings
Retrograde pyelography	Radiopaque dye is injected into ureters through a catheter in cystoscope inserted into urinary bladder; x-ray films are then taken	Detects obstructions of ureter caused by tumors, calculi, or strictures; visualizes renal pelvis and ureters without relying on renal filtration (useful if renal function is impaired)
Renal biopsy	Using ultrasound as a guide, biopsy needle is inserted through back and into kidney. Specimen is then removed for analysis	Determines cause of renal disease; detects rejection of transplanted kidney; used to perform tumor biopsy
Intravenous pyelography (IVP)	Dye injected intravenously is filtered at kidney and excreted into urinary tract; x-rays are then taken to view kidneys, ureters, and urinary bladder	Determines size of kidney, obstructions such as calculi or tumors, or anatomical abnormalities; relies on renal filtration of contrast medium
Cystography	Dye is inserted through catheter placed in urethra and threaded into urinary bladder; x-rays are then taken	Identifies tumors of urinary bladder and rupture of bladder by trauma; if x-rays are taken as patient voids, detects reflux of urine from urinary bladder to ureters
Laboratory Test	**Normal Values**	**Significance of Abnormal Values**
Urinalysis		
pH	4.5–8.0	Alkaline or acidic urine may indicate increased or decreased blood pH; may vary with diet
Color	Pale yellow amber	Color may change with certain drugs and foods; dark red-brown urine may indicate bleeding from kidney; bright red blood comes from lower urinary tract; some bacterial infections cause green tint
Appearance	Clear	Clouded urine may result from bacterial infection or from certain foods
Odor	Aromatic	Acetone odor may occur in diabetic ketoacidosis; asparagus in the diet gives a distinctive odor
Specific gravity	1.003–1.030	Increased in dehydration, increased ADH production, heart failure, glycosuria, or proteinuria; decreased in diabetes insipidus or renal failure
Protein	≤150 mg/24 h	Increased protein loss occurs in kidney infections or inflammation and after strenuous exercise
Glucose	None	Glucose appears in urine in the hyperglycemia of diabetes mellitus or Cushing's disease and after corticosteroid therapy
Ketones	None (unless fasting)	Appear in ketoacidosis or poorly controlled cases of diabetes mellitus, during dehydration, and after several hours of fasting
Urine Electrolytes		
Sodium	40–220 mEq/day	Increased levels occur with diuretic use and Addison's disease; decreased levels occur with dehydration or kidney, liver, or congestive heart failure
Potassium	25–100 mEq/day	Elevated levels occur with diuretics, dehydration, and starvation; decreased levels occur with kidney failure, some drugs
Blood Tests		
Electrolytes (serum)		
Sodium	Adults: 135–145 mEq/L	Increased levels occur with severe dehydration; decreased levels occur with SIADH, heart and kidney failure, and use of some diuretics
Potassium	Adults: 3.5–5.0 mEq/L	Increased levels occur with acute renal failure, acidosis; decreased levels can occur in some renal diseases that affect tubules and after use of some diuretics
Renin (plasma)	Adults: 0.1–4.3 ng/mL per hour	Elevated levels occur in hypovolemia and Addison's disease; decreased levels occur with ADH therapy
Acid phosphatase (plasma)	Adults: 0.11–5.5 U/L	Elevated levels occur with prostate cancer and bone cancer
Antistreptolysin O titer (ASO titer)	≤160 Todd units/mL	Increased levels occur in glomerulonephritis, rheumatic fever, bacterial endocarditis, and scarlet fever

Urinary System

(continued)

Table 39 Examples of Tests Used in the Diagnosis of Urinary System Disorders *(Continued)*

Diagnostic Procedure	Method and Result	Representative Uses
Blood Tests		
Bicarbonate (serum)	Adults: 24–28 mEq/L	Elevation or reduction of bicarbonate levels is important in diagnosis of acid–base disorders
BUN (blood urea nitrogen) (serum)	5–25 mg/dL	Estimates GFR; values increase in renal disease, dehydration, gastrointestinal bleeding, liver disease, and gout
Creatinine	0.6–1.2 mg/dL	Estimates GFR as follows: $\dfrac{(140 - age) \times \text{lean body weight (kg)}}{\text{plasma creatinine (mg/dL)} \times 72}$
Creatinine clearance	Male: 97–137 mL/min Female: 88–128 mL/min	Increased in renal failure; decreased by low muscle mass (starvation, aging). Both blood and urine are collected, comparisons reveal result
Uric acid (serum)	Adults: 2.5–7 mg/dL	Elevated levels occur with renal failure, gout, increased metabolism of nucleotides, leukemias, and lymphomas. Decreased levels occur with drug treatment for gout

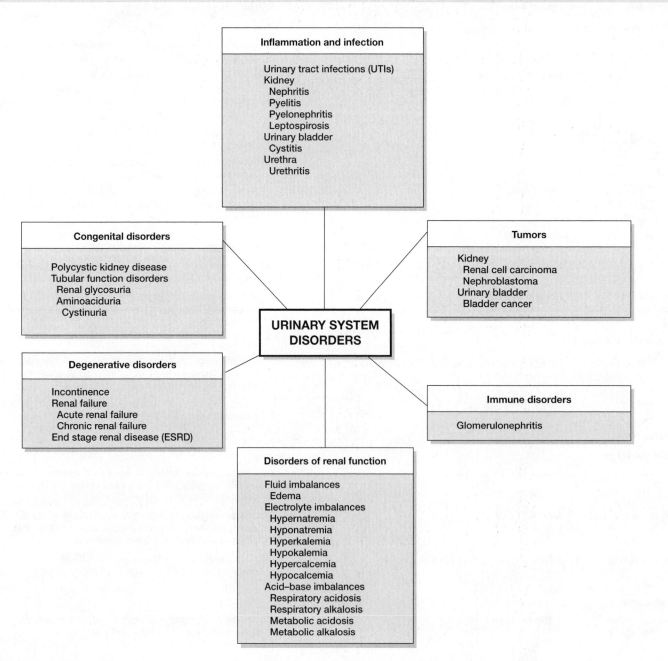

▶ **Figure 86** Disorders of the Urinary System.

Urinary System

the rest consists of formed elements. To have a plasma flow of 750 mL/min, the blood flow must be considerably greater. The patient's hematocrit is 45, which means that plasma accounts for 55 percent of the whole-blood volume. To calculate the renal blood flow, we must multiply the plasma flow by 1.8 (each 100 mL of blood has 55 mL of plasma, and 100/55 = 1.8):

$$750 \text{ mL/min} \times 1.8 = 1350 \text{ mL/min}$$

This value, 1350 mL/min, is the estimated tubular blood flow. The final step is to adjust the estimate to account for blood that enters the kidney, but flows to the renal pelvis, the capsule, or other areas not involved with urine production. This value is usually estimated as 10 percent of the total blood flow. We can therefore complete the calculation for this example as follows:

$$1350 \text{ mL/min} = 90 \text{ percent of total blood flow}$$
$$10 \text{ percent} = 1350/9 = 150 \text{ mL/min}$$
$$\text{Total blood flow} = 1350 + 150 = 1500 \text{ mL/min}$$

Urinary System Disorders

The kidneys filter the blood and remove metabolic waste products and excess water to maintain homeostasis by regulating blood volume and composition. During an average day a normal adult filters about 200 liters of blood and eliminates waste products in about 2 liters of urine. The glomerular filtration rate (GFR) is the amount of filtrate the kidneys produce each minute and is a usual way of assessing kidney function. The clearance of creatinine from the blood, the creatinine clearance, is commonly used to estimate the GFR. Most precisely, this involves measuring urine creatinine excretion over 24 hours, dividing this by the serum creatinine level and doing the calculations to express the results in mL/minute. As this involves collecting urine for 24 hours, often in clinical situations where perhaps drug doses need to take into account kidney function, an estimate of the creatinine clearance is made from entering measurements of the serum creatinine, patient age, lean body weight, and sex (if female, multiply the C-G result by 0.85 to reflect lower muscle mass) into the *Cockcroft-Gault equation*, which is used for calculating creatinine clearance:

$$\text{Creatinine clearance (mL/min)} = \frac{(140 - \text{age}) \times \text{weight (kg)}}{72 \times \text{serum creatinine (mg/dL)}}$$

Current laboratory tests often report an estimate of GFR, the *eGFR*, calculated from the serum creatinine, age and sex, adjusted if African-American, and using a standardized adult body surface area instead of body weight. This value is reported along with the serum creatinine level.

Conditions Affecting Filtration

Changes in *net filtration pressure (NFP)* can result in significant alterations in kidney function. Factors that can disrupt normal filtration rates include physical damage to the filtration apparatus and interference with normal filtrate or urine flow.

Physical Damage to the Filtration Apparatus

The kidneys can be injured by mechanical trauma, bacterial infection, circulating immune complexes, or exposure to metabolic poisons (such as mercury). The usual result is a sudden increase in the permeability of the glomerulus. When damage is

severe, plasma proteins and even blood cells enter the capsular spaces. Blood cells entering the filtrate will not be reabsorbed. The presence of blood cells in the urine is called **hematuria** (hē-ma-TOO-rē-uh).

The loss of plasma proteins has two immediate effects: (1) It reduces the osmotic pressure of the blood, and (2) it increases the osmotic pressure of the filtrate. The result is an increase in both the net filtration pressure and the rate of filtrate production.

Small amounts of protein can be reabsorbed, but when glomeruli are severely damaged, the nephrons are unable to reabsorb all the plasma proteins that enter the filtrate. Plasma proteins then appear in the urine, a condition termed **proteinuria** (prō-tē-NOO-rē-uh). Proteins and RBCs can form masses within the renal tubules; abnormal masses or precipitates in urine are called *casts*. Protein or RBC casts in urine indicate glomerular kidney damage.

Interference with Filtrate or Urine Flow

If the tubule, collecting duct, or ureter becomes blocked and urine cannot flow, capsular pressures gradually rise. When the capsular hydrostatic pressure and blood osmotic pressure equal the glomerular hydrostatic pressure, filtration stops completely. The severity of the problem depends on the site of the blockage. If it involves a single nephron, only a single glomerulus will be affected. If the blockage occurs within the ureter, filtration in that kidney will come to a halt. If the blockage occurs in the urethra, both kidneys will become nonfunctional. Recovery of renal function occurs if the blockage is recognized and removed within four to six weeks.

Elevated capsular pressures can also result from inflammation of the kidneys, a condition called **nephritis** (nef-RĪ-tis). A generalized nephritis may result from bacterial infections, exposure to toxic or irritating drugs, or autoimmune disorders. One of the major problems in nephritis is that the inflammation causes swelling, but the renal capsule prevents the kidney from increasing in size. The result is an increase in the hydrostatic pressure in the peritubular fluid and filtrate. This pressure opposes the glomerular hydrostatic pressure, lowering the net filtration pressure and the GFR.

Inherited Problems with Tubular Function

The tubular absorption of specific ions or compounds involves many different carrier proteins. Some individuals have an inherited inability to manufacture one or more of these proteins, so they experience impaired tubular function. For example, in **renal glycosuria** (glī-kō-SOO-rē-uh), a defective carrier protein makes it impossible for the proximal convoluted tubule (PCT) to reabsorb glucose from the filtrate. Although urine glucose levels are abnormally high, blood glucose is normal, which distinguishes this condition from diabetes mellitus. Affected individuals generally do not have any clinical problems, except when demand for glucose is high, as in starvation, acute stress, or pregnancy.

There are several types of **aminoaciduria** (a-mi-nō-as-i-DŪ-rē-uh), differing according to the identity of the missing carrier protein. Some of these disorders affect the reabsorption of an entire class of amino acids; others involve individual amino acids, such as lysine or histidine. **Cystinuria** is the most common disorder of amino acid transport, occurring in approximately 1 person in 12,500. Individuals with this condition have difficulty reabsorbing cystine and amino acids with similar

Urinary System

chemical structures, such as lysine, arginine, and ornithine. The most obvious and painful symptom is the formation of kidney and bladder stones that contain crystals of these amino acids. In addition to removing the stones, treatment for cystinuria involves drinking enough fluids to dilute the urine so that amino acid concentrations remain too low to precipitate into stones. Reducing urine acidity also helps.

Any of these problems with tubular absorption and secretion will have a direct effect on urinary volume. Urine cannot be concentrated past 1200 mOsm/L. Hence, the greater the number of solutes in the tubular fluid that enter the collecting ducts, the less water can be extracted and returned to the bloodstream by osmosis, and more urine is produced.

Diuretics

Diuretics (dī-ū-RET-iks) are drugs that promote the loss of water in the urine. Diuretics have several different mechanisms of action, but each affects transport activities or water reabsorption along the nephron and collecting system. (Some affect both.) Important diuretics in use today include the following:

- **Osmotic diuretics.** Osmotic diuretics are metabolically harmless substances that are filtered at the glomerulus and ignored by the tubular epithelium. Their presence in the urine increases its osmolarity and limits the amount of water reabsorption possible. **Mannitol** (MAN-i-tol) is an osmotic diuretic. It is used to accelerate fluid loss (usually to reduce brain edema), speed the removal of toxins from the blood, and elevate the GFR after severe trauma or other conditions have impaired renal function.

- **Drugs that block sodium and chloride transport.** A class of drugs called **thiazides** (THĪ-a-zīdz) reduces sodium and chloride transport in the proximal and distal tubules. Thiazides such as *hydrochlorothiazide* or *chlorthalidone* are often used to accelerate sodium and fluid losses in the treatment of hypertension and peripheral edema.

- **High-ceiling, or loop, diuretics.** The **high-ceiling diuretics**, such as *furosemide* inhibit transport along the nephron loop, reducing the osmotic gradient and the ability to concentrate the urine. They are called high-ceiling diuretics because they produce a much higher degree of diuresis than do other drugs. They are fast acting and are commonly used in a clinical crisis—for example, in treating acute pulmonary edema. In both the thiazide and the furosemide diuretics, water, Na^+, and K^+ are lost in the urine.

- **Aldosterone-blocking agents.** Blocking the action of aldosterone prevents the reabsorption of sodium along the distal convoluted tubule (DCT) and the collecting tubule and so increases fluid losses. The drug *spironolactone* is this type of diuretic. While not a particularly powerful diuretic, it is often used in conjunction with other diuretics because blocking the aldosterone-activated exchange pumps helps reduce the potassium ion loss. Aldosterone-blocking agents are also known as *potassium-sparing diuretics*. The natriuretic peptides could, in theory, also be used as diuretics, because they counteract the effects of both aldosterone and ADH at the kidneys.

- **Drugs with diuretic side effects.** Many drugs prescribed for other conditions promote diuresis as a side effect. For example, **ACE** (*angiotensin-converting enzyme*) **inhibitors** prevent the conversion of angiotensin I to angiotensin II by ACE.

In turn, that prevents the stimulation of aldosterone production and promotes water loss. Drugs that block carbonic anhydrase activity, such as acetazolamide (*Diamox*), have an indirect effect on sodium transport. Although they cause diuresis, these drugs are seldom prescribed with that effect in mind. (Because carbonic anhydrase is also involved in aqueous humor secretion, Diamox is used to reduce intraocular pressure in glaucoma patients.) Two nonprescription drugs that are more familiar, caffeine and alcohol, have pronounced diuretic effects. Caffeine produces diuresis directly by reducing sodium reabsorption along the tubules. Alcohol works indirectly by suppressing the release of ADH at the posterior lobe of the pituitary gland. After a hot, sweaty day in the sun, beer might *taste* refreshing, but if you drink beer, drink water as well, or you won't rehydrate—the fluid gains will be offset by increased urine production.

Urinalysis

There are a variety of sophisticated laboratory tests that can be performed on urine samples (see **Table 39**). There are also several basic screening tests that can be performed by recording changes in the color of test strips that are dipped in the sample. Urine pH and approximate urinary concentrations of glucose, ketones, bilirubin, urobilinogen, plasma proteins, and hemoglobin, as well as the density, or *specific gravity*, of the urine, can be monitored by this technique. The specific gravity (which is a measure of urine concentration) may also be determined by floating a simple device known as a **urinometer** (ū-ri-NOM-e-ter) or **densitometer** (den-si-TOM-e-ter) in a urine sample. A urine sample may also be spun in a centrifuge, and any sediment examined under the microscope. Mineral crystals, bacteria, red or white blood cells, and deposits, known collectively as **casts**, can be detected in this way. ▶**Figure 87** provides an overview of the major categories of urinary casts. During a urinary tract infection, bacteria may be cultured to determine their identities. New test strips can detect WBCs and nitrites found in urine infections.

More comprehensive analyses can determine the total osmolarity of the urine and the concentration of individual electrolytes and minor metabolites, metabolic wastes, vitamins, and hormones. A test for one hormone in the urine, *human chorionic gonadotropin* (hCG), provides an early and reliable proof of pregnancy. Urine drug tests are used to detect the presence of various drugs (including alcohol) in sports and employment, and to determine potential causes for impaired performance.

The information provided by urinalysis can be especially useful when correlated with the data obtained from blood tests. The term **azotemia** (a-zō-TE-mē-uh) refers to the presence of excess metabolic wastes in the blood. This condition may result from the overproduction of urea or other nitrogenous wastes by the liver ("prerenal azotemia syndrome"), or from decreased renal function. **Uremia** (ū-RĒ-mē-uh) (where the blood urea nitrogen and creatinine are elevated) describes the state of decreased renal function.

The total volume of urine produced in a 24-hour period may also be of interest. **Polyuria** (pol-ē-Ū-rē-uh) refers to excessive production of urine—well over 2 liters per day. Polyuria most commonly results from endocrine disorders—such as the various forms of diabetes—metabolic disorders, or damage to the filtration apparatus, as in glomerulonephritis. **Oliguria** (o-li-GŪ-re-uh) refers to inadequate urine production (50–500 mL/day). In

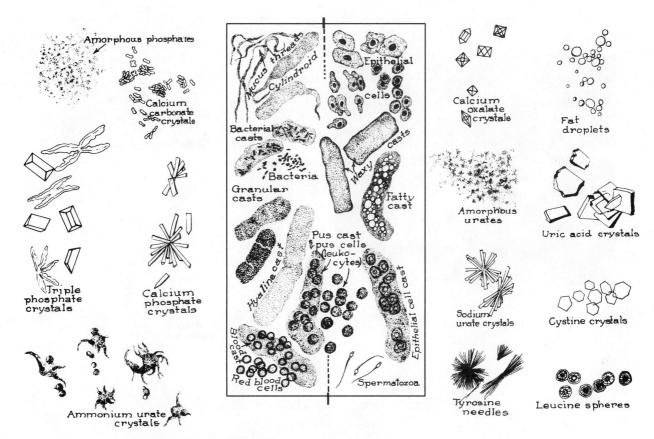

▶**Figure 87** **Microscopic Examination of Urine Sediment.**
Based on Clinical Diagnosis by Laboratory Methods by James Campbell Todd and Arthur Hawley Sanford, W. B. Saunders.

anuria (a-NŪ-rē-uh), a negligible amount of urine is produced (0–50 mL/day); this is usually a sign of renal failure and, unless resolved, a fatal problem. The simple, ongoing measurement of fluid intake and urine output over 24 hours (the "Ins and Outs" of nursing notes) is useful in assessing many patients.

Advances in the Treatment of Renal Failure

One normal kidney is sufficient to filter the blood and maintain homeostasis. As a result, renal failure will not develop unless both kidneys are significantly damaged. A normal adult glomerular filtration rate (GFR) is above 90 mL/min. Chronic renal failure is defined as a GFR less than 60 mL/min for over three months. The management of chronic renal failure typically involves restricting water and salt intake, controlling hypertension and diabetes, if present, and minimizing dietary protein intake. This combination reduces strain on glomerular filtration by minimizing the volume of urine produced and preventing the generation of large quantities of nitrogenous wastes. Acidosis, a common problem in persons with renal failure, can be countered by ingesting bicarbonate ions.

If drugs and dietary controls cannot stabilize the composition of blood, more drastic measures are taken. In **hemodialysis** (hē-mō-dī-AL-i-sis), a *dialysis machine* containing an artificial membrane is used to regulate the composition of blood (▶**Figure 88a**). The basic principle involved in this process, called **dialysis**, is passive diffusion across a selectively permeable membrane. The patient's blood flows past an artificial *dialysis membrane*, which contains pores large enough to permit the diffusion of small ions, but small enough to prevent the loss of plasma proteins. On the other side of the membrane flows a special **dialysis fluid**.

The composition of typical dialysis fluid is indicated in **Table 40**. As diffusion occurs across the membrane, blood composition changes. Potassium ions, phosphate ions, sulfate ions, urea, creatinine, and uric acid diffuse across the membrane into the dialysis fluid. Bicarbonate ions and glucose diffuse into the bloodstream. In effect, diffusion across the dialysis membrane replaces normal glomerular filtration, and the characteristics of the dialysis fluid (which can be modified for each patient) ensure that important metabolites remain in the bloodstream rather than diffusing across the membrane as metabolic wastes and excess fluid are removed.

For temporary kidney dialysis, a silicone rubber tube called a *shunt* is inserted into a medium-sized artery and vein (▶**Figure 88b**). (The typical location is the forearm, although the lower leg is sometimes used.) The shunt can be used like a tap in a wine barrel, to draw a blood sample or to connect the individual to a dialysis machine. For chronic dialysis, an artery and vein (usually in the forearm) are surgically connected to create an arteriovenous shunt.

While connected to the dialysis machine, the individual sits quietly as blood circulates from the shunt, through the machine, and back through the shunt. In the machine, the blood flows within a tube composed of dialysis membrane, and diffusion occurs between the blood and the surrounding dialysis fluid.

The use of a dialysis machine is suggested when a patient's *blood urea nitrogen (BUN)* exceeds 100 mg/dL (the normal value is up to 20 mg/dL) and the serum creatinine is 8 mg/dL or more (the normal value is up to 1.2 mg/dL). These results usually

Urinary System

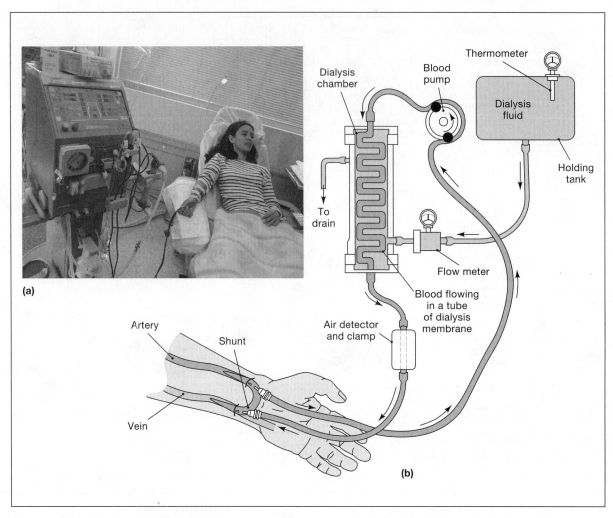

▶ **Figure 88 Hemodialysis. (a)** A patient is hooked up to a dialysis machine. **(b)** The path of blood during dialysis. Preparation for hemodialysis typically involves the surgical creation of a shunt that permits normal blood flow when the patient is not hooked up to the machine.

Table 40	The Composition of Dialysis Fluid	
Component	**Plasma**	**Dialysis Fluid**
Electrolytes (mEq/L)		
Potassium	4	3
Bicarbonate	27	36
Phosphate	3	0
Sulfate	0.5	0
Nutrients (mg/dL)		
Glucose	80–100	125
Nitrogenous wastes (mg/dL)		
Urea	20	0
Creatinine	1	0
Uric acid	3	0

Note: Only the significant variations are given; values for other electrolytes are usually similar. Although these values are representative, the precise composition can be tailored to meet specific clinical needs. For example, if plasma potassium levels are too low, the dialysis fluid concentration can be elevated to remedy the situation. Changes in the osmotic concentration of dialysis fluid can also be used to adjust an individual's blood volume, generally by adjusting the glucose content of the dialysis fluid.

reflect a creatinine clearance of 10 mL per minute or less and define **end stage renal disease (ESRD)**. Dialysis techniques can maintain patients who are awaiting a transplant, as well as those whose kidney function has been temporarily disrupted. Hemodialysis does have drawbacks, however: The patient must sit by the machine about 15 hours a week; between treatments, the signs and symptoms of uremia gradually appear; hypotension can develop as a result of fluid loss during dialysis; air bubbles in the tubing can cause an embolus to form in the bloodstream; there is more exposure to bloodborne infections such as hepatitis B and C; and the shunts can serve as sites of recurring infections. Because of the technical challenges, hemodialysis is usually done in dialysis centers by specially trained staff.

One alternative to the use of hemodialysis is **peritoneal dialysis**, in which the peritoneal lining is used as a dialysis membrane. Dialysis fluid is introduced into the peritoneum through a catheter in the abdominal wall, and the fluid is removed and replaced at intervals. One procedure, for example, involves cycling 2 liters of fluid in an hour—15 minutes for infusion, 30 minutes for exchange, and 15 minutes for fluid reclamation. This procedure may be performed in a hospital or at home. An interesting variation is **continuous ambulatory peritoneal dialysis (CAPD)**, in which patients self-administer 2 liters of dialysis fluid through the catheter and then continue normal activity until 4–6 hours later, when the fluid is removed and replaced with fresh dialysis fluid.

In 2009, 871,000 U.S. residents were under treatment for end stage renal disease. Probably the most satisfactory solution, in terms of overall quality of life, is *kidney transplantation*. This procedure involves the implantation of a new kidney obtained from a living donor or a cadaver. Of the 14,029 kidneys transplanted in 2013, 4715 were obtained from living (usually related) donors. In many cases, the failing kidneys are removed. An arterial graft is inserted to carry blood from the iliac artery or the aorta to the transplant, which is placed in the pelvic or lower abdominal cavity. In 2009, 398,861 U.S. patients were being treated by some form of dialysis, and 172,553 patients had a working transplanted kidney.

The success rate for kidney transplantation varies, depending on how well matched the recipient and donor are in MHL tissue antigens, which affects how aggressively the recipient's T cells attack the donated organ, whether infection develops, and whether the source is a living donor or cadaver. The one-year success rate for kidney transplantation is now 85–95 percent. The 10-year survival of a kidney transplant is now 45 percent. The use of kidneys taken from close relatives significantly improves the chances that the transplant will succeed for five years or more. Immunosuppressive drugs are administered to reduce tissue rejection, but unfortunately, this treatment also lowers the individual's resistance to infection.

Bladder Cancer

In the United States in 2013 there will be an estimated 72,500 new cases of **bladder cancer**, and 15,200 deaths. The incidence among males is three to four times that among females, and over 95 percent of patients are older than age 55. Environmental factors, especially tobacco use and exposure to *2-naphthylamine* or related compounds, are responsible for many and possibly most bladder cancers. For this reason, the bladder cancer rate is highest among cigarette smokers (with twice the risk of nonsmokers) and employees of chemical and rubber companies. The mechanism responsible appears to involve damage to tumor suppressor genes (such as *p53*) that regulate cell division. The prognosis is reasonably good for localized superficial cancers (five-year survival: 94 percent), but it is poor for persons with severe metastatic bladder cancer (five-year survival: 6 percent). Treatment of metastasized bladder cancer is very difficult, because the cancer spreads rapidly through adjacent lymphatic vessels and through the bone marrow of the pelvis.

Urinary Tract Infections

Urine is normally sterile (bacteria-free). **Urinary tract infections (UTIs)** result from the growth of bacterial or fungal invaders. The intestinal bacterium *Escherichia coli* is most commonly involved. Women are particularly susceptible to UTIs because the female urethra is relatively short and bacteria present around its external opening may reach the urethra and bladder where they can multiply and infect the mucosal lining. Sexual intercourse can push bacteria into the urethra resulting in a UTI. Most bacteria double in number every 20 minutes, so emptying the bladder after intercourse, and several times during the day as a routine, may reduce the number of infections.

If inflammation of the urethral lining occurs, the condition is termed **urethritis**; inflammation of the lining of the bladder is called **cystitis**. Many infections, including sexually transmitted diseases (STDs) such as gonorrhea, cause a combination of urethritis and cystitis. Both conditions cause painful urination—a symptom known as **dysuria** (dis-Ū-rē-uh)—and the urinary bladder becomes tender and sensitive to pressure. Despite the discomfort produced, the affected individual feels the urge to urinate frequently. Cystitis and urethritis generally respond to antibiotic therapies, although reinfections can occur. The condition may have no symptoms, but it can be detected by the presence of bacteria and blood cells in urine. After treatment, retesting urine for treatment failure may be indicated.

In untreated cases, the bacteria may proceed along the ureters to the renal pelvis. The resulting inflammation of the walls of the renal pelvis produces **pyelitis** (pī-e-LĪ-tis). If the bacteria invade the renal cortex and medulla as well, **pyelonephritis** (pī-e-lō-nef-RĪ-tis) results. Signs and symptoms of pyelonephritis include a high fever, intense pain on the affected side, vomiting, diarrhea, and the presence of blood cells and pus in the urine. More prolonged and intensive antibiotic therapy is required to treat this condition.

▶ Disorders of Fluid, Electrolyte, and Acid–Base Balance

Water and Weight Loss

The safest way to lose weight is to reduce the intake of food while ensuring that all dietary essentials are available in adequate quantities. Water must be included on the list of essentials, along with the amino acids, fatty acids, vitamins, and minerals. Because nearly half of our normal water intake comes from food, a person who eats less becomes more dependent on drinking fluids and on whatever water is generated metabolically.

At the start of a diet, the body conserves water and catabolizes lipids. That is why the first week of dieting may seem rather unproductive. Over that week, the level of circulating ketone bodies gradually increases. In the weeks that follow, the rate of water loss at the kidneys increases in order to excrete waste products, such as the hydrogen ions released by ketone bodies and the urea and ammonia generated during protein catabolism. The rate of fluid intake must also increase, or else the individual risks dehydration. While a person is dieting, dehydration is especially serious, because as water is lost, the concentration of solute in the extracellular fluid (ECF) climbs, further increasing the concentration of waste products and acids generated during the catabolism of energy reserves. This soon becomes a positive feedback loop: These waste products enter the filtrate at the kidneys, and their excretion accelerates urinary water losses.

Hypokalemia and Hyperkalemia

When the plasma concentration of potassium falls below 2 mEq/L, extensive muscular weakness develops, followed by eventual paralysis. We discussed this condition, called *hypokalemia* (*kalium*, potassium), in connection with ion effects on cardiac function. Causes of hypokalemia include the following:

- **Inadequate dietary K^+ intake.** If K^+ gains from the diet do not keep pace with the rate of K^+ loss in urine, K^+ concentrations in the ECF will drop.

Urinary System

- **The administration of diuretic drugs.** Several diuretics, including *Lasix*, can produce hypokalemia by increasing the volume of urine produced. Although the concentration of K$^+$ in urine is low, the greater the total volume, the more potassium is lost.

- **Excessive aldosterone secretion.** The condition of *aldosteronism*, characterized by excessive aldosterone secretion, results in hypokalemia, because the reabsorption of Na$^+$ is tied to the secretion of K$^+$.

- **An increase in the pH of the ECF.** When the H$^+$ concentration of the ECF declines, producing alkalosis, cells exchange intracellular H$^+$ for extracellular K$^+$. This ion swap helps stabilize the extracellular pH, but it gradually lowers the K$^+$ concentration of the ECF.

Treatment for hypokalemia generally includes increasing the dietary intake of potassium by salting food with potassium salts (KCl) or by taking potassium liquids or tablets, such as *Slow-K*. Severe cases are treated by cautious intravenous infusion of a solution containing K$^+$ at a concentration of 40–60 mEq/L.

High K$^+$ concentrations in the ECF produce an equally dangerous condition known as **hyperkalemia**. Severe cardiac arrhythmias appear when the K$^+$ concentration exceeds 8 mEq/L. Hyperkalemia results under the following circumstances:

- **Renal failure.** Kidney failure prevents normal K$^+$ secretion, thereby producing hyperkalemia.

- **The administration of diuretic drugs that block Na$^+$ reabsorption** (also called potassium-sparing diuretics). When sodium reabsorption slows, so does potassium secretion. Hyperkalemia can result.

- **A decline in the pH of the ECF.** When the pH of the ECF declines, producing acidosis, hydrogen ions move into the ICF in exchange for intracellular potassium ions. In addition, potassium secretion at the kidney tubules slows, because hydrogen ions are secreted instead of potassium ions. The combination of increased K$^+$ entry into the ECF and decreased K$^+$ secretion can produce a dangerous hyperkalemia very rapidly.

Treatment for hyperkalemia includes (1) the elevation of ECF volume with a solution low in K$^+$; (2) the stimulation of K$^+$ loss in urine by using appropriate diuretics, such as the loop diuretic *furosemide*; (3) the cautious administration of buffers (generally sodium bicarbonate) that can control the pH of the ECF; (4) restriction of dietary K$^+$ intake; and (5) the administration of enemas or laxatives containing compounds, that promote K$^+$ loss across the digestive lining (for example, *kayexalate*). In cases resulting from renal failure, kidney dialysis may also be required.

Diagnostic Classification of Acid–Base Disorders

The Anion Gap

Under normal circumstances, sodium (Na$^+$) ions are the primary cations in the ECF (potassium ion levels are much lower), and their positive charges are roughly balanced by the negative charges of the major anions (chloride and bicarbonate), minor anions (phosphate and sulfate), and plasma proteins. The concentrations of Na$^+$, Cl$^-$, and HCO$_3^-$ are relatively easy to determine. The concentration of Na$^+$ is greater than the concentration of Cl$^-$ plus that of HCO$_3^-$. The difference is called the anion gap:

$$\text{anion gap} = [\text{Na}^+] = ([\text{HCO}_3^-] + [\text{Cl}^-])$$

The anion gap in healthy individuals is 10–12 mEq/L. The calculation of the anion gap is useful in diagnosis, because it can be used to distinguish among different types of metabolic acidosis. The ECF of an individual in metabolic acidosis will have a low pH, and the P$_{CO_2}$ and concentration of HCO$_3^-$ will be reduced as the carbonic acid–bicarbonate system attempts to buffer the excess H$^+$.

If the anion gap is normal, the person's problem results either from the generation or ingestion of HCl or from the loss of bicarbonate:

1. HCl is a strong acid, and it dissociates completely into H$^+$ and Cl$^-$. The chloride ion gained is balanced by the loss of a bicarbonate ion, which buffers the hydrogen ion. The net result is that the anion gap remains unchanged, although the HCO$_3^-$ level drops.

2. In diarrhea, the body loses the HCO$_3^-$ contained in the buffers secreted into the intestinal tract. The movement of a bicarbonate ion across a cell membrane involves a countertransport mechanism that exchanges the bicarbonate ion for a chloride ion. Thus, for every bicarbonate ion secreted into the digestive tract and lost, a chloride ion is absorbed and retained. Again, the net result is that the anion gap remains unchanged, although the HCO$_3^-$ level drops.

If the individual in metabolic acidosis has an increased anion gap, the problem must reside either with the production or ingestion of organic acids or toxins or with renal failure:

1. Conditions such as *ketoacidosis* and *lactic acidosis* are caused by organic acids that release H$^+$ and various anions upon dissociation. (These anions are not considered in the calculation of the anion gap.) As a result, when the hydrogen ions are buffered by bicarbonate ions, HCO$_3^-$ levels decline. Because this decline is not accompanied by an elevation in Cl$^-$ concentrations, the anion gap increases.

2. In renal failure, acids that are normally excreted in the urine, such as sulfuric acid and phosphoric acid, are retained. The anions released by the dissociation of these acids are not considered in the calculation of the anion gap. Again, bicarbonate levels decline as they buffer the H$^+$ and the anion gap increases.

The Nomogram

When reviewing blood test results, clinicians may use a graphical representation of bicarbonate, carbon dioxide, pH, and P$_{CO_2}$ values. This graph, called a *nomogram*, is shown in ❱ **Figure 89**. A nomogram provides a visual summary of the information provided by blood gas analysis.

In the nomogram, the horizontal axis represents blood pH, and the vertical axis represents plasma HCO$_3^-$ concentration. The curving lines indicate the relationship between pH and bicarbonate levels at a specific value of P$_{CO_2}$. For example, at a P$_{CO_2}$ of 30 mm Hg (in the upper right-hand corner of the nomogram), the pH and bicarbonate values must lie somewhere along the indicated curve. The area at the center of the graph corresponds to the normal range of pH, bicarbonate levels, and P$_{CO_2}$.

Urinary System

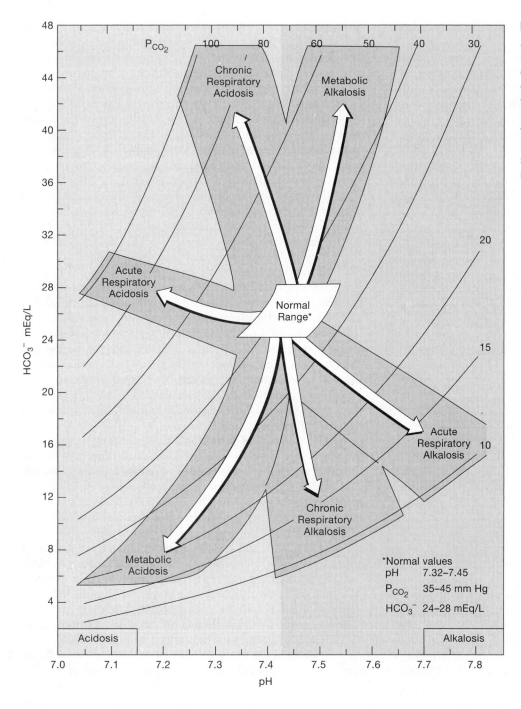

Figure 89 **An Acid–Base Nomogram.**
A nomogram is a graphical summary of the relationships among blood gases and pH. The central white box indicates the normal range of values for pH—the values in individuals with normal respiratory and acid–base balance. The black curving lines indicate the possible relationships between pH and HCO_3^- at a specific P_{CO_2} value. The most darkly shaded areas indicate the values observed in representative acid–base disorders.

With acid–base disorders, these values change. The new values, when plotted on the nomogram, will typically fall into one of the areas that are most darkly shaded in the diagram.

Fluid, Electrolyte, and Acid–Base Balance in Infants

Infants have very different requirements for the maintenance of fluid, electrolyte, and acid–base balance than adults have. A fetus obtains the water, organic nutrients, and electrolytes it needs from the maternal bloodstream. Buffers in the fetal bloodstream provide short-term pH control, and the maternal kidneys eliminate the H^+ generated. A newborn's body water content is high: At birth, water accounts for roughly 75 percent of body weight, compared with 50–60 percent in adults. Several factors contribute to this difference, including (1) the proportionately larger infant blood volume (10.0–12.5 percent of body weight, versus 6–7 percent in adults), (2) a proportionately larger volume of cerebrospinal fluid, (3) a proportionately larger interstitial fluid, and thus ECF, volume (due in part to existing for nine months under weightless conditions; the body water content also changes in orbiting astronauts), and (4) differences in the water content and in the proportions (in terms of body mass) of organs and tissues; for example, compared with adults, newborn infants have a proportionately larger heart and brain, which are 75–79 percent water, but less than half as much adipose tissue, which is 10 percent water.

During the first two to three days after delivery, roughly 6 percent of the infant's excess water is lost. Thereafter, the loss

Urinary System

is more gradual; the body water content typical for adults appears after age 2.

A newborn's distribution of body water is also different from that of an adult. In a newborn, the ECF accounts for roughly 35 percent of total body weight, versus 40 percent for the ICF. Because the ECF and the ICF in infants are similar in volume, the ICF is less effective than it is in adults at buffering changes in the ECF volume. As a result, less water must be lost from the ECF before the ICF volume is reduced enough to damage cells. As growth occurs, the ICF volume (as a percentage of the total body weight) remains relatively stable, whereas the ECF volume gradually decreases. The relative decline in the ECF volume becomes evident after a few months; the adult relationship between the ECF and ICF volumes (roughly 1:2) is reached after approximately two years.

Basic aspects of electrolyte balance are the same in newborns as in adults, but the effects of fluctuations in the diet are much more immediate in newborns because reserves of minerals and energy sources are much smaller. The problem is compounded by the fact that the metabolic rate (per unit body mass) of infants is twice that of adults. Thus, infants have an elevated demand for nutrients, and that demand must be met promptly. This is one reason that infants require frequent feedings; another is that infants have a much higher demand for water than adults have.

The elevated metabolic rate also means an accelerated production of waste products that must be eliminated at the kidneys. But because the kidneys of a newborn are unable to produce urine with an osmotic concentration above 450 mOsm/L, a newborn must produce more urine to eliminate metabolic waste products. Water losses at other sites are higher as well. Because the surface-to-volume ratio and the respiratory rate are relatively high, the rate of insensible perspiration is much higher in infants than in adults. To keep pace with rates of water loss in urine and insensible perspiration, newborns must consume, on a proportional basis, seven times as much water as adults do. If water intake is inadequate, waste products accumulate, and metabolic acidosis may develop. Although the kidneys become effective at concentrating urine after about one month, the elevated metabolic rate and water loss remain. As a result, infants continue to consume (and lose) roughly twice as much water as do adults per unit body weight. Infants are therefore at greater risk of dehydration; they can survive for only one to three days without water, whereas adults can survive for a week or more.

The increased metabolic rate of infants also means an accelerated demand for oxygen and more rapid generation of CO_2. The respiratory rate in newborns is relatively high—roughly 40 breaths per minute. And the breaths are deep: In proportion to their body weight, newborns must move twice as much air as adults do. As a result, the functional residual capacity of the lungs (the amount of air in the lungs after one quiet respiratory cycle) is about half that of adults. In adults, the functional residual capacity is large relative to the tidal volume (the amount of air moving into or out of the lungs in one respiratory cycle), and this proportion helps prevent rapid changes in the P_{O_2} and P_{CO_2} of alveolar air. In infants, as soon as the respiratory rate changes, the alveolar air composition changes. Because changes in the respiratory rate can cause sudden changes in the P_{O_2} and P_{CO_2} of arterial blood, newborns are at greater risk of developing respiratory acidosis or respiratory alkalosis.

Many newborns have some degree of respiratory acidosis caused by the interruption of placental circulation during labor. This condition is known as *fetal stress*. It is therefore important that newborns begin breathing as soon as possible after delivery so that the excess CO_2 can be eliminated. Many full-term newborns are given oxygen briefly to improve their blood oxygenation and to help them eliminate excess CO_2. (Prolonged oxygen administration to premature infants can have unwanted side effects.) The pH of infants in fetal stress rarely falls below 7.25; in most cases, no treatment besides clearing the airway and administering oxygen is required.

If the fetus struggles during labor or the delivery is prolonged, *fetal distress* can develop. This condition results from the combination of respiratory acidosis due to decreased placental circulation, and metabolic acidosis due to lactic acid production. The pH can fall as low as 7.0, with potentially fatal effects on the heart and central nervous system. Because fetal distress is one of the dangers of labor and birth, the fetus is commonly monitored during labor and delivery. Early in labor, the heart rate is checked; once the membranes are ruptured and the fetal scalp is accessible, blood samples can be taken from the scalp to monitor blood pH. If the fetal heart rate becomes too rapid and irregular early in labor, or if the arterial pH drops below 7.2, prompt delivery (often by cesarean section) may be recommended. Treatment of a newborn in fetal distress typically includes respiratory assistance, the administration of oxygen, and possibly the intravenous infusion of a sodium bicarbonate solution.

1. Jose, a 50-year-old office worker, is brought to the ER by his wife. Jose complains of shortness of breath on exertion over the last few months; it has become so severe that he now needs to rest after walking only a short distance. He recently almost fainted in the front yard after he checked the mailbox. Jose has been a heavy smoker for 25 years; he has a persistent cough with heavy mucus production. He has had several respiratory infections over the last few months.

   Jose is placed on oxygen. A chest x-ray and the following blood test results were obtained:

   RBC count: 6.2 million/mm^3

   Hct: 55

   Hb: 18 g/dL

   MCV: 98 mm^3

   MCH: 30 pg

   WBC count: 10,000/mm^3

   Platelets: 250,000/mm^3

   Arterial blood gases (obtained before oxygen administration):

   pH 7.3

   P_{CO_2} 48 mm Hg

   P_{O_2} 55 mm Hg

   HCO_3^- >30 mEq/L

   Chest x-ray: normal heart and expanded lungs with small areas of probable fibrotic changes

   What could cause Jose to suffer from shortness of breath? What are possible diagnoses?

2. Faith brings her 5-month-old infant to the pediatrician. The baby has a persistent deep cough with thick mucus, and the problem is only temporarily improved by antibiotic treatment. The cough started with a runny nose two months ago, but it has persisted and the infant has not gained weight over the interim. The infant has diarrhea with bulky, foul-smelling, oily-looking stools. On the basis of the following test results, what is a preliminary diagnosis? Fecal occult blood, stool pathogen culture, and parasite examinations are negative.

   Fecal fat test: 7 g/24 h (normal <7 g/24 h)

   Chest x-ray: bronchial wall thickening

   A Sweat chloride test is scheduled.

3. A 20-year-old woman lost consciousness while in a shopping mall. She had no identification with her. In the emergency room, the physician assessed her comatose state, noting among other physical findings, rapid respirations, and then ordered blood and urine studies. The physician obtained the following laboratory test results:

Blood Studies

   Complete blood count (CBC): within normal limits

   Serum electrolytes:

   Sodium: 135 mEq/L

   Potassium: 4.1 mEq/L

   Chloride: 100 mEq/L

   Blood glucose: 800 mg/dL

   Blood pH: 7.23

$PaCO_2$: 30 mm Hg
HCO_3^-: 12 mEq/L
Blood urea nitrogen (BUN): 9 mg/dL

Urinalysis

pH: 4.6
Glucose: (+)
Protein: (–)
Blood: (–)
Ketones: (+)

The physician ordered intravenous fluids containing another medication; the patient soon regained consciousness.

What was the diagnosis, and what was given intravenously? (NOTE: To answer these questions, you will need to check on the normal lab values in your text, in actual practice most lab results include the normal values alongside the patient's results to avoid errors. A clinician's expertise is A&P knowledge and understanding of patterns leading to logical diagnosis and treatment, not memorized tables or lists. Of course, knowing where to find and how to use accurate tables and lists is important.)

NOTES:

Answers to these problems can be found on pages 241–242.

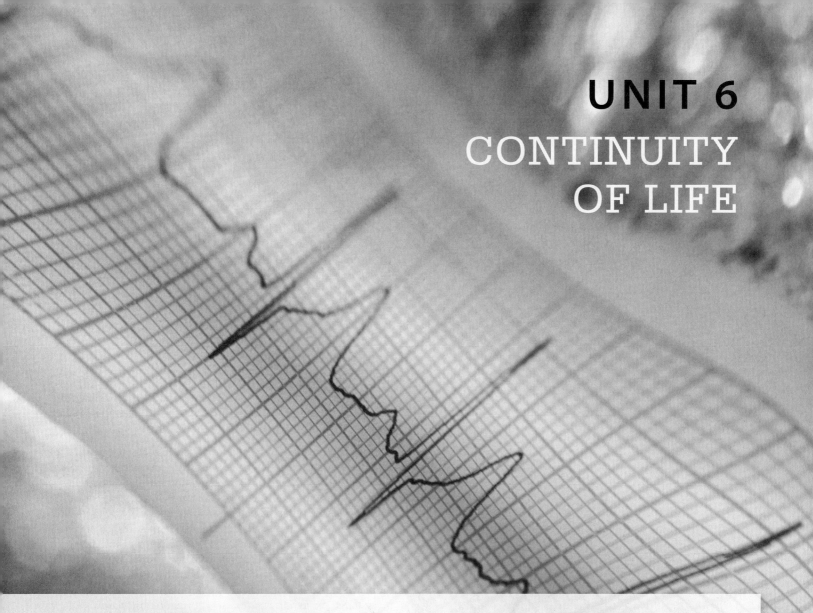

UNIT 6
CONTINUITY OF LIFE

An individual life span can be measured in decades, but the human species has survived for over a million years due to the activities of the reproductive system. Reproductive organs, called gonads, provide specialized reproductive cells, called gametes, and store them for variable lengths of time. Sperm form in the reproductive systems of males, and immature eggs (*oocytes*) in those of females. The gonads also secrete hormones that play a major role in the maintenance of normal sexual function and that affect a variety of other tissues, many not directly involved with reproduction. Finally, the male and female reproductive systems must ensure that complementary gametes meet under conditions that promote their successful interaction. Accomplishing this involves the integration of physiological, anatomical, and behavioral responses to appropriate stimuli.

Development begins with the union of sperm and oocyte and continues through maturity. In the first nine months, the period of prenatal (before birth) development, the fertilized oocyte forms a new and unique individual fetus composed of billions of cells. But developmental processes do not stop at birth, and postnatal development occurs as the individual progresses from infancy through childhood and adolescence and into maturity. These periods are characterized by growth and the appearance of increasingly complex structures and diverse functional capabilities. But the attainment of physical maturity does not mean an end to change, for the interactions between genetic programming and environmental factors gradually produce the characteristic signs of aging.

This unit considers the diagnosis and treatment of disorders affecting the reproductive system and factors that interfere with or complicate normal development. We will also consider some of the problems associated with aging.

MAJOR SECTIONS INCLUDED WITHIN THIS UNIT:

The Reproductive System and Development

END-OF-UNIT CLINICAL PROBLEMS

The Reproductive System and Development

▶ The Reproductive System and Its Disorders

The Physical Examination and the Reproductive System

The male reproductive system consists of the gonads (testes), a series of specialized ducts (the epididymis, ductus deferens, ejaculatory duct, and urethra), accessory glands (the seminal vesicles, prostate gland, and bulbourethral glands), and the external genitalia (penis and scrotum). The female reproductive system consists of the gonads (ovaries), derivatives of an embryonic system of ducts (the uterine tubes, uterus, and vagina), accessory glands (the greater and lesser vestibular glands), the external genitalia (the clitoris, labia majora, and labia minora), and secondary sexual organs (the mammary glands of the breasts).

Assessment of the Male Reproductive System

An assessment of the male reproductive system begins with a physical examination. Common signs and symptoms of male reproductive disorders include the following:

- **Testicular pain** may result from *testicular torsion* (twisting of the spermatic cord, which contains the vascular supply to the testis with resulting ischemia and intense pain). Significant pain may also result from various infections, including *gonorrhea* or other sexually transmitted diseases (STDs), and mumps, or the presence of an inguinal hernia. Testicular cancer and cryptorchidism are less likely to cause pain initially. Pain may be perceived as coming from the testicle while actually originating elsewhere in the reproductive tract, such as along the ductus deferens or within the prostate gland, or in other systems, as in *appendicitis* or a urinary obstruction (e.g., a ureteral kidney stone).

- **Urethral discharge** and associated pain on urination (*dysuria*) are commonly associated with STDs. These symptoms may also accompany disorders such as urinary tract infections, *epididymitis*, or *prostatitis* that may or may not be infectious or related to sexual contact.

- **Impotence**, or *erectile dysfunction* (*ED*), is the failure to achieve and maintain a sexually satisfactory erection. It can occur as a result of psychological factors, such as fear, anxiety, medications, or alcohol and other drugs. It can also develop secondarily to cardiovascular, hematological, hormonal, or nervous system problems that affect blood pressure or blood flow to the penile arteries.

- **Infertility**, which may be related to hormonal or other testicular disorders affecting sperm production, or a variety of anatomical problems along the reproductive tract.

Physical examination of the male reproductive system normally involves the examination of the external genitalia and palpation of the prostate gland. Inspection of the external genitalia entails the following observational steps:

1. The inspection of the penis and scrotum for skin lesions, such as vesicles, chancres, warts, and condylomas (wartlike growths). For example, painful small vesicles in clusters appear with the herpes

simplex virus infections. A chancre is a painless ulceration associated with early-stage *syphilis* (p. 222). In the examination of uncircumcised males, the foreskin is retracted to observe the lining of the prepuce. **Phimosis**, an inability to retract the foreskin in an adult uncircumcised male, generally indicates an inflammation of the prepuce and adjacent tissues.

2. The palpation of each testis, epididymis, and ductus deferens to detect the presence of abnormal masses, swelling, or tumors. Possible abnormal findings include the following:

- **Scrotal swelling** and distortion of the scrotal cavity due to the formation of *varicose veins* (p. 139) within the *pampiniform plexus* of testicular veins (a condition known as a *varicocele*), or the local accumulation of lymph (a *chylocele*), or serous fluid (a *hydrocele*).

- **Testicular swelling** due to an enlargement of the testis or the presence of a nodular mass. **Orchitis** is a general term for inflammation of the testis. Inflammation can result from an infection, such as syphilis, mumps, or tuberculosis. Testicular swelling may also accompany testicular cancer or testicular torsion.

- **Epididymal swelling** due to cyst formation (*spermatocele*), tumor formation, or infection. **Epididymitis** is an acute inflammation of the epididymis that may indicate an infection of the reproductive or urinary tract. The condition may also develop from irritation caused by the backflow, or *reflux*, of urine into the ductus deferens.

- **Swelling of the spermatic cord** may indicate (1) an inflammation of the ductus deferens (*deferentitis*), (2) an accumulation of serous fluid in a pocket of serous membrane (a *hydrocele*), (3) bleeding within the spermatic cord, or (4) testicular torsion.

3. A **digital rectal examination (DRE)** can screen for prostate enlargement, prostatitis, tumors (including prostate cancer), or an inflammation of the seminal vesicles. In this procedure, a gloved, lubricated finger is inserted into the rectum and pressed against the anterior rectal wall to palpate the posterior walls of the prostate gland and seminal vesicles.

If urethral discharge is present or if discharge occurs in the course of any of these procedures, the fluid can be cultured to check for the presence of pathogens. **Table 41** summarizes information about pathogens that are responsible for infections of the reproductive system. Other potentially useful diagnostic procedures and laboratory tests (for both males and females) are given in **Table 42**.

Assessment of the Female Reproductive System

Important signs and symptoms of female reproductive disorders include the following:

- **Acute pelvic pain**, a symptom that may accompany disorders such as pelvic inflammatory disease (PID), ruptured tubal

Table 41	Examples of Infectious Diseases of the Reproductive System	
Disease	**Organism(s)**	**Description**
Bacterial diseases		
Bacterial vaginitis	Varied, but often *Gardnerella vaginalis*	Vaginitis not associated with STDs, caused by imbalance of resident bacteria; symptoms include watery discharge, odor
Chlamydia	*Chlamydia trachomatis*	Chlamydial infections cause PID, nongonococcal urethritis
Gonorrhea	*Neisseria gonorrhoeae*	Infection of the male and female reproductive tracts; majority of females show no symptoms, but others may develop PID; majority of males develop painful urination (dysuria) and produce a viscous urethral discharge
Pelvic inflammatory disease (PID)	*Neisseria gonorrhoeae* *Chlamydia trachomatis*	An infection of the uterine tubes (salpingitis); symptoms include fever, abdominal pain, and elevated WBC counts; can cause peritonitis in severe cases; infertility can result from formation of scar tissue in the uterine tubes
Syphilis	*Treponema pallidum*	STD with a long period of chronic illness; symptoms of primary syphilis include chancres and enlarged lymph nodes; symptoms of secondary syphilis involve a reddish skin rash, fever, and headaches; tertiary syphilis affects CNS and cardiovascular system. Infection during pregnancy may cause severe fetal infection and malformations
Viral diseases		
Genital herpes	Herpes simplex viruses (HSV-1 and HSV-2)	Most cases caused by HSV-2; ulcers develop on external genitalia, heal, and recur
Genital warts	Human papillomavirus	Warts appear on external genitalia, perineum, and anus, of both sexes and on vagina and cervix of females; associated with cervical cancer. Many cases preventable with HPV vaccine Gardasil prior to infection
Fungal diseases		
Candidiasis	*Candida albicans*	Yeast infection that causes vaginitis; symptoms include itching, burning, and lumpy white discharge. Also called moniliasis; not associated with STDs
Parasitic diseases		
Trichomoniasis	*Trichomonas vaginalis*	Flagellated protozoan parasite of both male and female urinary and reproductive tracts; infection produces white or greenish-gray discharge in both sexes and intense vaginal itching in females; may be asymptomatic

Reproductive System

Table 42	Examples of Tests Used in the Diagnosis of Reproductive Disorders	
Diagnostic Procedure	**Method and Result**	**Representative Uses**
FEMALES		
Mammography	X-ray films are taken of breasts	Detects cysts or tumors of breast; effective in detecting early breast cancer
Laparoscopy	Fiber-optic tubing is inserted through incision in abdominal wall to view pelvic organs, remove tissue for biopsy, or perform surgical procedures	Detects pelvic organ abnormalities such as cysts or adhesions; determines cause of pelvic pain; enables diagnosis of pelvic inflammatory disease and endometriosis
Papanicolaou (Pap) smear	Cells from cervix are removed for cytological analysis	Screens for cervical cancer; reports signs of intraepithelial abnormalities of squamous or glandular cells
Colposcopy	Special instrument is used to view cervical tissue microscopically *in situ* and to guide removal of tissue for biopsy	Detects areas of dysplasia and malignancies of cervix; follow-up to abnormal Pap smear
Cervical biopsy	Tissue is removed from cervix for examination	Detects dysplasia and malignancy
Pelvic or transvaginal sonography	An ultrasonic probe is used suprapubically or inserted into the vagina	Obtains high-definition echograms of the ovaries, uterus, endometrium; detects tumors, cysts
MALES		
Transrectal ultrasonography	Ultrasound transducer is inserted rectally and scan is performed	Detects prostatic tumor and nodules or abnormalities of seminal vesicles and surrounding structures; used to guide biopsy of nodules
Laboratory Test	**Normal Values in Blood Plasma or Serum**	**Significance of Abnormal Values**
FEMALES		
Estrogen (serum)	Early uterine cycle: 60 to 400 pg/mL Middle: 100–600 pg/mL Late: 150–350 pg/mL Postmenopausal: <59 pg/mL	Detects hypofunctioning ovaries and helps determine timing of ovulation. Note the large range of normal values, due to variations in the levels of several different estrogens
Estradiol (serum)	Follicular phase: 20–150 pg/mL Ovulation: 100–500 pg/mL Luteal phase: 60–260 pg/mL	Decreased levels occur in ovarian dysfunction and in amenorrhea. Estradiol is the primary estrogen produced by ovaries
FSH (serum)	Before and after ovulation: 4–20 mIU/mL Midcycle: 10–30 mIU/mL Luteal phase: 1–12 mIU/mL Postmenopausal: 25–150 mIU/L	Helps determine the cause of infertility and menstrual dysfunction; increased levels occur in absence of estrogens (as during menopause); decreased levels occur with anorexia nervosa or hypopituitarism
LH (serum)	Follicular phase: 3–15 mIU/mL Midcycle: 22–105 mIU/mL	Determines timing of ovulation; LH is increased in ovarian hypofunction and polycystic ovary syndrome
Progesterone (serum)	Before ovulation: <1.0 ng/mL Midluteal: 3–20 ng/mL	Levels are increased after ovulation and in early pregnancy
Prolactin (serum)	Nonlactating females: 0–23 ng/mL	Values >100 ng/mL in a nonlactating female may indicate pituitary tumor
MALES		
Semen analysis		
Volume Sperm count Motility Sperm morphology	2–5.0 mL 20–150 million/mL 60–80% are motile 70–90% normal structure	Decreased sperm count causes infertility; infertility could also result if >40% of sperm are immotile or >30% of sperm are abnormal
Testosterone (serum)	Adult male: 270–1070 ng/dL Adult female: 10–75 ng/dL	Decreased level could indicate testicular disorder, or pituitary hypofunction; increased levels may cause virilization in females
Prostatic Acid Phosphatase (PAP) (serum)	Adults: 0.0–0.8 U/L at 37°C	Present in seminal fluid; increased serum levels occur with carcinoma of prostate gland
Prostate-specific antigen (PSA)	<4 ng/mL	Increased levels occur with prostate cancer, benign prostatic hypertrophy, and increasing age. Tables of age-specific normal ranges and the percentage of free PSA are used to interpret the results

(continued)

Reproductive System

Table 42	Examples of Tests Used in the Diagnosis of Reproductive Disorders (*Continued*)	
Laboratory Test	**Normal Values in Blood Plasma or Serum**	**Significance of Abnormal Values**
FEMALES AND MALES		
Serologic test for syphilis	Negative	Presence of antibodies indicates past or present infection with syphilis
Gonorrhea culture	Negative	Positive test indicates gonorrheal infection
Herpes simplex virus culture	Negative	Positive test indicates active Herpes simplex infection
***Chlamydia* culture**	Negative	Positive culture indicates Chlamydia infection

pregnancy, a ruptured ovarian cyst, or inflammation of the uterine tubes (*salpingitis*).

- **Bleeding outside normal menses**, which can result from tumors, hormonal fluctuation (including oral contraceptive use), PID, or endometriosis.

- **Amenorrhea (lack of menstruation)**, which may occur in women with anorexia nervosa, women who overexercise and are underweight, extremely obese women, postmenopausal women, and stressed or pregnant women.

- **Abnormal vaginal discharge**, which may be the result of a bacterial, fungal, or protozoan infection, including some sexually transmitted diseases (STDs).

- **Dysuria**, which may accompany an infection of the reproductive system due to the migration of the pathogen to the adjacent urethral entrance, even though the female reproductive and urinary tracts are distinct.

- **Infertility**, which may be related to hormonal disturbances, a variety of ovarian disorders, or anatomical problems along the reproductive tract.

A physical examination generally includes the following steps:

- The inspection of the external genitalia for skin lesions or anatomical abnormalities. Swelling of the labia majora can result from (a) regional cellulitis with lymphedema, (b) a *labioinguinal hernia* (rare), (c) bleeding within the labia as the result of local trauma, or (d) a *bartholin gland infection*, where an abscess develops in one of the greater vestibular glands (*Bartholin's glands*).

- The inspection or palpation of the perineum, vaginal opening, labia, clitoris, urethral meatus, and vestibule to detect lesions, abnormal masses, or discharge from the vagina or urethra. Samples of any discharge present can be tested to detect and identify any pathogens involved.

- The inspection of the vagina and cervix by using a speculum—an instrument that retracts the vaginal walls to permit direct visual inspection. The amount and appearance of vaginal discharge, and changes in the color of the vaginal walls may be important diagnostic clues. For example:

  - Cyanosis of the vaginal and cervical mucosa normally occurs during pregnancy, but it may also occur when a pelvic tumor exists or in persons with congestive heart failure.

  - Reddening of the vaginal walls occurs in vaginitis, caused by bacterial infections such as gonorrhea, protozoan infection by *Trichomonas vaginalis*, and yeast infections. It can also appear postmenopausally in some women (a condition known as *atrophic vaginitis*).

- The cervix is inspected to detect ulceration, polyps, or cervical discharge. A spatula and brush are then used to collect cells from the cervical os and to transfer them to a glass slide or preservative solution. After the sample is fixed with a chemical spray or dip, or is centrifuged, cytological examination is performed. This technique is the best-known example of a *Papanicolaou (Pap) test* (**Table 2**, p. 9), and the sampling process is commonly called a *Pap smear*. The test screens for the presence of cervical cancer. Ratings of Pap smear results are given in **Table 42**. Swabs of the cervix allow testing for infectious agents such as gonorrhea, *Chlamydia*, Herpes simplex, and human papillomavirus.

- A bimanual examination *for the palpation of the uterus, uterine tubes, and ovaries*. The physician inserts two fingers vaginally and places the other hand against the lower abdomen to palpate the uterus and surrounding structures. The contour, shape, size, and location of the uterus can be determined, and any swellings or masses will be apparent. Abnormalities in other reproductive organs, such as ovarian masses, endometrial growths, or tubal masses, can also be detected in this way.

Normal and Abnormal Signs Associated with Pregnancy

Pregnancy imposes many stresses on maternal body systems. Several clinical signs may be apparent in the course of a physical examination, including the following: **Chadwick's sign** is a normal cyanosis of the vaginal wall and is seen with the increase in pelvic blood flow of pregnancy.

- The size of the uterus increases dramatically during pregnancy; at full term, the uterus extends almost to the level of the xiphoid process.

- Significant uterine bleeding, causing vaginal discharge of blood, most commonly occurs in *placenta previa*, in which the placenta forms near the cervix. Subsequent cervical stretching leads to tearing and bleeding of the vascular channels of the placenta. Vaginal bleeding may also occur prior to miscarriage, or if the placenta suddenly separates from the uterine wall (*abruptio placentae*, p. 227). In all these cases there is significant risk of maternal and/or fetal death.

- Nausea and vomiting tend to occur in pregnancy, especially during the first three months.

- Edema of the limbs, especially the legs, is common, because the increased total blood volume and the weight of the uterus compress the inferior vena cava and its tributaries. As venous pressures rise in the lower limbs and inferior trunk, varicose veins (p. 139) and hemorrhoids (p. 187) may develop.

Reproductive System

- Back pain due to increased stress on muscles of the lower back is common. These muscles are strained as the weight of the uterus accentuates the lumbar curvature.

- A weight gain of 10 to 12.5 kg (22 to 27.5 lb) is now considered desirable, although 20 years ago weight increases of 20 to 25 kg (44 to 55 lb) were considered acceptable, and 45 years ago, before the linkage between prematurity with low maternal weight gain was recognized, weight gains of 7 to 10 lb were sometimes considered desirable. Except for women who start pregnancy very overweight (who with careful monitoring may benefit from little or no weight gain), failure to gain adequate weight during pregnancy can indicate or lead to serious neonatal problems.

- The combination of obesity, weight gain, estrogen, progesterone, prolactin, human placental lactogen (hPL), and other hormones that are elevated during pregnancy appears to promote the development of *insulin resistance*, a decrease in target cell sensitivity to insulin. As a result, diabetic women who become pregnant are at an increased risk of ketoacidosis and other diabetic complications. During diabetic pregnancies, glucose levels must be monitored and stabilized to prevent maternal problems as well as the increase in fetal deaths and developmental defects that can occur. *Gestational diabetes* develops in 3 to 8 percent of pregnancies in previously nondiabetic women in the United States. Women who are obese, have a family history of diabetes, or have higher than normal blood or urine glucose levels are at higher risk for gestational diabetes. For most women, the gestational diabetes resolves when pregnancy ends, but an increased risk for type 2 diabetes later in life persists.

- In some cases, a dangerous combination of hypertension, proteinuria, edema, and seizures occurs. We will consider this condition, called *eclampsia*, in a later section (p. 227).

Disorders of the Reproductive System

Representative disorders of the reproductive system are diagrammed in ▶ Figure 90.

Testicular Cancer

Testicular cancer occurs at a relatively low rate: about 3 cases per 100,000 males per year. Although only about 7900 new cases are reported each year in the United States, with less than 400 deaths, testicular cancer is the most common cancer among males ages 15 to 35. The incidence among Caucasian American males has more than doubled since the 1930s, but the incidence among African American males has remained unchanged. The reason for this difference is not known.

More than 95 percent of testicular cancers result from abnormal spermatogonia or spermatocytes, rather than abnormal sustentacular cells, interstitial cells, or other testicular cells. Treatment generally consists of a combination of orchiectomy and chemotherapy. The survival rate increased from about 10 percent in 1970 to about 95 percent in 1999, primarily as a result of earlier diagnosis and improved treatment protocols.

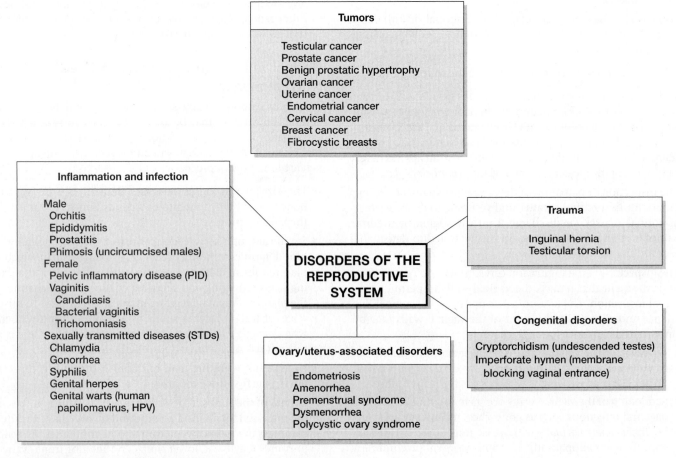

▶ Figure 90 Representative Disorders of the Reproductive System.

Ovarian Cancer

A woman in the United States has a lifetime risk of 1 chance in 70 of developing ovarian cancer. In 2010 an estimated 22,000 ovarian cancers will be diagnosed, and about 14,000 women will die from this condition. Ovarian cancer is the third most common reproductive cancer among women, after breast and uterine cancer. It is also the most dangerous, because it is seldom diagnosed in its early stages. The prognosis is relatively good for the rare cancers that originate in the general ovarian tissues or from abnormal oocytes. These cancers respond well to some combination of chemotherapy, radiation, and surgery. However, 85 percent of ovarian cancers develop from epithelial cells, from the epithelium that covers the ovary and sustained remission can be obtained in only about one-third of these cases. Early diagnosis greatly improves the chances of successful treatment, but as yet there is no standardized, effective screening procedure. (*Transvaginal sonography* can detect ovarian cancer at Stage I or Stage II, but there is a high incidence of false positive results. The blood test *CA 125* is elevated in many cases of ovarian cancer, but more benign conditions may also produce elevated results.)

A family history of ovarian cancer significantly increases the risk of the disease. Having a pathogenic mutation in the *BRCA1* or *BRCA2* genes markedly increases a woman's risk for ovarian and breast cancers, and increases a man's risk for prostate, breast, and pancreatic cancers. Perhaps 1 in 400 persons in the general population have one of these mutations, but 2 percent of breast and 10 percent of ovarian cancer patients have the abnormalities. With a positive family history and a *BRCA1* or *2* mutation, individual lifetime risk for breast cancer may be above 70 percent, and 50 percent for ovarian cancer. Prophylactic oophorectomy and/or bilateral mastectomy in high-risk patients can significantly reduce this risk.

The treatment required at Stage I may involve the unilateral removal of an ovary and uterine tube (a *salpingo-oophorectomy*) or a *bilateral salpingo-oophorectomy* (BSO) and *total hysterectomy* (removal of the uterus). The treatment of the more advanced stages or more dangerous forms of early-stage ovarian cancer includes radiation and chemotherapy, in addition to more extensive surgery.

Treatment of Stage III or Stage IV ovarian cancer, in which the cancer has spread into the peritoneal cavity, commonly involves the removal of the omentum, in addition to a BSO, a total hysterectomy, and aggressive chemotherapy. Some chemotherapy agents are introduced into the peritoneal cavity, where higher concentrations can be administered without the systemic effects that would accompany the infusion of these drugs into the bloodstream. This procedure is called *intraperitoneal therapy*.

Pelvic Inflammatory Disease

Pelvic inflammatory disease (PID) is the term used for infection of the uterus, uterine tubes, and ovaries. It starts by the spread of bacteria in the vagina through the cervix into the uterus and uterine tubes. Signs and symptoms of PID include fever, lower abdominal pain, and elevated white blood cell counts. In severe cases, the infection can spread to other visceral organs or produce a generalized peritonitis. Sexually active women ages 15 to 24 have the highest incidence of PID. Most cases are thought to be caused by sexually transmitted diseases (STDs), most often gonorrhea or chlamydia. The use of an oral contraceptive or condom decreases the risk of infection, whereas a previous episode of PID and the practice of douching increase the risk.

In 2011 there were an estimated 100,000 doctor visits for PID and over 50,000 hospitalizations.

Treatment with antibiotics can stop the infection, but uterine tube scarring can still occur, often following a *Chlamydia* infection that had few, if any symptoms and may not have been diagnosed and treated. Chronic abdominal pain may also persist. In 2011 over 1,400,000 cases of chlamydia were reported, out of an estimated 2,800,000 total cases. The damage and scarring of the uterine tubes can cause up to 75,000 cases per year of female infertility in the United States by preventing the passage of sperm to the egg or of a zygote to the uterus. Uterine tube scarring is also the leading cause of ectopic pregnancies. In **ectopic pregnancies** the zygote implants in the uterine tube. As the embryo grows, it can erode the wall of the uterine tube and cause massive bleeding requiring emergency surgery.

Uterine Tumors and Cancers

Uterine tumors are common tumors in women. It has been estimated that 40 percent of women over age 50 have benign uterine tumors involving smooth muscle and connective-tissue cells. When they are small, these *leiomyomas* (lē-ō-mi-Ō-maz), or *fibroids*, generally cause no problems. Stimulated by estrogens, however, they can grow quite large, reaching weights as great as 13.6 kg (30 lb). Menstrual abnormalities, occlusion of the uterine tubes, distortion of adjacent organs, and compression of blood vessels may then lead to complications. In symptomatic young women, observation or conservative treatment with drugs or restricted surgery may be utilized to preserve fertility. In older women, a decision may be made to remove the uterus.

Benign epithelial tumors in the uterine lining are called *endometrial polyps*. Roughly 10 percent of women probably have polyps, but because the polyps tend to be small and cause no symptoms, the condition passes unnoticed. If bleeding occurs, if the polyps become excessively enlarged, or if they protrude through the cervical os, they can be removed.

Uterine cancers are less common. In 2013 in the United States over 51,000 new cases will be reported, and 12,000 of these will result in death. There are two types of uterine cancers: (1) *endometrial* and (2) *cervical*.

Endometrial cancer is an invasive cancer of the endometrium. The condition most commonly affects women ages 50 to 70. Estrogen therapy, used to treat menopausal symptoms in recently postmenopausal women, increases the risk of endometrial cancer by 2 to 10 times. Adding progesterone therapy to the estrogen therapy seems to reduce this risk, but may also increase the risk of thrombotic events such as heart attack, pulmonary embolus, and stroke.

There is no perfect screening test for endometrial cancer, although measuring endometrial thickness by ultrasound, and endometrial biopsy may be used. The most common symptom is irregular postmenopausal bleeding, and diagnosis typically involves the examination of tissue from a biopsy of the endometrium obtained by suction or scraping. The prognosis varies with the degree of spread. Fortunately, endometrial cancer is usually diagnosed at an early stage. The one-year survival rate is 92 percent, and if the cancer has not spread, the five-year survival rate is 95 percent. The treatment of early-stage localized endometrial cancer involves a hysterectomy, and is usually curative, perhaps followed by localized radiation therapy. In advanced stages, more aggressive radiation

Reproductive System

treatment is recommended. Chemotherapy has *not* proved to be very successful in treating endometrial cancers; only 30 to 40 percent of patients benefit from this approach.

Cervical Cancer and Human Papillomaviruses (HPV)

There are over 100 types of human papillomaviruses. Most HPV types (70 percent) cause the common warts that occur on the skin of the hands, feet, arms, and legs and are spread by casual contact. Most infections take weeks to appear as visible growths and at least 90 percent of infections are suppressed or cleared by the immune system within two years. There are about 40 types that affect the genitals and anal and mouth mucosa, and these are usually transmitted by sexual contact. Of these, 15 types are associated with cancer of the cervix, the lower entrance part of the uterus that projects into the vagina. HPV types 6 and 11 cause about 90 percent of visible external genital skin warts. HPV types 16 and 18 cause 70 percent of cervical cancers. However, only persistent infection that lasts for many years progresses to cancer (10 to 15 years in the case of cervical cancer). Usually even these high-risk HPV infections are cleared by the immune system.

This is fortunate, as the CDC says HPV is the most common STD in the United States, eventually infecting 50 percent of sexually active persons. An estimated 79 million Americans are currently infected with HPV, and there are 14 million new infections every year. Infection does not mean visible genital warts will occur. Although most people clear the infection within two years, some infections progress to cancer. HPV-associated cancer occurs most commonly in the cervix, but can occur in the genital skin, anus, mouth, and throat of both sexes. In 2013 an estimated 12,000 HPV associated cancer cases with 4000 deaths will occur in the United States.

Vaccination against four HPV types (6, 11, 16, and 18) was started in the United States in 2006. Because of unscientific scare-mongering politics, only 32 percent of eligible girls aged 13–17 were vaccinated in that year. However, even that relatively low immunization rate made a difference; there was a 59 percent reduction in vaccine-preventable HPV infection in 14- to 19-year-old girls between 2007 and 2010. Australia was able to vaccinate a higher percentage of its young girls, and this has resulted in even more impressive reductions in HPV infections. Infections among Australian men dropped as well, and vaccination for both sexes is now recommended to increase "herd immunity." Since other strains of HPV may cause cervical cancer, and currently infected persons may develop cancer years, even decades after infection, screening by pap smears is still recommended for women over 30. (This test detects persistent HPV infection.)

Most women with cervical cancer develop no symptoms until late in the disease. At that stage, vaginal bleeding (especially after intercourse), pelvic pain, and vaginal discharge may appear. Early detection has reduced the mortality rate 70 percent since 1950 when the Pap/cervical smear was first introduced. This standard screening test is named for its developer, Dr. George Papanicolaou, a Greek-born, naturalized-American anatomist and cytologist. The cervical epithelium normally sheds its superficial cells, and a sample of cells scraped or brushed from the epithelial surface can be examined for abnormal or cancerous cells. Screening is recommended by one to three yearly Pap smear tests usually starting at age 21 until age 65. About 1,250,000 precancerous cervical lesions are detected yearly in the United States. Early treatment of

abnormal, but not cancerous, lesions detected by mildly abnormal Pap smears may prevent the progression to cancer. The treatment of localized, noninvasive cervical cancer involves the removal of the affected portion of the cervix. The treatment of more advanced cancers typically involves a combination of radiation therapy, hysterectomy, removal of lymph nodes, and chemotherapy. With HPV vaccination, many cases of cervical cancer will probably be prevented.

Endometriosis

Some conditions interfere with the normal menstrual cycle. In **endometriosis** (en-dō-mē-trē-Ō-sis), an area of endometrial tissue begins to grow outside the uterus, usually within the peritoneal cavity. The cause is unknown. Because this condition is most common in the inferior portion of the peritoneum, one possibility is that pieces of endometrium sloughed off during menstruation in some way travel through the uterine tubes into the peritoneal cavity, where they reattach. The severity of the condition depends on the size and location of the abnormal mass. Cyclic or persistent abdominal pain, bleeding, pressure on adjacent structures, and infertility in up to 40 percent of patients are common problems. As the island of endometrial tissue enlarges, the condition becomes more severe.

Endometriosis can generally be diagnosed by inserting a laparoscope through a small opening in the abdominal wall. Using this device, a physician can inspect the outer surfaces of the uterus and uterine tubes, the ovaries, and the lining of the pelvic cavity. Treatment of endometriosis may involve hormonal therapy to suppress uterine cycles, or surgical removal of the endometrial mass. Most symptoms improve after menopause.

Polycystic Ovary Syndrome (PCOS)

The adrenal glands in both sexes produce several hormones, including androgens. These hormones are chemical precursors to testosterone, which males produce mainly in the testes in large amounts. Females also produce testosterone; the ovaries normally produce some androgens and testosterone but much more estrogen. In **polycystic ovary syndrome (PCOS)** for uncertain, but partially inherited reasons, there is an excess of androgens. This hormone imbalance may also include increased insulin levels and is associated with weight gain, hirsutism, irregular menstruation, and decreased fertility. The oocytes may not mature in their follicles, and ovulation does not regularly occur, leaving small cysts to accumulate in the ovaries, which become enlarged and enclosed in a thickened capsule. Continued unsuccessful ovulation may result in infertility.

Weight loss is difficult to achieve but seems to reduce the hormonal imbalances, as does the diabetes drug metformin. Symptomatic treatment with cyclic progesterone or combination estrogen/progesterone contraceptive pills control menstrual irregularities, while mild anti-androgens and various cosmetic procedures can control hirsutism. The ovulation-inducing medication clomiphene, or other fertility drugs including gonadotropin-releasing hormone (GnRH), may induce ovulation. Surgically incising or resecting part of the ovary may temporarily restore ovulation. When pregnancy is achieved, it is usually normal.

Vaginitis

There are several forms of vaginitis, and mild cases are relatively common. **Candidiasis** (kan-di-DĪ-a-sis) results from a fungal

Reproductive System

(yeast) infection. The organism responsible appears to be a normal component of the vaginal environment in 30–80 percent of healthy women. Antibiotics, immunosuppression, pregnancy, and other factors that change the local environment can stimulate the growth of the fungus. Symptoms include itching and burning, and a lumpy white discharge may also be produced. Antifungal medications taken orally or applied topically are used to treat this condition. It may increase the risk of premature birth.

Bacterial (nonspecific) vaginitis (*BV*) results from the combined action of several bacteria. The bacteria involved are normally present in about 30 percent of adult women, but in BV the relative numbers change and this imbalance leads to more discharge and odor. In this form of vaginitis, the vaginal discharge contains epithelial cells and large numbers of bacteria. The discharge has a homogeneous, watery texture and a characteristic odor sometimes described as fishy or aminelike. Topical or oral antibiotics are effective in controlling this condition.

Trichomoniasis (trik-ō-mō-NĪ-a-sis) involves infection by the parasite *Trichomonas vaginalis*, and is spread by sexual contact with a carrier. Because trichomoniasis is a sexually transmitted disease, both partners must be treated to prevent reinfection. A foamy, green discharge that is intensely itchy and watery is characteristic. Both men and women can be asymptomatic carriers.

PMS

In many women, hormonal fluctuations associated with the reproductive cycle produce a variety of unpleasant effects. Several physical and physiological changes may occur 7 to 14 days before the start of menses. Fluid retention, breast enlargement, headaches, pelvic pain, and bloating are common. These physical changes may be associated with psychological changes, including irritability, anxiety, and depression. This combination of symptoms has been called **premenstrual syndrome (PMS)**.

The mechanism responsible for PMS has yet to be identified. Changes in sex hormone levels may be involved, either directly (by acting on peripheral organ systems) or indirectly (by modifying the release of neurotransmitters in the CNS). There are currently no laboratory tests or procedures for diagnosing PMS, but tracking the appearance of symptoms over a two- to three-month period can reveal characteristic patterns. Treatment may involve exercise, dietary change, and/or medication, depending on the nature of the primary symptom. For example, if a headache is the major problem, analgesics are prescribed, whereas diuretics may be used to combat bloating and fluid retention. For severe PMS, drugs that block GnRH secretion and stop uterine cycles for six months or more can be administered. This suppresses progesterone levels. During the interim, estrogens can be administered to prevent symptoms of premature menopause. Paradoxically, some women feel better with supplemental progesterone premenstrually. The antidepressant *Prozac* helps in some cases; underlying depression is common in PMS sufferers, but whether it is a cause or an effect is not known.

Birth Control/Contraception Strategies

For physiological, logistical, financial, or emotional reasons, most sexually active adults worldwide practice some form of conception control during their reproductive years. Between 1990 and 2005, increased use of contraceptives decreased total fertility and lowered pregnancy-related maternal deaths by 15 percent. In the United States, 99.9 percent of sexually experienced women have used contraception at some point in their lives. Two out of three U.S. women ages 15–44 currently employ some method of contraception. When the simplest and most obvious method, sexual abstinence, is unsatisfactory, another method or combination of methods of contraception must be used to avoid unintended pregnancies. Many methods are available, so the selection process can be quite involved. Because each method has specific strengths and weaknesses, the potential risks and benefits must be carefully analyzed on an individual basis. We will consider only a few of the available contraception methods here. See ❯**Figure 91** for comparative data on contraceptive methods and effectiveness. A mixture of social and technical factors hinder the consistent and correct use required for successful use of contraceptives. Fifty percent of pregnancies in the United States are unintended. A variety of contraceptive methods are available. Effective **hormonal contraceptives** manipulate the female hormonal cycle so that ovulation does not occur. The contraceptive pills produced in the 1950s used combined estrogen and progestins sufficient to suppress pituitary production of GnRH, so FSH was not released and ovulation did not occur. Most of the oral contraceptives developed subsequently contain much smaller amounts of estrogens, or only progesterone. Current *combination hormone contraceptives* are administered in a cyclic fashion, using medication for three weeks, then no medication during the fourth week when withdrawal bleeding usually occurs, mimicking a menstrual cycle. For user convenience, monthly injections, weekly skin patches, and insertable vaginal rings containing combined estrogen/progesterone hormone products are available. Some women may prefer or benefit from blocking menstruation completely, and preparations with continuous daily hormones are available.

At least 20 brands of combination oral contraceptives are now available, and more than 200 million women are using them worldwide. In the United States, 28 percent of women under age 45 use a combination pill to prevent conception. The failure rate for combination oral contraceptives, when used as prescribed, is 0.24 percent over a two-year period. (*Failure* for a birth control method is defined as a pregnancy.) Birth control pills are not risk free: Combination pills can worsen problems associated with severe hypertension, diabetes mellitus, epilepsy, gallbladder disease, heart trouble, and acne. Women taking oral contraceptives are also at increased risk of venous thrombosis, strokes, pulmonary embolism, and (for women over 35) heart disease. However, pregnancy has similar or higher risks.

Hormonal postcoital contraception, or the emergency "morning after" pill involves taking either combination estrogen/progesterone birth control pills or progesterone-only pills in two large doses 12 hours apart within 72 hours of unprotected sexual intercourse. Particularly useful when barrier methods malfunction or coerced intercourse occurs, it reduces expected pregnancy rates by up to 89 percent. The progesterone-only version is considered safe for nonprescription use and is available to everyone without prescription at drug stores.

Progesterone-only forms of birth control are now available: Depo-provera and the progesterone-only pill. *Depo-provera* is injected every three months. Uterine cycles are initially irregular, and

Reproductive System

Effectiveness of Contraceptive Methods

More Effective

Less than 1 pregnancy per 100 women in a year

Implant

0.05 %*

Intrauterine Device (IUD)

LNG - 0.2 % Copper T - 0.8 %

Sterilization

Male (Vasectomy)

0.15 %

Female (Abdominal, Laparoscopic, Hysteroscopic)

0.5 %

How to make your method most effective

After procedure, little or nothing to do or remember.

Vasectomy and hysteroscopic sterilization: Use another method for first 3 months.

6-12 pregnancies per 100 women in a year

Injectable

6 %

Pill

SUN MON TUES WED THUR FRI SAT

9 %

Patch

9 %

Ring

9 %

Diaphragm

12 %

Injectable: Get repeat injections on time.

Pills: Take a pill each day.

Patch, Ring: Keep in place, change on time.

Diaphragm: Use correctly every time you have sex.

---

18 or more pregnancies per 100 women in a year

Male Condom

18 %

Female Condom

21 %

Withdrawal

22 %

Sponge

24 % parous women
12 % nulliparous women

Condoms, sponge, withdrawal, spermicides: Use correctly every time you have sex.

Fertility awareness-based methods: Abstain or use condoms on fertile days. Newest methods (Standard Days Method and TwoDay Method) may be the easiest to use and consequently more effective.

Fertility-Awareness Based Methods

JANUARY

24 %

Spermicide

Spermicide

28 %

Less Effective

* The percentages indicate the number out of every 100 women who experienced an unintended pregnancy within the first year of typical use of each contraceptive method.

U.S. Department of Health and Human Services
Centers for Disease Control and Prevention

CONDOMS SHOULD ALWAYS BE USED TO REDUCE THE RISK OF SEXUALLY TRANSMITTED INFECTIONS.

Other Methods of Contraception

Lactational Amenorrhea Method: LAM is a highly effective, *temporary* method of contraception.
Emergency Contraception: Emergency contraceptive pills or a copper IUD after unprotected intercourse substantially reduces risk of pregnancy.

Adapted from WHO's Family Planning: A Global Handbook for Providers (2001) and Trussell et al (2011).

▶ **Figure 91 Birth Control Guide.**
[Centers for Disease Control and Prevention. Division of Reproductive Health, National Center for Chronic Disease Prevention and Health Promotion. http://www.cdc.gov/reproductivehealth/UnintendedPregnancy/PDF/effectiveness_of_contraceptive_methods.pdf]

eventually cease with continued use in roughly 50 percent of women using this product. The most common problems with this contraceptive method are (1) a tendency to gain weight and (2) a slow return to fertility (up to 18 months) after injections are discontinued. The progesterone-only pill is taken daily and may cause irregular uterine cycles. Skipping just one pill may result in pregnancy. One high dosage progesterone pill, available over the counter and with no age restrictions, is known as "Plan B," or the "morning after pill." This pill, when taken within 72 hours of un-protected sex, prevents 7 out of 8 unwanted pregnancies. Plan B is not intended for regular use as a birth control method.

LARC, or *Long-Acting Reversible Contraception*, using either Contraceptive Implants or Intrauterine Devices are available, but not widely used in the United States. The American Congress of OB-GYNs recommends increased use to reduce unintended pregnancies. These methods can be used by most women, including teenagers, the contraceptive effects last from 3 to 10 years, and they do not require any attention after insertion. The *Implanon implant*

is a single plastic rod that contains progesterone and is inserted under the skin. It is effective for three years, and fertility returns after removal.

An **intrauterine device (IUD)** consists of a small plastic T that releases either a form of progesterone or a small amount of copper into the uterine lining. Cervical mucus thickens, and the uterine lining is suppressed. The resulting changes affect sperm motility and implantation. Depending on the type, they provide birth control for 1 to 10 years. In the United States, IUD use has increased eightfold to 6 percent between 1995 and 2006–2010, and 18 percent of female OB-GYNs use them. Earlier fears of in-creased PID infections or infertility involved a brand of unusual design that was removed from the market. Currently, IUDs with progesterone are used to treat heavy menstrual bleeding prob-lems. IUDs are widely used in Europe, Scandinavia, and else-where around the world.

Barrier contraceptive methods include male **condoms**, also called *prophylactics* or "rubbers," which cover the glans and shaft of the penis during intercourse and keep spermatozoa from reaching the female reproductive tract. Latex condoms also reduce the spread of sexually transmitted diseases, such as syphilis, gonorrhea, HPV, and AIDS.

Vaginal barriers, such as the female condom, vaginal sponge, *diaphragm*, and *cervical cap*, rely on similar principles. A diaphragm consists of a dome of latex rubber with a small metal hoop supporting the rim. Because vaginas vary in size, women choosing this method must be individually fitted. Before intercourse, the diaphragm is inserted so that it covers the cervical os. The diaphragm must be coated with a small amount of spermicidal (sperm-killing) jelly or cream to be an effective contraceptive. The failure rate of a properly fitted and used diaphragm is estimated at 5–6 percent. The cervical cap is smaller and lacks the metal rim. It, too, must be fitted carefully, but unlike the diaphragm, it can be left in place for several days.

"Natural Family Planning," also called "Fertility Awareness" and the **rhythm method**, involves abstaining from sexual activity on the days ovulation might be occurring. The timing is estimated on the basis of previous patterns of menstruation, monitoring changes in body symptoms of ovulation including basal body temperature, cervical mucus texture, and, for some, urine tests for LH. Because of the irregularity of many women's menstrual cycles, this method of contraception has a high failure rate, although some women use it successfully.

Sterilization is a surgical procedure that makes an individual unable to provide functional gametes for fertilization. Either sexual partner may be sterilized. In a **vasectomy** (vaz-EK-to-mē), a segment of the ductus deferens is removed, making it impossible for spermatozoa to pass from the epididymis to the distal portions of the reproductive tract. The surgery can be performed in a physician's office in a matter of minutes. The spermatic cords are located as they ascend from the scrotum on either side; after each cord is opened, the ductus deferens is severed. A 1-cm section is removed, and the cut ends are tied shut. The cut ends cannot reconnect, and in time, scar tissue forms a permanent seal. Alternatively, the cut ends of the ductus deferens are blocked with silicone plugs that can later be removed. This more-recent vasectomy procedure may make it possible to restore fertility at a later date. After vasectomy, men experience normal sexual function, because the secretions of the epididymis and testes normally account for only about 5 percent of the volume of semen. Spermatozoa continue to develop, but they remain within the epididymis until they degenerate.

The uterine tubes can be blocked by a surgical procedure known as a **tubal ligation**. The failure rate for this procedure is estimated at 0.45 percent. Because the surgery requires that the abdominopelvic cavity be entered, most commonly by laparoscopy, general anesthetic is required and complications are more likely than with vasectomy. As in a vasectomy, attempts may be made to restore fertility after a tubal ligation, but as successful reversal may not be possible, both forms of contraception should be considered permanent.

Hormonal contraceptives, condoms, IUDs, and sterilization are the primary contraception methods for all age groups. But the proportion of the population using a particular method varies by age group. Sterilization, for example, is most popular among older men and women, who may already have had children. Relative availability also plays a role. For example, a sexually active female under age 18 can buy a condom more easily than

she can obtain a prescription for an oral contraceptive. But many of the differences are attributable to the relationship between risks and benefits for each age group.

Although pregnancy is a natural phenomenon, it has risks. In 1915 the maternal death rate in the United States was 608 per 100,000 live births. This rate fluctuated for several decades, but later dropped dramatically from 400 in 1939 to 13 in 1975. It has remained low (10–12 per 100,000 live births) since then. The long-term decline was associated with improvements in the use of antisepsis in childbirth and antibiotics if infection did occur. Monitored labor and delivery with hospital care to treat complications such as excessive bleeding also reduced deaths. The "average" maternal mortality rate encompasses a broad range: The rate is 7 per 100,000 women under 20, but 40 per 100,000 among women over 40. Although these risks are small, for pregnant women over age 35 the chances of dying from pregnancy-related complications are almost twice as great as the chances of being killed in an automobile accident and are many times greater than the risks associated with the use of oral contraceptives. For women in developing nations, the comparison is even more striking: The pregnancy-related mortality rate for women in parts of Africa is approximately 1 per 150 pregnancies. In addition to preventing pregnancy, combination birth control pills have been shown to reduce the risks of ovarian and endometrial cancers and fibrocystic breast disease.

Before age 35, the risks associated with oral contraceptive use are lower than the risks associated with pregnancy. The notable exception involves women who take the pill and also smoke cigarettes. Younger women are more fertile, so despite a lower mortality rate for each pregnancy, they are likely to have more pregnancies as the result of birth control failures.

After age 35, the risks of complications associated with oral contraceptive use increase, but the risks of using other methods remain relatively stable. Women over age 35 (smokers) or 40 (nonsmokers) are therefore often advised to use other forms of contraception. Because each contraceptive method has its own advantages and disadvantages, research on contraception continues.

Sexually Transmitted Diseases

Close physical contact can spread infectious diseases from person to person, and sexual contact is as close as two individuals can get. Some infections are spread almost exclusively by sexual contact. These infections are called **sexually transmitted infections/diseases**, or **STIs** or **STDs**. A variety of bacterial, viral, parasitic, and fungal infections are included in this category. At least two dozen STDs/STIs are currently recognized, and roughly 15 million people become infected each year in the United States. All STDs are unpleasant, and some are deadly. Here, we will discuss five of the most common STDs: *chlamydia, gonorrhea, syphilis, herpes,* and *genital warts (HPV)*. The deadliest, acquired immunodeficiency syndrome (HIV/AIDS), was discussed on pages 155–157.

Chlamydia

As diagnostic procedures improve, infections by the bacterium *Chlamydia trachomatis* are proving to be the most frequent cause of STDs. 1.4 million infections were reported in 2011 in the United States, most in the 15 to 24 year age group. Chlamydial infection can have a variety of clinical effects. It is responsible for the majority of cases of pelvic inflammatory disease PID (p. 217), and even asymptomatic infections may result in uterine tube blockage and infertility. In males it is probably the most common cause of *nongonococcal urethritis*. Infection during pregnancy can lead to premature birth and eye and lung infections in newborns. Chlamydial infections can be treated with antibiotics (erythromycin, tetracycline, and ceftriaxone are most often used), but sexual partners must be treated to prevent reinfection.

Gonorrhea

The bacterium *Neisseria gonorrhoeae* is responsible for gonorrhea, one of the most common STDs in the United States. Nearly 2 million cases were reported in the early 1970s. In 2011, over 350,000 cases were reported. The bacteria normally invade epithelial cells that line the male or female reproductive tract. In relatively rare cases, they will also colonize the pharyngeal or rectal epithelium.

The symptoms of genital infection differ according to the sex of the individual. It has been estimated that up to 80 percent of women infected with gonorrhea experience no symptoms or symptoms so minor that medical treatment is not sought. As a result, these women act as carriers, spreading the infection through their sexual contacts. An estimated 10 to 15 percent of women infected with gonorrhea experience more acute symptoms, because the bacteria invade the epithelia of the uterine tubes. This infection probably accounts for many of the cases of pelvic inflammatory disease (PID) in the U.S. population. As many as 80,000 women become infertile each year as a result of the formation of scar tissue along the uterine tubes after gonorrheal or chlamydial infection.

Seventy to 80 percent of infected males develop symptoms in 2 to 30 days painful enough to make them seek antibiotic treatment. The asymptomatic 20–30 percent are male carriers who unknowingly spread the infection. The urethral invasion is accompanied by pain on urination (dysuria) and, typically, by a viscous urethral discharge. A sample of the discharge can be cultured to permit the positive identification of the organism involved.

Syphilis

Syphilis (SIF-i-lis) results from infection by the bacterium *Treponema pallidum*. The first reported syphilis epidemics occurred in Europe during the 16th century, possibly introduced by early explorers returning from the New World. The death rate from the "Great Pox," as syphilis was called then, was appalling, far greater than it is today, even after we take into account the absence of antibiotic therapies at that time. It appears likely that the syphilis bacterium has mutated during the interim, to a form that reduces the immediate mortality rate, but prolongs the period of chronic illness and increases the likelihood of successful transmission. Syphilis still remains a life-threatening disease.

Primary syphilis begins as the bacteria cross the mucous epithelium and enter the lymphatic vessels and bloodstream. At the site of the invasion, the bacteria multiply. After an incubation period of 1.5 to 6 weeks, their activities produce a painless raised ulcerated lesion, or **chancre** (SHANG-ker) (**▶ Figure 92**). The chancre is infectious and persists for several weeks before fading away, even without treatment. In heterosexual men, the chancre tends to appear on the penis; in

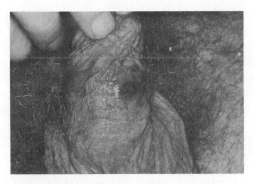

▶ Figure 92 **A Syphilitic Chancre.**

women, it may develop on the labia, vagina, or cervix. Lymph nodes in the region often enlarge and remain swollen even after the chancre has disappeared.

Symptoms of *secondary syphilis* appear roughly six weeks later. The skin lesions of secondary syphilis are also infectious. It generally involves a diffuse, reddish skin rash. Like the chancre, the rash fades over a period of two to six weeks. These symptoms may be accompanied by fever, headaches, and malaise. The combination is so vague that the disease may easily be overlooked or misdiagnosed. In a few instances, more serious complications, such as meningitis, hepatitis, or arthritis, develop. During primary and secondary syphilis the bacterium circulates in the blood and can be spread by unscreened transfusion.

The individual then enters the *latent phase*, which is noninfectious. The duration of the latent phase varies widely. Fifty to seventy percent of untreated individuals with latent syphilis do not develop the symptoms of *tertiary syphilis*, or *late syphilis*, although the bacterial pathogens remain within their tissues. Those who develop tertiary syphilis may do so 10 or more years after primary infection.

The most severe symptoms of tertiary syphilis involve the central nervous system and the cardiovascular system. **Neurosyphilis** may result from a bacterial infection of the meninges or the tissues of the brain or spinal cord. **Tabes dorsalis** (TĀ-bēz dor-SAL-is) results from the invasion and demyelination of the posterior columns of the spinal cord and the sensory ganglia and nerves. In the cardiovascular system, the disease affects the major vessels, leading to aortic stenosis, aneurysms, or focal calcification.

Equally disturbing are the effects of transmission from an infected mother to fetus across the placenta. In those cases where infection does not lead to fetal death, congenital syphilis is marked by infections of the developing bones and cartilages of the skeleton and progressive damage to the spleen, liver, bone marrow, and kidneys. The risk of fetal transmission may be as high as 95 percent, so maternal blood testing is recommended early in pregnancy. Blood donations are screened to prevent transfer through transfusion. The treatment of syphilis involves the administration of *penicillin* or other antibiotics.

Herpes

Genital herpes results from infection by herpes viruses. Two different viruses are involved. Eighty to ninety percent of genital herpes cases are caused by the virus known as HSV-2 (herpes simplex virus Type 2), which usually attacks the skin of the external genitalia. The remaining cases are caused by HSV-1, the virus that most commonly causes "cold sores" or "fever blisters" around the mouth. Typically, within a week of the initial herpes infection, the individual develops multiple small, painful, infectious ulcerated lesions on the affected skin or mucous membrane, with associated lymphadenopathy. In women, ulcerations may also appear on the cervix. These ulcerations gradually heal over the next two to three weeks. Recurring lesions are common, although subsequent incidents are less severe.

During delivery, infection of the newborn with herpes viruses present in the mother's vagina can lead to serious illness, because the infant has few immunological defenses. The antiviral agent *acyclovir* has helped in treating initial infections and in reducing recurrences.

Human Papillomavirus (HPV)/Genital Warts

Genital warts, or *condyloma acuminata*, result from infection by one of a number of strains of *human papillomavirus* (HPV) discussed on page 218. Several of these strains are responsible for many cases of cervical, anal, vaginal, penile, and possibly even oral cancer. Roughly 14 million Americans become infected with genital warts each year and approximately 79 million are currently infected. There is no satisfactory treatment for this problem. Most people will eventually clear the infection, but this can take years and reinfection with another strain is possible. The traditional treatments of individual warts have included cryosurgery, erosion by caustic chemicals, surgical removal, and laser surgery to remove the warts. These treatments expunge the visible signs of infection, but the virus remains within the epidermis. Barrier contraceptives reduce the risk of cervical cancer and may reduce infection, but vaccination, currently available for four strains of HPV, holds the most promise.

▶ Problems with Development and Inheritance

Disorders of Development

Fetal development is a complex process, and developmental disorders are extremely diverse. ▶ Figure 93 surveys representative disorders of development.

Teratogens and Abnormal Development

Teratogens (TER-a-tō-jenz) are stimuli that disrupt normal fetal development by damaging cells or altering their chromosomal structure or that interfere with normal induction. **Teratology** (ter-a-TOL-o-jē)—literally, the "study of monsters"—deals with extensive departures from the pathways of normal development. Teratogens that affect the embryo in the first trimester can disrupt cleavage, gastrulation, or neurulation. The embryonic survival rate will be low, and most survivors will have severe anatomical and physiological defects that affect all the major organ systems. Errors introduced into the developmental process during the second or third trimester are likely to affect specific organs or organ systems, for the major organizational patterns are already established. Nevertheless, the alterations reduce the chances for long-term survival of the infant.

We encounter many powerful teratogens in everyday life. The location and severity of the resulting defects vary with the nature of the stimulus and the time of exposure. Radiation is a powerful teratogen that can affect all cells. Even the x-rays used in diagnostic procedures can break chromosomes and produce developmental errors; thus, nonionizing procedures such as ultrasound are used to track embryonic and fetal development. Fetal exposure to the microorganisms responsible for syphilis or rubella ("German measles") can also produce serious developmental abnormalities, including congenital heart defects, mental retardation, and deafness.

Some chemical agents are teratogenic only if they are present at a time when embryonic or fetal targets are vulnerable to their effects. Thousands of critical inductions are under way during the first trimester, initiating developmental sequences that will produce the major organs and organ systems of the body. In almost every case, the nature of the inducing agent remains unknown, and the effects of unusual compounds within the maternal circulation cannot be predicted. As a result, virtually any unusual chemical that reaches an embryo has the potential for producing developmental abnormalities. For example, during the 1960s, many Europeans used **thalidomide**, a drug that is effective in promoting sleep and preventing nausea. Thalidomide was commonly prescribed for women in early pregnancy, with disastrous

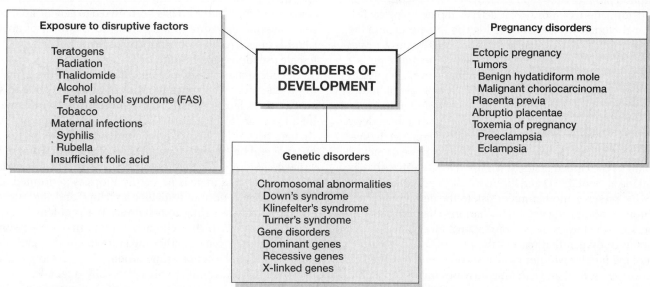

▶ Figure 93 **Disorders of Fetal Development.**

results. The drug crossed the placenta and entered the fetal circulation, where, for a few days, it interfered with the induction process responsible for limb development. Many infants exposed to the drug were born without limbs or with drastically deformed ones. Thalidomide was not approved by the U.S. Food and Drug Administration (FDA) then, and it could not be sold legally in the United States. Although the FDA is often criticized today for its approval process, in this case the combination of rigorous testing standards and complex bureaucratic procedures protected the public. Thalidomide is valuable in treating some forms of leprosy and multiple myeloma and is now used carefully to treat these conditions.

However, even when extensive testing is performed with laboratory animals, uncertainties remain because the chemical nature of the inducer responsible for a specific process may vary from one species to another. For example, thalidomide produces abnormalities in humans and monkeys, but developing mice, rats, and rabbits are unaffected by the drug. More powerful teratogens have an effect regardless of the time of exposure. Pesticides, herbicides, and heavy metals are common around agricultural and industrial environments, and these substances can contaminate the drinking water in the area. A number of prescription drugs, including certain antibiotics, tranquilizers, sedatives, steroid hormones, diuretics, anesthetics, and analgesics, also have teratogenic effects. Pregnant women should read the "Caution" label before using any drug without the advice of a physician. Most "natural" herbs and substances have not been tested, and their chemical composition may vary from source to source, so their effects during pregnancy are unknown, and they should not be considered uniformly safe to use during pregnancy. (We do know that some plants produce teratogens and store them in their leaves, presumably as a defense against herbivorous animals.)

The term **Fetal Alcohol Spectrum Disorder (FASD)** is used to refer to a broad range of developmental problems associated with maternal alcohol consumption. FASD may affect up to 12,000 infants born each year in the United States. The most severe cases of FASD are called **Fetal Alcohol Syndrome (FAS)**, which accounts for up to 8000 cases each year. FAS occurs when maternal alcohol consumption produces developmental defects in the fetus, such as skeletal deformation, cardiovascular defects, and neurological disorders. Fetal mortality rates can be as high as 17 percent, and the survivors have problems in later development. The most severe cases involve mothers who binge drink (4 or more drinks per occasion) or do regular heavy drinking, consuming the alcohol content of at least 7 ounces of hard liquor, 10 beers, or several bottles of wine each day. However, because the effects produced are directly related to the degree of exposure, there is probably no level of alcohol consumption that can be considered completely safe. Fetal alcohol syndrome may be the number-one cause of mental retardation in the United States. It is important to realize that by not drinking alcohol, FASD and FAS is 100 percent preventable.

Smoking presents another major risk to the developing fetus. In addition to introducing potentially harmful chemicals, such as nicotine, smoking lowers the P_{O_2} of maternal blood and reduces the amount of oxygen that reaches the placenta. A fetus carried by a smoking mother will not grow as rapidly as one carried by a nonsmoking mother, and smoking increases the risks of spontaneous abortion, prematurity, and fetal death. The rate of infant mortality after delivery is also higher when the mother smokes, and postnatal development can be adversely affected as well.

Nutrition and Abnormal Development

Low maternal levels of the B vitamin folate/folic acid are associated with increased risk of the neural tube birth defects spina bifida (cleft spine) and anencephaly (absent brain). The early embryo forms a tube that closes 14 days after conception (day 29 of pregnancy as measured from the first day of the last menstrual cycle) to form the brain at one end and the spinal cord at the other. This is before most women even know they are pregnant. If the top doesn't close, anencephaly results. If the bottom doesn't close, spina bifida occurs. Anencephaly is fatal, and depending on the extent of the spinal opening, spina bifida can be very disabling.

Woman taking 400 micrograms of folate daily starting before conception have folate blood levels that significantly reduce the risk of neural tube defects in pregnancy. Any woman planning pregnancy should start taking prenatal vitamin supplements that include this amount of folate daily at least one month before conception. Because the defect occurs so early in pregnancy, and the reality is that 50 percent of pregnancies in the United States are unplanned, it is prudent that all women of child-bearing age (15 to 44) take 400 micrograms of folate daily. To help achieve this, in 1996 the FDA mandated folate enrichment of many grain products including flour, rice, cereals, pasta, and breads. By 2006 there was a 31 percent reduction in neural tube birth defects in the United States.

Induction and Sexual Differentiation

The physical (phenotypic) sex of a newborn infant depends on the hormonal cues it receives during development, not on the genetically determined sex of the individual. If something disrupts the normal inductive processes, the individual's genetic and anatomical sexes may be different. These **disorders of sexual development (DSDs)** or **intersex situations** were previously called **pseudohermaphrodism**. The baby may be born with ambiguous genitalia, in which the outer genitalia vary from the genetic sex and may have both male and female elements. For example, if a female embryo becomes exposed to male hormones, the embryo will develop many of the sexual characteristics of a male. Such situations are relatively rare. The most common cause is the hypertrophy of the fetal adrenal glands and their production of androgens in high concentrations; in some cases, this condition has been linked to genetic abnormalities. Maternal exposure to androgens, in the form of anabolic steroids or as a result of an endocrine tumor, can also produce a female pseudohermaphrodite.

Genetic male intersex persons may result from an inability to produce adequate concentrations of androgens due to some enzymatic defect. In **testicular feminization syndrome**, the infant appears to be a normal female at birth. Typical physical changes occur at puberty, and the individual develops the overt physical and behavioral characteristics of an adult woman. Menstrual cycles do not begin, however, because the vagina ends in a blind pocket, and there is no uterus. Biopsies performed on the gonads reveal a normal testicular structure, and the interstitial cells are busily secreting testosterone. The problem apparently involves a defect in the cellular receptors that are sensitive to circulating androgens. Neither the embryo nor the adult tissues can respond to the testosterone produced by the gonads, so the person develops as, and remains physically, a female.

If detected in infancy, many cases of ambiguous genitalia can be treated with hormones and surgery to produce genitalia of male or female appearance. Depending on the arrangement of the internal organs and gonads, normal reproductive function may be more difficult to achieve. Sex hormones also affect the brain's development, and sex assignment must take into account behavioral as well as anatomic factors. In one instance, a genetic male who had been raised as a female after undergoing surgery for ambiguous genitalia later adopted a male name, attire, and behavior after hormonal changes occurred at puberty.

Intersex disorders are one type of a developmental problem caused by hormonal miscues or by an inability to respond appropriately to hormonal instructions. Another example is maternal exposure to *diethylstilbestrol* (*DES*), a synthetic steroid prescribed in the 1950s to prevent miscarriages. Daughters exposed to DES while *in utero* have higher-than-normal infertility and prematurity rates, due to uterine, vaginal, and uterine tube abnormalities, and they have an increased risk of developing vaginal cancer and possibly breast cancer. Use was discontinued after studies showed DES was ineffective in preventing miscarriages but there remains a cohort of women, now in their 50s, who are still at risk due to DES exposure *in utero*.

Technology and the Treatment of Infertility

Infertility (*sterility*) is usually defined as an inability to achieve pregnancy after one year of appropriately timed intercourse. Problems with fertility are relatively common: An estimated 10–15 percent of U.S. married couples of reproductive age are infertile, and another 10 percent are unable to have as many children as they desire. It is thus not surprising that reproductive physiology has become a popular field, and that the treatment of infertility has become a major medical industry. In 2011, 61,610 infants conceived by **assisted reproductive technologies (ART)** were born in the United States, and that represented over 1 percent of the number of live births that

year. Both sexes need to produce functional gametes, plus have correctly functioning reproductive anatomy and physiology to bring the sperm and oocytes together for fertilization. In addition, a woman must then support the developing embryo through fetal growth and development to birth.

An infertile woman may not produce functional oocytes, or may have anatomical or physiological barriers to fertilization, or be unable to maintain pregnancy and have recurrent miscarriages. An infertile man lacks a sufficient number of motile sperm capable of successful fertilization. Because the infertility of either sexual partner has the same result, the diagnosis and treatment of infertility must involve evaluations of both partners. Statistics vary, but approximately 33 percent of infertility cases are attributed to the female partner, 33 percent to the male partner, and the rest to both partners or an unknown cause. Primary evaluation for male infertility is the relatively simple semen analysis in which determining sperm number, appearance, and motility require a sample, a microscope, and a trained evaluator. As the female reproductive tract and oocyte(s) are internal, female infertility evaluation requires more technology.

Recent advances in our understanding of reproductive physiology are providing new solutions to fertility problems. The various problems, and the approaches to solving them—called assisted reproductive technologies (ART)—include the following (◗ Figure 94):

• **Low sperm count/abnormal spermatozoa.** In cases of male infertility due to low sperm counts or unexplained infertility, semen from one or several ejaculates can be concentrated, and the most motile sperm selected and then placed in the uterus. This technique, intrauterine insemination or artificial insemination, is often combined with hormonal treatment of the woman to induce ovulation. Normal fertilization and pregnancy may then occur. In cases in which a male's spermatozoa

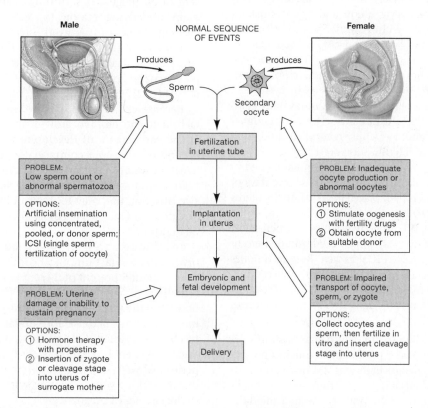

◗ **Figure 94 Infertility Problems and Treatment Options.**

are very few, or are unable to penetrate the oocyte, single-sperm fertilization (intracytoplasmic sperm injection or ICSI) can be accomplished by microscopically manipulating the sperm, oocyte, and the corona radiata. The fertilized egg is then used in an IVF (discussed later) procedure with essentially the same successful live birth rate as IVF cycles with no diagnosed male infertility. If the male cannot produce functional spermatozoa, donor sperm obtained from a *sperm bank* (which screens the donor and stores the sperm) can be introduced into the female reproductive tract (artificial insemination). In this case intrauterine insemination or IVF may not be necessary.

- **Hormonal problems.** If the problem involves the woman's inability to ovulate because her gonadotropin or estrogen levels are low, or if she is unable to maintain adequate progesterone levels after ovulation, these hormones can be provided to induce ovulation and/or maintain the luteal endometrium.

- **Problems with oocyte production.** *Fertility drugs*, such as clomiphene (Clomid), stimulate ovarian oocyte production. Clomiphene blocks the feedback inhibition of the hypothalamus and pituitary gland by estrogens. As a result, circulating FSH levels rise, so more follicles are stimulated to complete their development. Injected purified gonadotropins, such as Pergonal (FSH and LH) and Metrodin (FSH), are also used to accelerate ovum development. The chance that a single oocyte will be fertilized through well-timed sexual intercourse is about 1 in 3. Increasing the number of oocytes released raises the odds of fertilization and therefore the odds of a pregnancy. It is not easy to determine just how much ovarian stimulation is needed, however, so treatment with fertility drugs commonly results in multiple mature eggs. Careful monitoring of follicle development and avoidance of fertilization if too many eggs are present reduces the chances of a multiple birth.

- **Blocked uterine tubes.** Blockage of or damage to the uterine tubes can interfere with oocyte, sperm, and zygote transport. Normally, fertilization occurs within the uterine tube. However, this site is not essential, and fertilization can also take place *in vitro*—in a test tube or petri dish (*vitro*, glass). The ovaries are stimulated with injected hormones, and a large "crop" of mature oocytes is "harvested" from tertiary follicles by inserting a long needle through the vaginal wall (with ultrasound guidance) into the follicles. The individual oocytes are examined for defects. Then sperm and a carefully controlled fluid environment are provided, and if fertilization occurs, early development will proceed normally. The zygote can be maintained in an artificial environment through the first two to five days of development. The embryo is then placed directly into the uterus through a tube introduced into the cervix. The process, called **in vitro fertilization**, or **IVF**, was involved in 99.9 percent of ART procedures in the United States in 2010. Almost 64 percent of these IVF procedures involved ICSI as well. Both techniques had about 33 percent overall live birth success. The success rate is strongly linked to maternal (source of oocyte) age: 40 percent under 35 years of age, 31 percent ages 35–37, 21 percent ages 38–40, 12 percent ages 41–42, and 5 percent ages 43–44.

- **Abnormal oocytes.** Unlike sperm, which have a short existence and are continually being produced, a woman's oocytes age with her. Age-related changes in the characteristics and quality of the oocytes, rather than changes in hormone levels or uterine responsiveness, seem to be the primary cause of infertility

in older women. Viable donor eggs can be obtained from younger women. Through treatment with fertility drugs, the donor's ovaries are stimulated to produce a large crop of oocytes, which are then collected and fertilized in vitro, generally by the spermatozoa of the recipient's husband. After cleavage has begun, the pre-embryo is placed in the recipient's uterus, which has been synchronized with the donor's menstrual cycle by ovulatory hormones and "primed" by progesterone therapy. Although the mother has no genetic relationship to the embryo, pregnancy proceeds normally. The live birth rate of this procedure in women over 40 with young donor eggs is over 55 percent.

- **An abnormal uterine environment.** If fertilization and transport occur normally, but the uterus cannot maintain a pregnancy, the problem may involve low levels of progesterone secretion by the corpus luteum. Hormone therapy may solve this problem. If a woman's uterus simply cannot support development, an IVF procedure can introduce a zygote or cleavage-stage embryo into the uterus of another woman (a surrogate mother or *gestational carrier*). If the surrogate woman becomes pregnant, fetal development will proceed normally even though the gestational carrier may have no genetic relationship to the embryo. The egg or sperm used may come from the couple wanting a child, the surrogate, or another egg or sperm donor.

Ectopic Pregnancies

Implantation normally occurs at the endometrial surface that lines the uterine cavity. The precise location within the uterus varies, although in most cases implantation occurs in the uterine body. By contrast, in an **ectopic pregnancy**, implantation occurs outside the uterus, 95 percent of the time within the uterine tube.

The incidence of ectopic pregnancies is approximately 2 percent of all pregnancies. Abnormal uterine tube function is the main underlying reason for ectopic pregnancy. Anatomical blockage from scarring from infection (PID) or blockage from endometriosis (p. 218) are among the main risk factors. Women who douche regularly have a higher risk of experiencing an ectopic pregnancy, presumably because the flushing action may lead to PID (p. 217).

Because extrauterine sites are not favorable for placental attachment and embryo growth, many ectopic pregnancies end at the early stages of development. The uterine tubes cannot expand enough to accommodate a developing embryo, and it usually ruptures during the first trimester. Abdominal pain and vaginal spotting are early warning signs of this event. The massive bleeding that may follow can be severe enough to be life-threatening. Once detected, laparoscopic or surgical repairs can be performed.

In a few instances, the ruptured uterine tube releases the embryo with an intact umbilical cord, so further development can occur. About 5 percent of these very rare abdominal pregnancies actually complete full-term development. Normal birth cannot occur, but the infant can be surgically removed from the abdominopelvic cavity. Because abdominal pregnancies are possible, it has been suggested that men as well as women could act as surrogate mothers if a zygote were surgically implanted into the peritoneal wall. It is not clear, though, how the endocrine, cardiovascular, nervous, and other systems of a man would respond to the stresses of pregnancy.

Problems with Placentation

In a **placenta previa** (PRĒ-vē-uh; "in the way"), endometrial implantation occurs in or near the cervix. This condition causes problems as the growing placenta approaches the internal os (internal cervical orifice). In a **total placenta previa**, the placenta extends across the internal os, whereas a partial placenta previa only partially blocks the internal os. The placenta is characterized by a rich fetal blood supply and the erosion of maternal blood vessels within the endometrium. Where the placenta passes across the internal os, the delicate complex hangs like an unsupported water balloon. As the pregnancy advances, even minor mechanical stresses can be enough to tear the placental tissues, leading to massive fetal and maternal bleeding.

Most cases of placenta previa can be diagnosed by ultrasound in the second trimester. As the uterus enlarges, the placenta may retract from covering the cervical os. If it does not, then, by the seventh month, when the placenta reaches its full size, the cervical canal is dilated, and the uterine contents push against the placenta where it spans the internal cervical os without the support of the uterine wall. Minor, painless bleeding may occur, but sudden catastrophic bleeding is a risk to both the mother and the fetus. The treatment of total placenta previa involves bed rest for the mother until the fetus reaches a gestational age and size at which cesarean delivery can be performed with a reasonable chance of survival of the neonate (newborn).

In an **abruptio placentae** (ab-RUP-shē-ō pla-SEN-tē), part or all of the placenta tears away from the uterine wall sometime after the fifth month of gestation. This is a serious and dangerous condition, and the bleeding and pain are usually sufficient to prompt an immediate visit to a physician. In severe cases, the bleeding leads to maternal anemia, shock, and kidney failure. Although maternal mortality is only 0.5 to 1 percent, the fetal mortality rate from this condition ranges from 20 to 35 percent, depending on the severity of the fetal blood loss.

Problems with the Maintenance of a Pregnancy

The rate of maternal complications during pregnancy is relatively high. Pregnancy stresses maternal systems, and can overwhelm homeostatic mechanisms. By late pregnancy maternal blood volume has doubled, and the heart workload is increased by 30 to 50 percent with similar increases in kidney filtration. Lung and digestive processes are compromised by uterine growth and fetal metabolic demands. Maternal cellular and humoral immune system responses are blunted as part of tolerance to fetal foreign antigens. Pregnancy complications are more common at the extremes of pregnancy statistics. In other words, there is increased risk with both younger (teenage) and older (over 35) mothers, first pregnancies, multiple fetal pregnancies (twins and above), obese or underweight mothers, and those with preexisting health problems.

Some problems are unique to pregnancy. The term **toxemia** (tok-SĒ-mē-uh) **of pregnancy**, also known as hypertension in pregnancy and **preeclampsia**, refers to disorders that affect the maternal cardiovascular and urinary systems, and the placenta. Affecting up to 8 percent of pregnancies after 20 weeks gestation, sustained hypertension and protein in the urine define the condition initially. The hypertension, fluid imbalances, swelling, and edema may progress to central nervous system (CNS) disturbances, leading to **eclampsia**, with maternal coma, convulsions, and death. Roughly 4 percent of individuals with preeclampsia develop eclampsia. The mother can be saved ("cured") only if the fetus is delivered immediately. Once the fetus and placenta have been delivered (removed from the uterus), symptoms of eclampsia disappear over a period of hours to days.

Severe cases of eclampsia account for 20 percent of maternal deaths and contribute to an estimated 10,500 fetal deaths in the United States each year. Prenatal care involves monitoring the mother's vital signs and urine to detect early signs of preeclampsia so that treatment can slow or prevent further progression of the condition. The cause is unknown, but immune system and vascular problems are implicated.

Complexity and Perfection

The expectation of prospective parents that every pregnancy will be idyllic and every baby will be perfect reflects deep-seated misconceptions about the nature of the developmental process. These misconceptions lead to the belief that when serious developmental errors occur, someone or something is at fault and that blame might be assigned to maternal habits (such as smoking, alcohol consumption, or improper diet), maternal exposure to toxins, infections or drugs, or the presence of other disruptive stimuli in the environment. The prosecution of women who give birth to severely impaired infants for "fetal abuse" (exposing a fetus to known or suspected risk factors) is an extreme example of this philosophy.

Although environmental stimuli can indeed lead to developmental problems, such factors are only one component of a complex system that is normally subject to considerable variation. Even if every pregnant woman were packed in cotton and confined to bed from conception to delivery, developmental accidents and errors would continue to occur with regularity.

Spontaneous mutations are the result of random errors in replication; such incidents are relatively common. At least 10 percent of fertilizations produce zygotes with abnormal chromosomes. Because most spontaneous mutations fail to produce visible defects, the actual number of mutations must be far larger. Most of the affected zygotes die before completing development, and only about 0.5 percent of newborns show chromosomal abnormalities that result from spontaneous mutations.

Due to the nature of the regulatory mechanisms, prenatal development does not follow precise, predetermined pathways. For example, much variation exists in the pathways of blood vessels and nerves, because it does not matter how blood or neural impulses get to their destinations, as long as they do get there. If the variations fall outside acceptable limits, however, the embryo or fetus fails to complete development. Very minor changes in heart structure can result in the death of a fetus, whereas large variations in venous distribution are common and relatively harmless. Virtually everyone can be considered abnormal to some degree, because no one has characteristics that are statistically average in every respect. An estimated 20 percent of your genes are subtly different from those found in the majority of the population, and minor defects or anomalies such as extra nipples or birthmarks are quite common, affecting perhaps 2 to 3 percent of infants.

Current evidence suggests that as many as half of all conceptions produce zygotes that do not survive the cleavage stage. These zygotes disintegrate within the uterine tubes or uterine cavity; because implantation never occurs, there are no obvious signs of pregnancy. This preimplantation mortality is commonly

Reproductive System

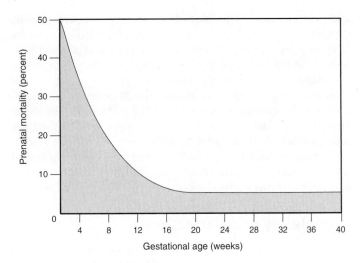

▶ **Figure 95 Prenatal Mortality.**

associated with chromosomal abnormalities. Of those embryos that do implant, roughly 20 percent fail to complete five months of development, with an average survival time of eight weeks. In most cases, severe problems affecting early embryogenesis or placenta formation are responsible. ▶ **Figure 95** graphically shows the relation of prenatal mortality to gestational age.

Prenatal mortality tends to eliminate the most severely affected fetuses. Those with less extensive defects may survive, completing full-term gestation or arriving via premature delivery. **Congenital malformations**, or **birth defects**, are structural abnormalities that are present at birth and that affect major systems. Spina bifida, hydrocephalus, anencephaly, cleft lip, and Down's syndrome are among the most common congenital malformations; we described those conditions in earlier chapters of the main text. The incidence of significant congenital malformations at birth averages about 3 percent. Of these 3 percent, only 10 percent are currently attributed to environmental factors (including infection, nutrition, and toxins). The remaining 90 percent have evidence of chromosomal abnormalities or genetic factors, including a family history of similar or related defects.

Medical technology continues to improve our abilities to understand and manipulate physiological processes. Genetic analysis of potential parents can now provide estimates of the likelihood of specific problems, although the problems themselves remain outside our control. But even with a better understanding of the genetic mechanisms involved, we will probably never be able to control every aspect of development and thereby prevent spontaneous abortions and congenital malformations. Too many complex, interdependent steps are involved in prenatal development, and malfunctions of some kind are statistically inevitable.

Monitoring Postnatal Development

Each newborn should be closely and rapidly scrutinized after delivery. The maturity of the newborn may be determined prior to delivery by means of ultrasound or amniocentesis (**Table 43**). Immediately on delivery, the newborn is checked and assigned an **Apgar rating** at one and five minutes of life, which evaluates the heart rate, respiratory rate, muscle tone, response to stimulation, and color at one and five minutes after birth. In each category, the

infant receives a score ranging from 0 (poor) to 2 (excellent), and the scores are then totaled. An infant's Apgar rating (0 to 10) has been shown to be an accurate predictor of newborn survival and of the presence of neurological damage. For example, newborn infants with cerebral palsy tend to have a low Apgar rating.

In the course of this examination, the newborn's breath sounds, depth and rate of respiration, color, and heart rate are noted. Both the respiratory rate and the pulse rate are considerably higher in infants than in adults. (See **Table 1**, p. 7.) Later, after parental contact and bonding with the new individual, a more complete physical examination of the newborn focuses on the status of its vital systems. Inspection of the infant normally includes the following:

- The head of a newborn may be misshapen after vaginal delivery, but it generally assumes its normal shape within the next few days. However, the size of the head must be checked to detect hydrocephalus, and the cranial vault is checked to ensure that cerebral hemispheres are present and anencephaly does not exist.

- The eyes, nose, mouth, and ears are inspected for reflex responses and for obstruction.

- The abdomen is palpated to detect abnormalities of internal organs.

- The heart and lungs are auscultated to check for breath sounds and heart murmurs.

- The external genitalia are inspected. The scrotum of a male infant is checked to see if the testes have descended.

- Cyanosis of the hands and feet is normal in newborns, but the rest of the body should be pink. A generalized cyanosis may indicate congenital cardiovascular disorders. Measurements of the neonate's body length, head circumference, and body weight are taken. A weight loss in the first 48 hours is normal, because fluid shifts occur as the infant adapts to the change from weightlessness (floating in amniotic fluid) to normal gravity. (Comparable fluid shifts occur in astronauts returning to Earth after extended periods in space.)

The nervous and muscular systems of newborns are assessed for normal reflexes and muscle tone. Reflexes commonly tested include the following:

- The **Moro reflex** is triggered when support for the head of a supine infant is suddenly removed. The reflex response consists of extension of the trunk and a rapid cycle of extension–abduction and flexion–adduction of the limbs. This reflex normally disappears at an age of about 3 months.

- The **stepping reflex** consists of walking movements triggered by holding the infant upright, with a forward slant, and placing the soles of the feet against the ground. This reflex normally disappears at an age of about 6 weeks.

- The **placing reflex** can be triggered by holding the infant upright and drawing the top of one foot across the bottom edge of a table. The reflex response is to flex and then extend that leg. This reflex also disappears at an age of about 6 weeks.

- The **sucking reflex** is triggered by stroking the lips. The associated *rooting reflex* is initiated by stroking the cheek, and the

Reproductive System

Table 43   Tests Performed during Pregnancy and on the Neonate

Diagnostic Procedure	Method and Result	Representative Uses
Amniocentesis	A needle, inserted through abdominal wall into uterine cavity, collects amniotic fluid for genetic diagnosis; performed in weeks 14–15 or closer to delivery if complications arise	Detects chromosomal abnormalities and birth defects such as spina bifida; late-pregnancy level of hemolysis in erythroblastosis fetalis; determines fetal lung maturity; evaluates fetal distress
Chorionic villi biopsy	Test performed during weeks 8–10 of gestation; small pieces of villi are suctioned into a syringe	Detects chromosomal abnormalities and biochemical disorders
Pelvic ultrasonography	Standard ultrasound	Detects multiple fetuses, fetal abnormalities, and placenta previa; estimates fetal age, growth, and sex
External fetal monitoring	Monitoring devices on maternal abdominal surface measure fetal heart rate and force of uterine contraction	Detects irregular heart rate or fetal stress
Internal fetal monitoring	Electrode is attached to fetal scalp exposed by dilated cervix to monitor heart rate; catheter is placed in uterus to monitor uterine contractions	As above

Laboratory Test	Normal Values	Significance of Abnormal Values
Amniotic fluid analysis		
Karyotyping	Normal chromosomes	Detects chromosomal defects such as those in Down's syndrome
Bilirubin	Traces only	Increased values may indicate amount of hemolysis of fetal RBCs by mother's Rh antibodies
Meconium	Not present	Present in fetal distress
Lecithin/sphingomyelin ratio (L/S ratio)	$\geq$2:1 ratio	Ratio below 2:1 indicates fetal lung immaturity
Creatinine	$\geq$2 mg/dL of amniotic fluid indicates week 36 of gestation	Less than 2 mg/dL indicates fetal immaturity (not as accurate as L/S ratio)
Alpha-fetoprotein (AFP)	Week 16 of gestation: 5.7–31.5 ng/mL (lowers with increasing gestational age)	Increased values indicate possible neural tube defect, such as spina bifida
Blood tests		
Human chorionic gonadotropin (hCG) (maternal serum or urine)	Nonpregnant females (serum): <0.005 IU/mL Nonpregnant females (urine): negative Pregnant females (urine): <500,000 IU over 24 hours	Determines pregnancy; used in home pregnancy tests
TORCH Toxoplasmosis Other (syphilis, *Varicella*) Rubella Cytomegalovirus (CMV) Herpes simplex 1 and 2 virus	Pregnant women: negative for IgM antibodies to these pathogens Neonates: negative for IgM antibodies to these pathogens	Pathogens causing these disorders can cross placenta and infect fetus; fetal infections cause mild to severe problems, such as stillbirth, and birth defects; mother should be free of active herpes lesions to deliver vaginally
Alpha-fetoprotein, AFP (serum)	Adults: <40 ng/mL	>500 ng/mL occurs in liver tumors; in pregnancy, levels peak at weeks 16–18; elevated levels occur with Down's syndrome, anencephaly, and spina bifida
Blood type, Rh factor	Rh$^+$ or Rh$^-$	A sensitized Rh$^-$ mother carrying an Rh$^+$ baby can result in erythroblastosis fetalis
Two-hour postprandial glucose test	Adults: 70–140 mg/dL (2 hours after glucose administration)	Increased level indicates possible gestational diabetes, confirmed by more detailed testing
Vaginal culture for group B beta-hemolytic strep	None present	Presence increases risk of prematurity and neonatal sepsis
Neonates		
Blood from umbilical cord	Blood typing	Screens for possible maternal/fetal Rh and ABO blood type incompatibility
Bilirubin	<12 mg/dL	Increased levels occur in jaundice due to immaturity of newborn's liver
Phenylalanine (serum)	1–3 mg/dL	Levels over >4 mg/dL occur after infant is fed in PKU, or phenylketonuria (more commonly tested in urine)
Galactose-1-phosphate	18.5–28.5 U/g hemoglobin	Decreased value indicates galactosemia

Reproductive System

response is to turn the mouth toward the site of stimulation. These reflexes persist until age 4–7 months.

- The **Babinski reflex** is positive, with fanning of the toes in response to stroking of the side of the sole of the foot. This reflex disappears at about age 3 years as descending motor neuron pathways become established.

These procedures check for the presence of anatomical and physiological abnormalities. They also provide baseline information that is useful in assessing postnatal development. In addition, newborn infants are typically screened for genetic or metabolic disorders discussed previously, such as phenylketonuria (PKU), congenital hypothyroidism, galactosemia, and sickle cell anemia.

The excretory systems of the newborn infant are assessed by examining the urine and feces. The first urination may be pink, owing to the presence of uric acid derivatives. The first bowel movement consists of a greenish-black mixture of epithelial cells and mucus called *meconium*.

Pediatrics is a medical specialty focusing on postnatal development from infancy through adolescence. Infants and young children cannot clearly describe the problems they are experiencing, so pediatricians and parents must be skilled observers. Standardized testing procedures are also used to assess developmental progress. In the **Denver Developmental Screening Test (DDST)**, infants and children may be checked repeatedly during their first five years. The test checks gross motor skills, such as sitting up or rolling over, language skills, fine-motor coordination, and social interactions. The results are compared with normal values determined for individuals of similar age. These screening procedures assist in identifying children who may need special teaching and attention.

Too often, parents tend to focus on a single ability or physical attribute, such as the age when the infant takes the first step or the growth rate compared against standardized growth charts. This kind of one-track analysis has little practical value, and parents can become overly concerned with how their infant compares with the norm. *Normal values are statistical averages*, not absolute realities. For example, most infants begin walking at 11 to 14 months of age. But about 25 percent start before then, and another 10 percent have not started walking by the 14th month. Walking early does not indicate true genius, and walking late does not mean that the infant will need physical therapy. The questions on screening tests such as the DDST are intended to identify *patterns* of developmental deficits. Such patterns appear only when a broad range of abilities and characteristics are considered.

Chromosomal Abnormalities and Genetic Analysis

Embryos that have abnormal autosomal chromosomes rarely survive. However, two types of abnormalities in autosomal chromosomes, translocation defects and trisomy, do not invariably kill the individual before birth.

In a **translocation defect**, crossing-over occurs between different chromosome pairs. A piece of chromosome 8, for example, may become attached to chromosome 14 rather than to the other chromatid of chromosome 8. In their new position, the translocated genes may function abnormally, becoming inactive or overactive.

In **trisomy**, something goes wrong in meiosis: At fertilization, one of the gametes contributes an extra copy of one chromosome, so the zygote has three copies of this chromosome rather than two. The location of the trisomy is indicated by the number of the chromosome involved. Zygotes with extra copies of chromosomes seldom survive. Individuals with trisomy 13 and trisomy 18 may survive until delivery but rarely live longer than a year. The notable exception is trisomy 21.

Trisomy 21, or **Down's syndrome**, is the most common viable chromosomal abnormality. There were 11.8 Down's syndrome births per 10,000 births between 1999 and 2003 in the United States. Affected individuals exhibit mental retardation and characteristic physical malformations, including a facial appearance that gave rise to the term *mongolism*, once used to describe this condition. The degree of mental retardation ranges from moderate to severe. Few individuals with Down's syndrome lead independent lives. Anatomical problems affecting the cardiovascular system may prove fatal during childhood or early adulthood. Although some individuals survive to moderate old age, many develop Alzheimer's disease while still relatively young (before age 40).

For unknown reasons, maternal age is linked to the risk of having a child with trisomy 21. For a maternal age below 25, the incidence of Down's syndrome approaches 1 in 2000 births, or 0.05 percent. For maternal ages 30 to 34, the odds increase to 1 in 900, and over the 10 years from age 35 to 44, the odds change from 1 in 290 to 1 in 46. These statistics are becoming increasingly significant, because many women have delayed childbearing until their mid-thirties or later.

Abnormal numbers of sex chromosomes do not produce effects as severe as abnormalities induced by extra or missing autosomal chromosomes. Individuals with **Klinefelter's syndrome** have the sex chromosome pattern XXY. The phenotype is male, but the extra X chromosome causes reduced androgen production. As a result, the breasts are slightly enlarged, the testes fail to mature, and affected individuals are sterile. The incidence of this condition among newborn males averages 1 in 750.

Individuals with **Turner's syndrome** have only a single, female sex chromosome; their sex chromosome complement is abbreviated XO. This kind of chromosomal deletion is known as **monosomy**; its incidence at delivery has been estimated as 1 in 10,000. The condition may not be recognized at birth, because the phenotype is normal female. But maturational changes do not appear at puberty. The ovaries are nonfunctional, and estrogens are produced only at negligible levels.

Fragile-X syndrome causes mental retardation, abnormal facial development, and increased testicular size in affected males. The origin is an X chromosome that contains a *genetic stutter*, which is an abnormal repetition of a single triplet. In this case, the problem is the repetition of the nucleotide sequence CCG at one site of the *FMR1* gene on the X chromosome. Usually this triplet is repeated from 5 to 40 times. Affected persons with Fragile X syndrome have more than 200 repetitions. (A genetic stutter on another chromosome is responsible for Huntington's disease, discussed on page 99.) The presence of the stutter in some way disrupts the normal functioning of adjacent genes and so produces the signs of the disorder. Female carriers of Fragile X syndrome, with one normal X chromosome and one affected X chromosome, are less affected, and those with fewer than 200 repetitions have even fewer problems.

Reproductive System

Many of these conditions can be detected prior to birth through the analysis of fetal cells. In **amniocentesis**, a sample of amniotic fluid is removed and the fetal cells it contains are analyzed. This procedure permits the identification of over 20 congenital conditions, including Down's syndrome and neural tube defects. The needle inserted to obtain a sample of fluid is guided into position by using ultrasound. Unfortunately, amniocentesis has two major drawbacks:

1. Because the sampling procedure represents a potential threat to the health of the fetus and mother, amniocentesis is performed only when known risk factors are present. Examples of risk factors include a family history of specific conditions or, in the case of Down's syndrome, a maternal age over 35.

2. Sampling cannot be performed safely until the volume of amniotic fluid is large enough that the fetus is unlikely to be injured during the sampling process. The usual time for amniocentesis is at a gestational age of 14 to 15 weeks; it can then take several more weeks to obtain results. By the time the results are received, the option of therapeutic abortion may no longer be available.

In an alternate procedure known as **chorionic villus sampling**, cells collected from the chorionic villi during the first trimester (usually at weeks 10 to 12) are analyzed. Although it can be performed at an earlier gestational age than can amniocentesis, chorionic villus sampling may have a higher risk of miscarriage than amniocentesis, and does not test for neural tube defects such as spina bifida. In 2014 it is possible to noninvasively detect cell free fetal DNA in a pregnant woman's blood and screen selected high risk pregnancies for chromosomal abnormalities. To confirm a diagnosis, invasive amniocentesis or chorionic villus sampling is needed.

Assisted reproductive technologies (ART) (pp. 225–226) result in zygotes. A single cell taken from the eight-cell blastomere can be tested for genetic abnormalities and, if absent, that screened zygote used in an IVF procedure. This Preimplantation Genetic Testing is increasing in popularity.

Death and Dying

Despite exaggerated claims, few cases of individuals who have reached an age of 120 years have been substantiated. Estimates for the average life span of individuals born in the United States during 2010 are 76.2 years for males and 81.1 for females. Interestingly enough, the causes of death vary with the age group. Consider the graphs shown in ▶**Figure 96** and the data in **Table 44**, which indicate the United States mortality statistics for various age groups. The major cause of death in young people is accidents; in adults over age 40 to 45, it is cardiovascular disease. More specific information about the major causes of death is given in **Table 44**. Many of the characteristic differences in mortality result from changes in the individual's functional capabilities, which are linked to development or senescence. The picture differs significantly for those living in countries and cultures with different genetic and environmental pressures.

The differences in mortality values for males and females are related to differences in the accident rates among young people and in the rates of heart disease and cancer among older individuals. For instance, more young males die from accidents and homicide than females, and lung cancer kills older men more often than

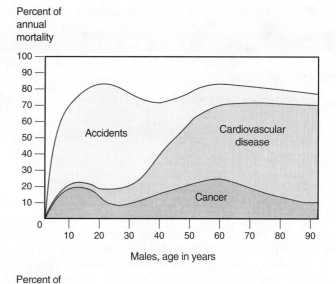

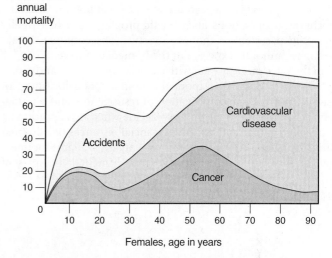

▶**Figure 96   Major Causes of Postnatal Mortality.**

it does older women. But among women, the incidence of lung cancer and related killers, including pulmonary disease, heart disease, and pneumonia, has been steadily increasing as the number of women smokers has risen. This change has narrowed the difference between male and female life expectancies.

Experimental evidence and calculations suggest that the human life span has an upper limit of about 150 years. As medical advances continue, research must focus on two related issues: (1) extending the average life span toward that maximum and (2) improving the functional capabilities of long-lived individuals. The first objective may be the easiest from a technical standpoint. It is already possible to reduce the number of deaths attributed to specific causes. For example, new treatments promote remission in a variety of cancer cases, and anticoagulant therapies reduce the risks of death or permanent damage after a stroke or heart attack. Many defective organs can be replaced with functional transplants, and the use of controlled immunosuppressive drugs increases the success rates of these operations. Artificial hearts have been used, although with limited success thus far, and artificial kidneys and endocrine pancreases are under development.

The second objective poses more of a problem. Few people past their mid-nineties lead active, stimulating lives, and many

Reproductive System

Table 44	The Five Major Causes of Death in the U.S. Population (2006–2010)				
Rank	**Age 1–14**	**Age 15–24**	**Age 25–44**	**Age 45–64**	**Age 65 and over**
1	Accidents	Accidents	Accidents	Cancer	Heart disease
2	Cancer	Homicide	Cancer	Heart disease	Cancer
3	Congenital anomalies	Suicide	Heart disease	Accidents	COPD*
4	Homicide	Cancer	Suicide	COPD*	Cerebrovascular disease
5	Heart disease	Heart disease	Homicide	Chronic liver disease and cirrhosis	Alzheimer's disease

*COPD = chronic obstructive pulmonary disease

may find the prospect of living another 50 years rather horrifying unless the quality of their lives could be significantly improved. Our abilities to prolong life now involve making stopgap corrections in systems that are on the brink of complete failure. Reversing the process of senescence would entail manipulating the biochemical operations and genetic programming of virtually every organ system. Although investigations continue, breakthroughs cannot be expected in the immediate future.

In the interim, we are left with some serious ethical and moral questions. If we could postpone the moment of death almost indefinitely with some combination of resuscitators and pharmacological support, how would we decide when it is appropriate to do so? How can medical and financial resources be allocated fairly? Who gets the limited number of organs available for transplant? Who should be selected for experimental therapies of potential significance? Should we take into account that health care of an infant or child may add decades to a life span, whereas the costly insertion of an artificial heart in a 60-year-old may add only months to years? How shall we allocate the costs of sophisticated procedures that can reach hundreds of thousands of dollars per individual over the long run? Are these societal, individual, or family responsibilities? Will only the rich be able to survive into a second century of life? Should the government provide the funds? If so, what will happen to tax rates as the baby boomers become elderly citizens? And what about the role of the individual? If you decline treatment, are you mentally and legally competent? Could your survivors bring suit if you were forced to survive or if you were allowed to die? These and other difficult questions will not go away. In the years to come, we will have to find answers we are content to live and die with.

1. Gayle is in her sixth month of pregnancy and notices that her face and hands are swelling. At her next checkup, she mentions this swelling to her obstetrician. Gayle's blood pressure is now 150/110. The obstetrician examines the results of Gayle's urinalysis:

   pH: 5

   Color: amber

   Appearance: cloudy

   Specific gravity: 1.040

   Protein: 2+ on dipstick (160 mg/day or 16 mg/dL)

   Glucose: trace

   Ketones: none

   Gayle is immediately hospitalized. What is the likely initial diagnosis?

   a. placenta previa

   b. preeclampsia

   c. ectopic pregnancy

   d. erythroblastosis fetalis

   e. gestational diabetes

2. Moira and Sam have been attempting to conceive for the last year. They have consulted a fertility specialist, who conducted lab and diagnostic tests. The lab results are as follows:

   Moira

   LH (serum)

   follicular = 15 mIU/mL

   midcycle = 80 mIU/mL

   Progesterone (serum)

   before ovulation (follicular) = 0.5 ng/mL

   after ovulation (luteal) = 4 ng/mL (within normal range)

   FSH (serum)

   pre- and post-ovulation = 12 mIU/mL

   midcycle = 30 mIU/mL

   What do these tests indicate about Moira's monthly hormonal cycle? What aspects of her reproductive status remain uncertain?

   Sam

   Semen Analysis

   Volume: 3.0 mL

   Sperm count: 10 million/mL

   Sperm motility: 65%

   Abnormal sperm: 20%

   What do these tests indicate about Sam's reproductive status? What else remains uncertain about his reproductive status?

   What are some treatment options available to this couple?

3. John, a 66-year-old man, reports to his physician that he has had trouble urinating for the last several months, with narrowed stream and difficulty initiating urination. For

the last two days, urination has become more frequent, painful, and produces only small amounts of urine. Urinalysis results are within normal limits, with the exception of the presence of bacteria on culturing. Other observations and lab tests are the following:

Digital rectal exam: enlarged prostate and "rock hard" prostatic nodule in one lobe

Transrectal ultrasonography: confirms abnormal mass present

Urine culture: bacteria present

Prostate-specific antigen: 12 ng/mL

What are likely diagnoses? Why is it difficult for John to urinate?

NOTES:

Answers to these problems can be found on page 242.

Case Studies

These case studies are examples of how knowledge of underlying anatomy and physiology and organ system interactions can produce the overlapping signs and symptoms of disease. What may at first appear to be a straightforward single health problem can be complicated by linked organ systems and past medical history. The assessment of older patients can be made more difficult because they may have several chronic health problems, each involving different systems and managed by a different suite of medications, before the latest medical crisis developed. When new health problems arise, the homeostatic responses that might otherwise compensate for the condition may be compromised by preexisting conditions. It is therefore important to recognize various possibilities and reevaluate the situation as new information becomes available. When solving an algebraic equation, there is only one correct answer. But a medical problem may have many possible solutions, and the correct answer can change over time. Nevertheless, as in mathematics, solving a health problem involves critical thinking and use of the scientific method. These exercises are designed to show you how a clinician would work through complex medical problems through a series of relatively simple steps. Pause where indicated to answer the associated questions before continuing to read the Case. Then at the end of the Case, review your answers and compare them with the answers presented on pages 243–246.

CASE STUDY 1 George's Demise

George P., a 65-year-old man, is brought to the emergency room by his family because he almost fainted when he stood up at home. George reports that he has felt tired and weak for the last few days. He is short of breath but attributes this symptom to his chronic bronchitis (he has smoked for 25 years). *(Questions 1, 2, 3)*

When George arrives at the emergency room (ER), his blood pressure is 100/70 while he is lying down, and his pulse is 110 bpm. While he is sitting, his blood pressure is 90/70 and his pulse is 115 bpm. On palpation, he has mild epigastric tenderness, and his stool guaiac test is positive for blood. *(Questions 4, 5, 6)*

A stat Hct/Hb is performed, and the results are Hct is 33, Hb is 10.0 g/dL. He is immediately given IV fluids and oxygen, and his blood is sent for typing and cross-match testing. *(Question 7)*

A nasogastric tube is inserted into George's stomach; suction reveals that the stomach contains black-brown material resembling coffee grounds. *(Question 8)* A guaiac test on this material is positive for blood. The stomach is flushed with water introduced through the tube, and the suction fluid gradually clears. *(Question 9)* Acid inhibitors are administered intravenously, and George is given antacids orally. A gastroenterology consult is requested with diagnostic and potentially therapeutic endoscopy planned. *(Question 10)*

During the administration of IV fluids, George's blood pressure stabilizes at 120/70, and his pulse slows to 90 bpm. A second blood test reveals a hematocrit of 29 and a hemoglobin concentration of 9.0 g/dL. George now feels more comfortable and is breathing easier; he is transferred from the emergency room to the intensive care unit (ICU) for further observation. An hour later, his Hct is 30 and his Hb is 9.1.

While talking with his family, George suddenly experiences anterior chest pain that radiates into his neck and left shoulder, with diaphoresis (cold sweats) and shortness of breath. *(Question 11)* Nitroglycerin and antacids have no effect. *(Question 12)* His blood pressure falls (hypotension), and IV fluids, blood transfusion, and further medication cannot stop the trend. *(Question 13)* Four hours after the first chest pain, George's heart stops (cardiac arrest), resuscitation attempts prove futile, and he dies. *(Question 14)*

Discussion Questions

1. What is the likely explanation for fainting on standing? What is this phenomenon called?

2. Disorders affecting which systems may be responsible for sensations of tiredness and weakness?

3. What are some of the long-term effects of smoking on systems other than those discussed in Question 2?

4. What is the significance of these postural changes in blood pressure and heart rate?

5. What might account for the combination of epigastric tenderness and a positive stool guaiac test?

6. How do these values compare with normal values?

7. Why is George treated with IV fluids and oxygen? Why does the hospital staff perform blood typing and cross-match testing?

8. What is a nasogastric tube? Why did the physician decide to perform this test?

9. Why use water? Is it significant that the fluid clears?

10. What is the preliminary diagnosis at this point? Why are acid inhibitors and antacids administered? What might be seen on endoscopy?

11. What disorders may account for these symptoms?

12. Why are nitroglycerin and antacids administered? What does their lack of effect mean?

13. What can account for the decline in blood pressure despite the therapies administered?

14. What are the final diagnoses? Discuss the mechanisms responsible for the progression of this case.

CASE STUDY 2 Christine's Workout

On Monday, Christine K., a 35-year-old woman, complains to her physician of bilateral anterior chest and arm pain. *(Question 1)* The pain started after Christine worked hard in the backyard the previous Saturday, lifting and carrying plants, fertilizer, and concrete blocks. (She had continued working despite arm pain that developed about midday.) Christine has no problem breathing, but she cannot raise her arms above her shoulders, and her shoulder and chest muscles are very sore to the touch (tender). *(Question 2)* On questioning, Christine reports that her hands are not numb, and her urine is not dark. *(Question 3)*

On physical examination, muscles around the elbows, arms, shoulders, and anterior chest are swollen and tender. Christine can move her fingers and wrists normally, but she has some pain in the proximal forearm muscles. The ranges of motion at the elbow and shoulder joints are reduced. No bruises are evident, and Christine has normal sensation in her fingertips. *(Question 4)*

On inspection, her urine is cloudy and light brown; on dipstick testing, it tests positive for protein and blood (the urine dipstick test cannot distinguish hemoglobin from myoglobin). Microscopically, there are no RBCs in the urine. *(Question 5)*

Blood tests are ordered to check for CPK (CK) and LDH, and kidney function tests are performed to assay BUN and creatinine. *(Question 6)* Treatment initially consists of pain medication, ice applied to the sore muscles, and a high fluid intake (at least 4 quarts every 24 hours). *(Question 7)*

Christine is asked to return the next day and to call the physician sooner if her urine remains cloudy, if the volume decreases, or if her hands or forearms feel worse. *(Question 8)* Christine returns the next day to have her urine rechecked and to get the results of her lab tests. Her arms and chest are still sore and swollen, her hands still have normal sensation, and her urine is clear and dilute, but urine tests show traces of protein and hemoglobin. Her lab tests from the previous day show CK and LDH levels 10 times normal values, but kidney tests were normal. *(Question 9)*

Christine continues a high fluid intake for the next week, and her sore muscles gradually became less swollen and tender. During this period, her urine and blood enzyme tests returned to normal. *(Question 10)*

Discussion Questions

1. On the basis of the case history, discuss the possible causes of these symptoms, and rank them from high probability to low probability.

2. What is significant about breathing and arm elevation?

3. Why ask about her hands and her urine?

4. Discuss the potential importance of these observations.

5. What do these findings suggest?

6. Why are these tests selected? What is the physician trying to determine?

7. What is the preliminary diagnosis?

8. What problems might be present if Christine's urine remains cloudy or declines in volume or if her peripheral sensation changes?

9. What is the significance of these lab results?

10. What is the final diagnosis?

CASE STUDY 3 Charlene's Pain

Charlene, a 75-year-old woman, arrives at her physician's office reporting pain in her right back, thigh, and leg. *(Question 1)* This pain, which is continuous, was initially dull and aching but has gotten progressively more intense, especially over the last few days. The pain seems unrelated to a position change or to eating, and Charlene has had no problems with urination or defecation. *(Question 2)* Her general medical health is poor, with severe COPD (probably from her 50 years of smoking). *(Question 3)* She has needed continuous low-flow oxygen administration, bronchodilators, and oral prednisone (corticosteroid) treatment for the last 18 months. Despite this, she is chronically dyspneic, with labored respirations at 20–30 breaths per minute. *(Questions 4, 5, 6)*

The physician notes that the physical examination reveals a thin, pale, weak elderly woman with a round, moon-shaped face, breathing with pursed lips and getting oxygen from nasal prongs. *(Question 7)* Her vital signs are normal except for the increased respiratory rate. She is not tender over her back or vertebral column, right thigh, or leg and has a normal range of motion in her right hip, knee, and ankle. She can walk normally. Her skin is clear, and her abdominal exam is normal. Her breath sounds are decreased, but she has no increased cough or sputum production. A urinalysis is performed, with normal results. *(Question 8)*

X-rays of her lower back and hip show osteoporosis with DJD, a possible compression fracture at L_3 (of uncertain age), but no acute problems. *(Question 9)* The physician reaches a tentative diagnosis of musculoskeletal pain possibly associated with the lumbar spine compression fracture. *(Question 10)* Blood samples are sent to the lab for analysis, and pain medications are prescribed.

Discussion Questions

1. What are the most likely sources of pain that is distributed in this way?

2. What do these observations tell you about the nature of the problem?

3. Discuss the pattern of symptoms characteristic of COPD.

4. On the basis of this history, what else might be responsible for Charlene's pain?

5. How do bronchodilators work? What are the systemic effects of prednisone?

6. Relate these observations to what you now know about Charlene's medical history.

7. What do these observations reveal about the nature of the problem?

8. Why is it significant that there has been no change in coughing or sputum production and that the urinalysis was normal?

9. What was not found on the x-rays that could have accounted for the reported symptoms?

10. Explain the proposed linkage between the compression fracture and musculoskeletal pain.

11. Would you expect this, on the basis of the tentative diagnosis?

12. Why is the bone scan ordered?

The blood test results are normal. However, over the next three days, the pain intensifies. *(Question 11)* There are no new signs or symptoms, and Charlene continues her pain medications. A bone scan is ordered for the following day, but that night her friends bring Charlene to the emergency room with severe mental confusion. *(Question 12)*

The consulting internist wonders about (1) a toxic reaction to the pain medications, (2) a stroke, or (3) an infection masked by chronic steroid use. *(Question 13)* Laboratory tests are then performed; the results are normal except the chest x-ray and the hypoxia consistent with the preexisting COPD. A lumbar puncture is performed, and the results suggest meningitis. Charlene is admitted to the hospital and started on antibiotics pending the results of bacterial culture of blood, urine, sputum, and CSF samples.

The next morning, Charlene's family physician reexamines her in the hospital and notes the appearance of a new, red, papular rash along the side of Charlene's right ankle and the top of her right foot. *(Question 14)* Red papules are also scattered diffusely across her trunk and arms. During the next day, the areas involved develop distinctive clear vesicles typical of a herpes zoster (shingles) infection.

Charlene is started on IV acyclovir, and during the next two weeks, her rash fades and her mental status improves. *(Question 15)* The leg pain, diagnosed as postherpetic neuralgia, persists for three months but then subsides. There is no recurrence. *(Question 16)*

13. Discuss each of these possibilities. What tests might be performed to determine which of these factors (if any) is responsible for Charlene's condition?

14. Does the rash affect the diagnosis, or is it an interesting but relatively insignificant secondary problem?

15. What is acyclovir used for?

16. What was the final diagnosis? Explain the linkage between (a) the pattern and timing of pain development, (b) the meningitis and its effects on Charlene's mental state, and (c) the distribution of the rash.

CASE STUDY 4 Hal's Athletic Adventure

Discussion Questions

Hal, a 50-year-old politician, was a college basketball player and stayed fairly active with sports and weekend pickup games through his thirties and early forties. *(Question 1)* Having a members-only gym near his office facilitated this practice. However, intense campaigning and successful election to the U.S. Congress meant extra travel, fried chicken dinners, hotel rooms, and prolonged sitting, and weight gain.

After years of relative inactivity, with only sporadic visits to the gym, Hal participates in an intense basketball game. Midway through the game, he collapses suddenly with a sharp pain behind the right ankle. *(Question 2)* While he is on the floor, the pain eases, and he is able to move his ankle and foot reasonably well. But when he tries to stand, his ankle is very weak and painful. He proceeds (hopping) to a physician.

On examination, there is normal and painless range of motion in his right hip, knee, and toes. With his right leg dangling, he can flex (dorsiflex) and extend (plantar flex) the ankle with only minimal pain. *(Questions 3, 4)* The region posterior to the ankle and proximal to the heel is swollen and tender. *(Question 5)* When he lies prone, with the right knee bent, there is a palpable depression posterior to the ankle, extending toward the knee. Grasping and compressing the calf muscles of the normal left leg produces an ankle extension reflex (plantar flexion), but doing this with the right leg has no motion of the ankle joint. *(Question 6)*

1. What are the effects of (a) aging and (b) reduced levels of activity on the systems critical to the performance of athletic activities? How do these effects interact?

2. List at least three possible causes of the ankle problem, and relate them to relevant factors detailed in your answer to Question 1.

3. What is the significance of the reduction in pain and Hal's ability to move the ankle and foot when he is not standing?

4. What does this tell the physician about the nature of the injury?

5. What structures are located in this region?

6. What does this mean, and what was the final diagnosis? Why was Hal still able to extend the ankle when it was not bearing weight?

CASE STUDY 5 Jim and That Darned Cat

Discussion Questions

Jim, 65 years old, visits his new physician on Friday afternoon. Jim has a headache that will not go away. *(Question 1)* He and his wife retired and moved into a new home about three months ago. A week ago he was looking for one of their cats, which he found under the steps to the porch. When backing out from beneath the porch, he banged his head. Jim thinks that the headache started either at that time or shortly thereafter during an argument with his wife. *(Question 2)* Jim wants to get rid of the cats, because he thinks they are too much trouble and because they make him sneeze, but his wife won't consider it. Their arguments have been heated, and he has often developed mild headaches afterward.

Jim describes this headache as continuous, all over his head, and extending into his posterior neck muscles. *(Question 3)* His neck muscles are tight. His vision is normal, as is his hearing, and he has no nausea. Tylenol and ibuprofen have helped slightly. *(Question 4)* He hasn't been sleeping well and remains upset about the cats.

1. Discuss the various possible causes of headaches.

2. List the possible causes for Jim's headache that match this history.

3. Discuss the significance of each of these symptoms.

4. Does the lack of substantial pain relief with these drugs suggest anything about the nature of the problem?

5. Discuss the significance of each of these physical exam findings. What possible problems were less likely because of these signs?

6. How does the situation on Monday match the preliminary diagnosis?

On physical examination, Jim is alert and oriented, and his vital signs are normal. His neck is supple, with normal range of motion. The skull is normal on palpation, and there are no signs of abnormal cranial nerve function. His tympanic membranes are normal, as are his pupil reflexes, eye motion, and retinae. His gait and deep tendon reflexes are also normal. The physician reaches a preliminary diagnosis of a muscle tension headache and gives Jim a mild sedative/muscle relaxant/pain medication. *(Question 5)* Jim has instructions to call or return if the headache does not improve.

On Monday, Jim calls the physician's office to report that the headache is worse. He is seen that morning at 11 a.m. His appearance and manner have changed markedly over the weekend. *(Question 6)* He could not drive, so his wife brought him to the office. He seems drowsy (which he attributes to the medication), and his headache is worse. It is a throbbing pain that increases if he lies down, presses against his head, or flexes his neck. His vision is still normal. Although he is not vomiting, he has not wanted to eat today. *(Question 7)*

On physical examination, Jim still has normal vital signs, and he has no fever. However, his blood pressure is 140/90, up from 120/80 on Friday. His neck motion is normal, although flexion increases the severity of the headache. His cranial nerve, eye, and gait/balance tests are normal and symmetrical. *(Question 8)* Because there has been no improvement and the symptoms have worsened, Jim is sent for an urgent CT scan of the head and a series of blood tests. The radiologist reports to the family physician that the CT scan shows large, bilateral subdural hematomas of uncertain age. There is no sign of a skull fracture, and the blood tests (CBC, WBC, diff. count) are normal. *(Question 9)* Jim is immediately referred to a neurosurgeon. After discussion, Jim signs informed consent papers and is taken to the operating room. The neurosurgeon draining the subdural hematomas reports that there is an initial high-pressure release of blood when the skull is pierced. *(Question 10)*

After recovering from anesthesia, Jim finds that his headache is virtually gone. He jokes with his family physician that he is glad that he didn't have to get his head shaved (he is already bald) and says he has decided to buy a big dog.

7. Discuss the significance of these symptoms.

8. Why is it important that the results are symmetrical? What would be the implications if they had been asymmetrical?

9. Is it significant that the blood tests are normal? How does this result affect the diagnosis? If results had been abnormal, what else could be involved?

10. What does the high-pressure release tell you about the status of the hematoma?

CASE STUDY 6 Joe's Abdominal Pains

Joe, a 70-year-old man, goes to the hospital in 1992 with severe upper abdominal pain and vomiting that began after his last meal. *(Question 1)* An x-ray and blood analysis show elevated liver function tests and indications of pancreatic and gallbladder problems. *(Question 2)* The initial diagnosis is acute cholelithiasis. *(Question 3)*

Joe is sent to surgery, and his inflamed gallbladder is removed. After surgery, he remains hospitalized for six weeks with continued epigastric pain and elevated serum levels of amylase. Over this period, he is given nutrients through an intravenous infusion. *(Question 4)*

During surgery, the surgeon noted that Joe's liver had a lumpy, scarred appearance, and a liver biopsy was taken. The pathologist reports signs of chronic inflammation and scarring, cirrhosis. The attending physician visits Joe and obtains more information about his history. Joe had vascular surgery for femoral artery occlusion in 1980 and surgery to repair an aortic aneurysm in 1985. *(Questions 5, 6, 7, 8)* During each of these procedures, he received blood transfusions. Joe was a "social drinker" (two or three drinks a day for 35 years), often at business lunches or dinners. He has never used injected drugs and has been heterosexual and monogamous with his wife for the last 40 years. *(Question 9)* Joe had been a POW in a German concentration camp during World War II, where unknown substances were injected into his chest as part of an experiment. (The experimental goals and methods are not known. The lack of medical ethics is appalling.) He has had hypertension and occasional symptoms of

Discussion Questions

1. What disorders may produce these symptoms?

2. What was probably seen on the x-ray? What blood tests were likely to have been abnormal?

3. Discuss the origin, symptoms, and treatment options for cholelithiasis (gallstones).

4. What condition does Joe have? What nearby organs may be affected? Is it serious? Why does he receive nutrients by infusion rather than by mouth?

5. What are the technical terms for liver inflammation and liver scarring? What are the most likely causes of these conditions?

6. What factors are apparent in this history that may have affected liver function?

7. Discuss (a) the condition responsible for femoral artery occlusion and (b) the goal of the surgery.

8. Describe the repair of an aortic aneurysm.

9. Which forms of hepatitis are unlikely, given this history?

10. What are the most likely routes of viral hepatitis infection?

11. Why was Joe given these instructions?

post-traumatic stress disorder since that time, but until his vascular surgery, his health was otherwise good. Blood tests on admission showed no signs of hepatitis A or B, but his blood contained antibodies to hepatitis C (this test became available in 1990). *(Question 10)* He is told to stop drinking and to restrict the use of drugs, such as tranquilizers. *(Question 11)*

In the three years after his release from the hospital, Joe develops ascites, peripheral edema, splenomegaly, lowered serum protein levels, a reduced platelet count, and elevated blood ammonia levels. *(Question 12)* He becomes somewhat disoriented and has difficulty speaking if he eats a high protein meal. *(Question 13)*

After another year, Joe reenters the hospital by ambulance. He is unresponsive, his hematocrit is 25, and he has vomited dark blood. His blood ammonia levels are above 500. A gastroscopic examination reveals ruptured blood vessels along the esophagus, with bleeding into the stomach. *(Question 14)* Epinephrine is injected into the bleeding vessels, and the bleeding stops. *(Question 15)* The gastric contents are aspirated, and the digestive tract evacuated. His hematocrit rises, his ammonia levels decline, and he awakens confused but relatively alert. *(Question 16)* He is discharged from the hospital after two weeks and placed on a low-protein diet. *(Questions 17, 18)*

12. Are these symptoms related to Joe's hepatitis? Discuss the linkage, if any.

13. Why is Joe experiencing disorientation and difficulty in speaking?

14. Discuss his status on admission and the implications of his symptoms.

15. Why was epinephrine injected at the bleeding site?

16. Why aspirate the gastric contents?

17. Discuss the factors that accounted for his hospitalization and the reason for the dietary restrictions.

18. What other treatments might be used to relieve Joe's symptoms? Are any of these options practical in this case?

CASE STUDY 7 Mrs. M.'s Summer Problems

Mrs. M. is a 47-year-old woman who spends three long, warm days in a wet swimming suit while she participates in canoe races. *(Question 1)* Afterward, she develops vulvar pruritus and notices a white, curdy ("cottage cheese") vaginal discharge. *(Question 2)* She consults her doctor for the first time in three years. While evaluating and treating her vaginitis, the physician notices that Mrs. M. has not had a Pap test or breast exam since her tubal ligation six years ago, and has never had a mammogram. She has no history of diabetes. Because Mrs. M. is usually healthy, active, and has trouble scheduling regular checkups, her physician convinces her to have those tests performed while she is still in the office. *(Questions 3, 4)*

The pelvic exam is normal except for inflammation of the vulva and vagina. A KOH prep of the vaginal discharge shows a yeast infection. On breast exam, a 2-cm, hard, painless lump that dimpled the overlying skin is found superior to the left nipple. There is no other breast abnormality or palpable axillary lymphadenopathy. Mrs. M. had felt the lump for several months, but she did not want to mention (or think about) it, because her aunt recently died of breast cancer. *(Questions 5, 6)*

A mammogram shows changes consistent with breast cancer. During the next few weeks, Mrs. M. and her husband consult surgeons, the Internet, and nutritionists. A lumpectomy is recommended, with axillary node biopsy followed by irradiation and/or chemotherapy to be determined by the extent of lymph node involvement. *(Question 7)* Mrs. M. decides to have the lumpectomy with no radiation. Because axillary lymph node biopsies are negative and the tumor cells are estrogen-sensitive, she agrees to have tamoxifen therapy, which reduces the risk of recurrence of breast cancer; tamoxifen is an anti-estrogen drug that blocks estrogen effects on breast tissue, including estrogen-sensitive breast cancer, but has some estrogen-like effects on the uterus. *(Question 8)*

The surgeon removes a large area around the tumor to avoid possible recurrence, as the area will not be irradiated. The surgery leaves a loss of tissue, including the nipple, but Mrs. M. declines breast augmentation surgery or a prosthesis. Six years later, there is no evidence of disease, and she remains on tamoxifen. Because recent studies on prolonged use of tamoxifen reveal an increased risk of uterine cancer with use longer than five years, she is switched to another anti-estrogen, raloxifene, that does not have an increased risk of uterine cancer.

Discussion Questions

1. Discuss the likely cause of Mrs. M.'s initial problem.

2. What organisms may be responsible?

3. What is the likely treatment for this condition?

4. As people age, their risk of developing various diseases increases. Early detection and treatment by screening tests improves outcomes. Discuss the relationship between normal good health, aging, and diseases such as cancer.

5. What is the likely evaluation for the breast lump?

6. Is it significant that her aunt had breast cancer?

7. What stage is her breast cancer? Discuss the pros and cons of aggressive treatment of breast cancer and other cancers.

8. Why is estrogen sensitivity important?

CASE STUDY 8 Cynthia's Bad Back

Cynthia, 28 years old, barely manages to make it to her physician's office. She has terrible right lower back pain. *(Question 1)* She has had a few twinges over the last few days that disappeared in seconds, but today she woke up at 3 a.m. with steadily increasing pain in her right back at about waist level. She at first attributed it to physical activity, but she remembered no injury and was able to move and bend her back, arms, and legs without any change in the level of pain. Cynthia has no previous history of back pain or chronic health problems. *(Questions 2, 3)* She has no cough or respiratory symptoms, no dysuria or dark urine, and no nausea, vomiting, or bowel problems. Her last menses, three weeks ago, was normal. *(Question 4)*

On entering the examination room, the physician finds Cynthia kneeling on the floor with her head resting on a chair seat. On request she is able to stand, walk, and sit on the exam table without limitation of movement. She does not have a fever, but her blood pressure and pulse are elevated. *(Question 5)* Her head, neck, and chest exams are normal. She is not tender over her kidneys, spinal vertebrae, or muscles. The neurological exam is normal. On abdominal examination, she is tender in her right lower abdomen but has no masses, guarding, or rebound. Neither the liver nor the spleen is enlarged, and bowel sounds are normal. A urinalysis is normal, and a pregnancy test is negative. *(Questions 6, 7)*

On pelvic examination, Cynthia's vulva, vagina, and cervix are normal in appearance, but on palpation the pain on the right side increases with movement of the cervix. Her uterus is small, firm, and nontender, but her adnexal area where the right uterine tube and ovary are located is very tender, with a 3-cm mass detected on palpation. Cynthia feels pain both in her abdomen and in her back when this area is palpated. Rectal examination confirms the pelvic findings; the stool guaiac test is negative. *(Question 8)*

After Cynthia is admitted to the hospital, blood tests, x-rays, and ultrasound exams rule out pneumonia, cholecystitis, ectopic pregnancy, and appendicitis but show an enlarged cystic right ovary. Abdominal surgery is performed, and the ovary is found to be twisted on its supporting ligaments. *(Question 9)* The twisting is reduced, the cyst removed, and the ovary stabilized to prevent recurrence of the problem. Cynthia recovers from the surgery and resumes a normal life.

Discussion Questions

1. Discuss the range of potential causes of acute back pain, including referred pain.

2. Is the history consistent with musculoskeletal injury?

3. What does the continuous nature of the pain imply about the condition?

4. Discuss the implications of these findings; what would have been likely diagnoses if there had been a cough, dysuria, dark urine, nausea, vomiting, bowel problems, or abnormal menstrual periods?

5. Why might her bp and pulse be elevated?

6. Discuss the relevance of each of these findings. What potential causes of acute back pain are being eliminated?

7. What results would you expect to have seen, if these conditions were not ruled out?

8. Why was the stool guaiac test performed?

9. What is a comparable condition that affects males?

ANSWERS TO END-OF-UNIT CLINICAL PROBLEMS

UNIT 2 ANSWERS

1. The patient is most likely suffering from rheumatoid arthritis (RA), although several other joint disorders could have been present. Osteoarthritis normally occurs in older individuals, is less symmetric, and weight-bearing joints are more commonly affected. Uric acid crystals in the joints cause gout, a form of arthritis (but note the absence of uric acid in the synovial fluid). Bacterial arthritis is ruled out because there are no bacteria in the synovial fluid. Systemic lupus erythematosus (SLE) also causes joint problems, but the patient did not have the characteristic skin lesions found in SLE.

2. The weakness of the pelvic girdle muscles in a young boy suggests a possible muscle disorder, such as muscular dystrophy (MD). Aldolase is an enzyme present in high concentrations in muscle fibers, and its appearance in serum indicates muscle damage. A muscle biopsy revealed histological changes characteristic of MD, including the absence of normal amount of dystrophin. The nerve conduction tests were normal.

3. Because his ankle has been supported by the cast, his calf muscles (soleus and gastrocnemius muscles) will probably have partially atrophied (shrunk) from disuse, reducing the diameter of the calf and reducing the muscle strength produced during plantar flexion. If his skin were even lightly tanned, the newly exposed skin would be paler, and it may be wrinkled due to the muscular atrophy that has occurred. Because the normal shedding of epidermal skin cells has been confined, clumped scaly patches will flake off, and there may be intense itching. While the cast was in place, the secretions of the integumentary glands were contained, and if the cast was not fiberglass and lined with a padding that permitted washing, there may be a pronounced smell when the cast is removed. Because the thigh and hip muscles have become accustomed to picking up that foot with a heavy cast on it, the first few steps may show unconscious "high stepping" around the exam room, while the patient scratches wildly, exclaims over how awful the leg looks and how terrible the cast smells. Fortunately, none of these changes will be permanent if Jeremy resumes his normal pattern of activities.

UNIT 3 ANSWERS

1. The occurrence of optic neuritis in a young adult, the plaques, and abnormal CSF values point to multiple sclerosis (MS). Clinically, a history of *multiple* neurological deficits at *multiple* times and involving *multiple* sites on the body is the classic pattern for MS. Because MS is an autoimmune problem, IgG antibody levels increase in the CSF.

2. The physical changes result from elevated levels of the adrenal hormone cortisol. This can be caused by a pituitary tumor overproducing ACTH or a primary tumor in the adrenal gland that makes too much cortisol. Taking large doses of corticosteroid medication for a long time is another cause. Answer (b) would be the preliminary diagnosis.

   The pituitary gland is anatomically seated in the sella turcica. Erosion or enlargement of the sella turcica indicates a possible pituitary gland tumor. The MRI confirmed the presence of an abnormal mass in the pituitary gland.

3. The unilateral distribution and pain suggest sensory nerve involvement following the contours of a single dermatome. The condition appears to be affecting the dermis and epidermis. The appearance and distribution of the rash and a childhood history of chicken pox (varicella) lead to the diagnosis of reactivated varicella zoster, or shingles. A viral culture could be taken of the blister fluid to detect the varicella zoster virus, but the rash is so distinctive that a clinical diagnosis can be made by appearance alone.

UNIT 4 ANSWERS

1. Melanie should have informed her dentist about her heart condition so that she could have taken antibiotics prior to the procedure. The antibiotics could have prevented infection on the abnormal areas of her heart. Brief bacteremia (bacteria in the blood) may happen when vascular areas that normally have some bacteria present, such as entrances to the respiratory, intestinal, and genitourinary systems, are manipulated. Minor bacteremia, such as from tooth brushing, is usually cleared without problem by the reticular-endothelial system within the liver and spleen. Some surgical or dental procedures can cause more significant bacteremia. After an oral procedure, such as a dental cleaning, many more bacteria than usual may enter the bloodstream. Cleaning the teeth with instruments forces bacteria, present in tartar and plaque, into the pocket of gum tissue that surrounds the teeth. The bacteria enter the blood through these areas, which are highly vascular. Moving through the heart, the bacteria are capable of lodging on the endocardium if a heart defect, particularly one previously treated with an artificial patch or valve, is present. The bacteria can then propagate there and cause multiple problems. Infective endocarditis can be fatal without treatment. In 2007 an American Heart Association review reduced the number of heart conditions for which antibiotic prophylaxis is recommended before invasive dental procedures.

2. CK (creatine phosphokinase) is an enzyme found within cardiac and skeletal muscle cells, and troponin T is an intracellular protein specific to cardiac muscle cells. Because these are intracellular proteins, elevated levels in the blood would indicate that the cells have been damaged, presumably as a result of blocked blood supply caused by a myocardial infarction. The membranes of these cells broke down, releasing intracellular proteins into the tissue fluid. Levels of CK cardiac isoenzymes begin to rise 3–6 hours after the myocardial infarction (MI). Troponin T levels begin to rise shortly after an MI. Troponin T and CK-MB are early indicators of an MI.

3. Teresa is suffering from hypovolemic shock secondary to acute intraabdominal hemorrhage. A ruptured spleen is the most likely bleeding site. The accumulation of blood in the peritoneal cavity causes irritation of the diaphragm and phrenic nerves, which in turn causes pain, and this leads to abdominal rigidity and referred pain to the top of the shoulder(s).

UNIT 5 ANSWERS

1. The primary problem is chronic bronchitis, a form of COPD. It has led to chronic hypoxemia and hypercapnia. The low P_{O_2} has led to erythrocytosis, and the increased blood viscosity, added to pulmonary vasoconstriction shunting blood away from fibrotic areas, may cause pulmonary hypertension. Any exertion can cause acute hypoxia; he may need home oxygen administration, bronchodilators, and other treatments.

2. The infant is diagnosed with cystic fibrosis. The major effect is an alteration of exocrine gland function due to an abnormal chloride transport mechanism. Thick mucus predisposes the infant to respiratory infections. Mucus plugs the pancreatic ducts, leading to problems with digestive exocrine enzyme release and signs of nutritional malabsorption syndrome.

3. The patient had an undiagnosed and untreated case of diabetes mellitus, which created the emergency situation. The markedly elevated blood glucose and the presence of glucose in the urine confirmed this diagnosis. Blood pH was low and indicated acidosis. The bicarbonate levels were decreased, indicating a metabolic cause. The patient was suffering from ketoacidosis, a potentially fatal complication of uncontrolled type 1 diabetes mellitus.

   Ketoacidosis is the consequence of excessive lipid catabolism for energy purposes. In the absence of insulin, the body's cells lack the accessibility to the glucose in the blood, and the body utilizes fats for energy instead. Glycosuria causes polyuria and may lead to dehydration. Excessive catabolism of fats causes the buildup of keto acids in the blood, decreasing the blood pH. Respiratory compensation with hyperventilation occurred, as shown by the decreased $PaCO_2$. Despite the compensation, the nervous system was adversely affected by the decreased pH, and loss of consciousness occurred.

The patient was given IV fluids, which treat dehydration, and IV insulin. The pH balance was restored over time as the glucose entered her cells for energy use.

UNIT 6 ANSWERS

1. Gayle has hypertension and proteinuria, the hallmarks of preeclampsia, answer (b).

2. Moira's test results show that she is ovulating normally and has regular hormonal cycles. However, problems with her uterine tubes or uterine lining may be present. Further tests, such as a hysterosalpingogram and endometrial biopsy, are performed. Sam's sperm count was low, and this is the most likely cause of the couple's infertility problem. If low sperm count is the only factor involved, his sperm can be collected, concentrated, and artificially introduced into the uterus immediately after ovulation. The sperm may also be incapable of normal fertilization; if this is the case, intracytoplasmic sperm injection (ICSI) of a single sperm into a harvested egg may be successful, or a sperm donor can be used. It is important to realize that completely normal babies who grow into normal, fertile adults result from successful IVF treatment.

3. John has two problems. First, he has a urinary tract infection that caused the dysuria and increased frequency of urination. This condition is probably related to chronic partial obstruction of the urethra by an enlarged prostate; the resulting incomplete emptying of the bladder during urination increases the likelihood of urinary tract infections. This urine infection can be treated with antibiotics. The second problem is the enlarged prostate, which appears to be the result of a malignancy in the prostate. Treatment will depend on the staging of the cancer, the type of prostate cells involved, and John's overall health.

CASE STUDY 1 ANSWERS George's Demise

1. The cause of fainting is a reduction in the blood flow to the brain. In this situation, George is probably experiencing orthostatic, hypovolemic, hypotension; his blood pressure drops and pulse rises when he goes from a sitting to a standing position.

2. Suggestion: Make a concept map similar to the one for muscle weakness (Figure 32, p. 63). Have branches for the nervous, cardiovascular, respiratory, and endocrine systems, then plug in a few representative possible disorders (muscular dystrophy, dehydration, hypovolemia, heart failure, anemia, pulmonary insufficiency, and so forth). Then rule out some of the possible disorders (such as muscular dystrophy) by comparing them with the patient's age and condition. This list then becomes a worksheet for the rest of the problem. Examples include:

    Circulatory: decreased blood flow to the brain due to (a) blockage of arteries; (b) ineffective heart pumping from irregular heart rate, weak heart muscle, or other factors **OR** decreased blood oxygen-carrying capacity due to (a) anemia associated with blood loss or impaired blood production; (b) low blood volume due to dehydration (decreased intake or increased output, as in diabetes or diarrhea); (c) hypovolemia from acute blood loss

    Respiratory: decreased blood oxygenation due to (a) poor airflow (asthma), (b) loss of alveolar space (COPD/emphysema or infection), (c) decreased diffusion across respiratory membrane (pulmonary fibrosis)

    Endocrine: metabolic disorders involving the adrenal or other glands

    Digestive: malnutrition or dehydration

    Urinary: dehydration

    Nervous: neuronal malfunction, as in autonomic dysfunction; psychological, as in anxiety

3. Smoking damages the lungs, promoting development of COPD, and damages arteries by promoting plaque formation and CAD. The combination increases the incidence of heart attacks, strokes, and peripheral vascular disease. Chronic hypoxemia can lead to elevated Hct, further increasing the cardiac workload.

4. Normally, the heart rate and blood pressure rise together, and blood pressure is stabilized when a person moves to a standing position by reflexes that elevate the heart rate and cause peripheral vasoconstriction. In this case, the neural controls are working properly (heart rate rises, so presumably vasoconstriction is under way), but the blood pressure falls. The most likely reason is that George's blood volume is so low that even these mechanisms cannot stabilize systemic blood pressure.

5. The epigastric tenderness could mean a stomach problem such as peptic ulcer or esophageal varices. The presence of blood in the stool (as indicated by the positive guaiac test) indicates bleeding somewhere along the digestive tract. In lower gastrointestinal (LGI) bleeding, recognizable blood is usually visible with the stool. In moderately rapid upper gastrointestinal (UGI) bleeding, blood that reaches the rectum is partially digested and is usually black and tarry. Slower, insidious bleeding may not visibly affect the stool and only be recognized by the occult blood stool guaiac test.

6. Both Hct and Hb are lower than normal, again indicating that there may be chronic as well as acute blood loss.

7. The increased fluid volume will elevate George's blood pressure. His shortness of breath could be due to the low oxygen-carrying capacity of the blood (anemia); giving him oxygen will improve oxygen delivery to the tissues. The typing and cross-match testing prepare him for a transfusion of whole blood or packed red blood cells.

8. A nasogastric tube is a flexible tube inserted into the nose that extends down the esophagus and into the stomach. The medical workers are checking for presence of blood in the stomach, whether it is fresh (bright red) or partially digested (coffee ground brown). This could be from a gastric ulcer or ruptured blood vessel.

9. The water flushes the stomach contents, and continued suction allows for continued monitoring and removes acidic fluid. If the suction fluid had not cleared, it would have indicated that the bleeding was continuing.

10. The preliminary diagnosis is a gastric or duodenal ulcer, and the drugs administered are intended to control acid production and permit healing of the epithelium. Endoscopy could confirm the diagnosis, locate the bleeding site, and in many cases take steps to control any ongoing blood loss.

11. The new symptoms may indicate acute angina or a heart attack, triggered by decreased oxygen to the myocardium as the result of anemia, hypotension, and preexisting coronary artery disease. Reflux of gastric acids into the esophagus can cause epigastric and chest pain, but not hypotension. The suite of symptoms in this case are most likely cardiovascular problems.

12. Nitroglycerin dilates coronary arteries. Antacids neutralize gastric acids, just in case cardiac problems aren't involved, and the symptoms are being caused by esophageal reflux of stomach acid.

13. If the heart muscle is too damaged to pump effectively, cardiogenic shock occurs, and blood pressure declines.

14. Death from heart attack is the probable cause of death. The progression was (1) underlying coronary artery disease aggravated by chronic bronchitis related to smoking, followed by (2) anemia and hypotension due to gastrointestinal bleeding probably from an ulcer, leading to (3) ischemia of the heart muscle and (4) fatal myocardial infarction.

CASE STUDY 2 ANSWERS Christine's Workout

1. You may want to begin with another concept map of possible causes. Include the following as a minimum:

    Cardiovascular: heart—low probability in this instance due to relatively young age, continued work, and description of pain (not crushing pain)

    Respiratory: lungs—low probability due to lack of shortness of breath

    Nervous: spinal cord—low probability due to absence of trauma to the region and no radiation of pain

    Musculoskeletal: high probability due to association with lifting and carrying, restriction of motion

2. The lack of difficulty in breathing indicates that the problem is not affecting lung function. The restriction of arm elevation indicates a probable musculoskeletal origin; heart/lung problems would not affect the range of motion.

3. Numb hands would indicate peripheral nerve compression or spinal nerve involvement or possible compartment syndrome where muscle swelling affects peripheral blood flow to peripheral nerves. Dark urine would indicate the presence of hemoglobin, red blood cells, or myoglobin in the urine.

4. Tenderness indicates focal damage to tissues rather than central circulatory or respiratory problems. Pulse and color assess peripheral circulation; sensation indicates normal neurological function. A normal range of motion in the limb indicates that the problem is probably muscular rather than skeletal.

5. Microscopic urine examination that showed RBCs would indicate that the dipstick test revealed hemoglobin. In the absence of RBCs, deciding what the urine dipstick test means is aided by the clinical exam and the blood test results. Myoglobin in the urine indicates acute muscle fiber damage leading to the leakage of myoglobin across the sarcolemma into the bloodstream.

6. CK and LDH are indicators of muscle damage; isozymes may confirm that the source is skeletal muscle rather than cardiac muscle. Kidney function tests assess whether the myoglobin has damaged the kidneys; myoglobin in high concentrations can clog the filtration membranes and inactivate glomeruli.

7. The likely diagnosis is traumatic myositis.

8. If Christine's urine remains cloudy or declines in volume, these signs may indicate acute renal failure due to interference with glomerular function. If her sensation diminishes, it may indicate compartment syndrome and compression of sensory nerves.

9. These results are consistent with the preliminary diagnosis.

10. Traumatic myositis due to overexertion. Christine should have responded to her body's warning signs and reduced her activity when pain developed.

CASE STUDY 3 ANSWERS Charlene's Pain

1. The most likely sources are (a) nerve compression (radicular pain) from the lumbar spine, possibly involving the sciatic nerve or its roots; or (b) muscle damage to muscles of the spine, buttocks, hip, or thigh. A compression fracture of the lumbar spine could result in pain via either mechanism.

2. Muscle pain is increased by the use of a muscle and worsens quickly with a change in position or gradually due to the stiffening of the muscle at rest. Pain due to nerve compression is most affected by a change in body position. Referred pain must be considered, as abdominal pain from the urogenital or gastrointestinal tracts can be referred to the back and hip. Normal urination and defecation indicate that neurological control is not impaired; incontinence and poor anal sphincter control would suggest damage to the nerve supply.

3. All COPD symptoms are related to poor oxygen delivery. The increased respiratory rate is an attempt at compensation; the upright posture lowers resistance along the airways and maximizes pulmonary gas exchange. Charlene cannot meet the increased oxygen demand accompanying activity, and this limits her ability to move around.

4. Some causes are (a) bony metastasis from lung or other cancer, (b) lumbar or hip fractures associated with osteoporosis from age and prednisone therapy, or (c) ischemia from vascular blockage to muscles (atherosclerosis).

5. Bronchodilators relax muscles that surround the bronchial airways, lowering resistance and improving airflow to the alveoli. Prednisone is a corticosteroid analog that reduces swelling of the lining of the airways, reducing resistance, but chronic high doses can cause osteoporosis, muscle weakness, suppression of the immune response, hypertension, diabetes, and other adverse side effects. Chronic systemic (as opposed to inhaled corticosteroids) prednisone use is a treatment of near last resort for severe COPD.

6. Thin: increased energy expenditure in breathing with COPD

   Pale: poor blood oxygenation from COPD, thin skin from prednisone

   Moon-faced: long-term prednisone administration, comparable to Cushing's syndrome

   Pursed lips: increases expiratory pressure and helps prevent alveolar collapse (see discussion of PEEP)

7. Broken bones or metastatic bone growths would be tender. Charlene's range of motion and walk would be affected if there were joint, muscle, or neural abnormalities.

8. Coughing and sputum production would increase if there were a lung infection. Urinalysis would be abnormal if there were a renal infection or kidney stone.

9. An abnormal lumbar disc could be present, but it would not show in a standard x-ray. However, disc problems were not indicated in the initial examination.

10. The fracture can produce pain from stimulation of sensory nerves. In addition, the change in bone shape and height can compress or distort spinal nerve roots or peripheral nerves, causing pain.

11. Not necessarily. Bony fractures are usually worst at the start and then show gradual improvement. Focal tenderness on palpation of the vertebral column in the area of the compression fracture is commonly present.

12. A bone scan measures blood flow to bone. It increases in new fractures or tumors but is normal in healed fractures.

13. Likely causes are (a) a toxic reaction to pain medications, which is more common in older individuals; (b) a stroke related to age, smoking history, and COPD; or (c) CNS infection. Useful tests include (1) drug levels in blood, (2) detailed neurological exam, (3) CT or MRI scan of the head, (4) blood culture, and (5) lumbar puncture.

14. The rash is a characteristic sign of infection by the virus herpes varicella–zoster, also known as herpes zoster or shingles. The lesions on the rest of the body are due to the same virus. In most cases, just one or two dermatomes are involved, but Charlene has disseminated Herpes zoster, including viral meningitis. Thus, all of her recent symptoms have a single, infectious, treatable cause.

15. Acyclovir is an antiviral agent.

16. The final diagnosis is disseminated herpes varicella–zoster infection probably related to immunosuppression from chronic prednisone treatment, with nerve pain preceding the rash and varicella meningitis causing confusion and a deteriorating mental state.

CASE STUDY 4 ANSWERS Hal's Athletic Adventure

1. Cardiovascular: (a) aging causes decreased cardiac reserve and decreased resilience of vessels; (b) decreased activity further reduces cardiac reserve

   Muscular: (a) aging causes decreased elasticity, decreased strength, and decreased range of motion; (b) decreased activity further reduces strength and range of motion

   Nervous: (a) aging causes decreased coordination and increased reflex response times; (b) decreased activity further reduces coordination

2. Fracture: related to poor coordination, reduced range of motion, and slow reflexes

   Ankle sprain: related to decreased elasticity of ligaments and tendons

   Calcaneal (Achilles) tendon tear: an acute sprain developing when the calf muscles apply extreme forces to the calcaneal tendon

3. The strength of ankle motion is reduced; smaller muscles whose tendons cross the joint can produce grossly normal movement when the workload is reduced.

4. The symptoms indicate that the bones are not broken and the joint capsule is not damaged.

5. This region normally contains the inferior portions of the gastrocnemius and soleus muscles and the base of the calcaneal tendon.

6. Compressing the normal calf exerts tension on the calcaneal tendon and produces plantar flexion. A rupture of the calcaneal tendon produced pain and trouble walking. Because the connection between the muscles and the calcaneus is lost, compression of the calf and the reflexive muscle contraction has no effect on the position of the foot. The final diagnosis is a complete rupture of the calcaneal tendon.

CASE STUDY 5 ANSWERS Jim and That Darned Cat

1. A concept map would be a good place to start. Without going into the subcategories, the map should include vascular (migraine, ischemia, hemorrhage), muscular, trauma, infection, tumor, and drug reactions.

2. A muscle tension headache would be consistent with psychological stress. Nasal congestion could be linked to sinusitis, another potential cause. There was trauma, which could indicate concussion, intracranial bleeding, or other brain injury.

3. A continuous headache is most often associated with muscle tension; migraine headaches are episodic. The generalized nature and tight muscles support a tension type headache. Migraines are usually localized or focal, more disabling, have associated light sensitivity and GI upset. Sinus headaches are primarily restricted to the facial region. Jim's normal vision, normal hearing, and the lack of nausea all indicate normal neurological function.

4. Both Tylenol and ibuprofen are analgesics; ibuprofen has anti-inflammatory effects. Their ineffectiveness provides few clues to the nature of the problem, because both are relatively weak pain relievers.

5. A normal mental state, normal vital signs, and lack of fever suggest that there is no CNS infection. The supple neck further supports the conclusion that an infection (such as meningitis) is not present. Other supporting evidence includes the normal skull (no fracture), normal tympanic membranes (no blood in middle ear), and normal cranial nerve function and retinal structure (no elevated intracranial pressure). This combination makes meningitis, a brain tumor, and epidural bleeding less likely. Physical exam findings are less sensitive than modern imaging techniques, but are more immediately available and can detect those patients more likely to have abnormalities.

6. These symptoms do not match the preliminary diagnosis. Tension headaches normally respond to medication and do not cause changes in mental state.

7. The pattern now is consistent with abnormal neurological function, in this case reduced mental function, possibly from elevated intracranial pressure and the distortion of the soft brain tissues.

8. Unilateral signs would imply the presence of a tumor or focal abnormality such as an aneurysm or stroke, or unilateral subdural mass.

9. A normal blood test tentatively rules out an infection as the cause. If blood tests were abnormal, encephalitis or meningitis could have been responsible for the signs and symptoms.

10. There was an acute onset of bleeding—probably at the time Jim bumped his head—followed by chronic leakage that further elevated intracranial pressure.

CASE STUDY 6 ANSWERS Joe's Abdominal Pains

1. Begin with a concept map that includes as a minimum:

   Digestive tract: anatomical blockage (gallstones or tumor), infection (viral or bacterial gastroenteritis), or hemorrhage (peptic ulcers)

   Accessory organs: gallbladder (gallstones) or pancreas (pancreatitis)

2. Gallstones are frequently radiopaque and visible on an x-ray. Among the blood test values, most likely candidates would be elevated alkaline phosphatase, SGOT, bilirubin, amylase, and SGPT values.

3. Gallstones originate as precipitates of cholesterol and bile salts within the gallbladder as bile is concentrated. Contraction of the gallbladder then pushes the stones into the bile duct. Peristalsis of the cystic duct and common bile duct around the stone produces acute, colicky pain. Treatment may include (a) surgical removal of the gallstone (and the gallbladder), (b) administration of IV fluids and bed rest until the stone is passed into the duodenum, (c) medications to promote the dissolution of gallstones, and (d) the use of a lithotripter to shatter the gallstones. Surgical removal of the gallbladder is the best bet to ensure that the condition does not recur.

4. Joe has pancreatitis (inflammation of the pancreas), probably as a result of blockage of the common bile duct where it meets the pancreatic duct. Taking intravenous nourishment avoids stimulation of the pancreas and the release of additional enzymes that would promote inflammation of this organ by essentially self-digestion. Pancreatitis is very painful, can be self-perpetuating, and may lead to malnutrition and diabetes.

5. Joe has cirrhosis (scarring) of the liver. Cirrhosis may be the end result of hepatitis (liver inflammation) caused by exposure to toxins (alcohol, drugs, chemicals), by chronic viral infection, or by congenital blockage of the biliary tract.

6. Blood transfusions received prior to screening tests of donated blood could carry viral hepatitis. Social drinking by itself might not cause cirrhosis but could accelerate liver damage from chronic viral hepatitis. Joe's POW experiences may indicate viral hepatitis exposure at a much earlier date.

7. (a) Atherosclerosis and hypertension lead to plaque formation in the lining of arteries, and in this case the plaques restricted blood flow to the lower limbs. (b) The goal of the surgery is to remove the constricted portion of the vessel and replace it with a vascular graft to restore normal blood flow.

8. See page 139 for details.

9. Hepatitis A, toxin exposure, or congenital biliary blockage are not likely candidates, given the absence of chronic hepatic inflammation with hepatitis A, Joe's age, and no known chemical exposure.

10. Hepatitis A is most often passed orally, while hepatitis B spreads through contact with blood or other body fluids. Hepatitis C is transmitted through blood transfusions (20 percent of cases prior to screening), intravenous drug use with needle sharing, or by unknown mechanisms.

11. The liver is responsible for the metabolism of alcohol and most other drugs. The goal is to reduce the workload of the liver.

12. The decreased liver function and scarring lead to (a) decreased amino acid and plasma protein synthesis, causing edema, ascites, and low serum protein levels; and (b) impaired blood flow and higher pressure in the portal blood system. This leads to splenomegaly with resulting low platelet counts. High ammonia levels may occur in the systemic circulation when portal blood bypasses the liver and or ammonia is not properly metabolized.

13. The high ammonia levels are related to dietary protein metabolism and portal hypertension causing shunting of blood to the systemic circulation. The neurotoxic effects of ammonia have led to disturbances in Joe's mental function (hepatic encephalopathy).

14. He is in a coma, probably due to high ammonia levels. His low hematocrit and the vomiting of blood indicate internal bleeding.

15. Epinephrine causes intense vasoconstriction and will normally slow or stop the bleeding.

16. The aspiration and gastric cleaning must be done to determine whether or not the bleeding has stopped. They also prevent recurrent bleeding, due to irritation by the gastric contents, and further acid production, through local reflexes.

17. Gastric bleeding and ruptured esophageal varices caused anemia. The large amount of blood proteins entering the GI tract led to the absorption of large quantities of amino acids. The amino acids were metabolized, but this process produced large amounts of ammonia, and the damaged liver was unable to convert the ammonia into urea. Ammonia levels rose, and Joe went into a coma. The low-protein diet will reduce the quantity of amino acids absorbed and will help control ammonia levels.

18. Treatment with interferon and an antiviral medicine could halt the progression of hepatitis C, but in this case the damage was already done before the condition was diagnosed. A liver transplant would potentially solve the problem, but Joe is too old and too sick to be considered for this procedure.

CASE STUDY 7 ANSWERS Mrs. M.'s Summer Problems

1. The likely cause is yeast growth in the vagina from increased moisture and heat over a prolonged period.

2. Probably a *Candida* species.

3. The application of a vaginal anti-yeast cream.

4. Some cancers appear to be related to genetics, infectious agents, or toxin exposure and can be asymptomatic (and potentially curable) for a long time before the general health of the individual is affected. A person can have a serious, life-threatening illness without feeling sick.

5. A breast tumor may indicate (a) breast cancer, (b) a benign cyst or fibroma, (c) a localized infection of a portion of the gland complex, or (d) the presence of scar tissue. Mammography (an x-ray of the breast), ultrasonography, and biopsy may all be needed to reach the diagnosis.

6. Mrs. M. may have an inherited increased risk of developing breast cancer. Perhaps equally important, her aunt's death frightened her into denying abnormalities in her own body and seeking early diagnosis and treatment.

7. Stage 1, an early stage prior to metastasis. Aggressive treatment may increase disfigurement and have side effects, but also lowers the chances for recurrence. The individual must consider (a) the stage of the cancer, (b) the rate of cancer development, (c) the success rate from surgery alone, (d) the age and life expectancy of the individual, and (e) the potential side effects for each level of possible treatment.

8. Tamoxifen blocks estrogen receptors, and these cancer cells are stimulated by estrogen. Thus tamoxifen use should suppress the development of any cancer cells that are not surgically removed.

CASE STUDY 8 ANSWERS Cynthia's Bad Back

1. Make another concept map that includes trauma (to skeleton, intervertebral discs, or muscles); infection (of skeleton, muscles, or discs); and referred pain from (a) vascular problems (aorta or other abdominal vessels), (b) visceral structures (pancreas, ovaries, liver, stomach, large intestine, urinary bladder, uterus, ovaries), or (c) kidneys.

2. No, because such injuries would affect range of motion and alter perceived pain. Referred pain from another anatomical source is suspected.

3. The continuous nature of the pain suggests that the problem is not gastrointestinal, renal (kidney stones), or gallbladder related (gallstones) because pain from these conditions undergoes frequent changes in intensity. Also described as colicky or crampy pain.

4. These findings suggest that we can rule out pneumonia (no cough and fever), urinary tract infection and kidney stones (no dysuria or dark urine), hepatitis and appendicitis (no nausea, vomiting, or bowel problems), and ectopic pregnancy (normal menstrual periods).

5. Pain, anxiety, and fear all cause an elevation in bp and heart rate.

6. Musculoskeletal problems and disc problems are unlikely, and nerve function is normal. Referred pain is the likely cause, and the exam is localizing the source.

7. You would expect a normal urinalysis.

8. This test checks for bleeding in the colon, which could indicate bowel tumors or other digestive tract problems.

9. The comparable condition in males is called testicular torsion.

Front Matter Author photos (p. viii, top to bottom): PK Martini; PK Martini; William C. Ober and Claire E. Ober; William C. Ober and Claire E. Ober.

Unit 1 Ocean/Corbis; 3a: Omikron/Science Source; 3b: BIO-PHOTO ASSOCIATES/Science Source; 3c: CMSP/Custom Medical Stock Photo, Inc.; 4b: Dr. Kathleen Welch; 4c: J. Croyle/Custom Medical Stock Photo, Inc.; 4d: Warrick G./Science Source; 5a: Philips Healthcare; 5b: Alexander Tsiaras/Science; 5c: Pascal Goetgheluck/Science Source; 6b: Southern Illinois University/Science Source; 6c: Pascal Goetgheluck/Science Source; 7: Pearson Education; 12a: Eric Grave/Science Source; 12b: Spike Walker/Science Source; 12c: Spike Walker/Science Source; 12d: Mae Melvin/CDC; 13a: Larry West/Science Source; 13b: Science Photo Library/Science Source; 13c: dcb/Shutterstock; 13d: James Gathany/CDC; 13e: CC-BY-SA photo: Luis Fernández García; 14a: P. Motta and A. Familiari/University "La Sapienza," Rome/Science Photo Library/Science Source; 14b: CNRI/Science Photo Library/Science Source; 15a: Douglas Chapman/University of Washington Department of Pathology; 15b: Courtesy of the Greenwood Genetic Center, Greenwood, SC.

Unit 2 Ocean/Corbis; 20: Doug Martin/Science Source; 21: Zuber/Custom Medical Stock Photo, Inc.; 26a: Science Photo Library/Science Source; 26b: Harold Chen, M.D.; 26c: REUTERS/Rick Wilking; 27: Will & Deni McIntyre/Science Source; 28a: Carters News Agency; 28b: Barbara Penoyar/Photodisc/Getty Images; 28c: SPL/Science Source; 29: Russell Illig/Photodisc/Getty Images; 30a: SIU Biomed Comm/Science Source; 30b: Courtesy of Eugene C. Wasson, III and staff of Maui Radiology Consultants, Maui Memorial Hospital; 30c: Courtesy of Eugene C. Wasson, III and staff of Maui Radiology Consultants, Maui Memorial Hospital; 31a: Courtesy of Smith & Nephew, Inc.; 31b: Courtesy of Smith & Nephew, Inc.; 31c: Courtesy of Smith & Nephew, Inc.; 34: Diego Cervo/Shutterstock; 38: Levent Konuk/Shutterstock.

Unit 3 Ocean/Corbis; 41a: Ralph T. Hutchings; 41b: Hinerfeld/Custom Medical Stock Photo, Inc.; 42: William C. Ober; 43: Jeff Widener/AP Images; 45: Southern Illinois University/Science Source; 47a: Lori Joyce/Anne Orsene; 47b: Michael Chorost; 49: From: Acromegaly. Philippe Chanson and Sylvie Salenave. Orphanet Journal of Rare Diseases 2008, 3:17, fig. 2. doi:10.1186/1750-1172-3-17; 50a: Medical-on-Line/Alamy; 50b: Scott Camazine/Science Source; 51a: Wellcome Image Library/Custom Medical Stock Photo, Inc.; 51b: Biophoto Associates/Science Source.

Unit 4 Ocean/Corbis; 53: MECKES/OTTAWA/Science Source; 55a: Biophoto Associates/Science Source; 55b: SIU/Custom Medical Stock Photo, Inc.; 55c: Medtronic, Inc.; 58: Jim Wehtje/Getty Images; 59a: Science Photo Library/Science Source; 59b: From: "Non-compacted Cardiomyopathy: Clinical-Echocardiographic Study" by Nilda Espinola-Zavaleta; M. Elena Soto; Luis Muñóz Castellanos; Silvio Játiva-Chávez; Candace Keirns in March 2007 by www.medscape.com; 59c: U. Joseph Schoepf; 62b: Jack Star/Getty Images; 65: Dr. P. Marazzi/Science Photo Library/Science Source; 67: James Gathany/CDC; 71: Wellcome Trust Library/Custom Medical Stock Photo, Inc.

Unit 5 Ocean/Corbis; 73: B. Luster/AP; 76a: CNRI/Science Source License; 76b: Susan Law Cain/Shutterstock; 79ab: God's Littlest Angels; 84: Philippe LeRoux/SIPA/Newscom; 88: Beranger/Science Source.

Unit 6 Ocean/Corbis; 92: Gavin Hart and N. J. Fiumara/CDC.

INDEX

Note: Italicized page numbers refer to figures; page numbers followed by "t" refer to tables.